VLSI DESIGN

A SHANTHI

A KAVITA

Publishing Globally

NEW AGE INTERNATIONAL (P) LIMITED, PUBLISHERS

LONDON • NEW DELHI • NAIROBI

Bangalore • Chennai • Cochin • Guwahati • Hyderabad • Kolkata • Lucknow • Mumbai

Visit us at **www.newagepublishers.com**

Published by New Age International (P) Ltd., Publishers
First Edition: 2006
Reprint: 2018

GLOBAL OFFICES

- **New Delhi** **NEW AGE INTERNATIONAL (P) LIMITED, PUBLISHERS**
  7/30 A, Daryaganj, New Delhi-110002, (INDIA)
  Tel.: (011) 23253771, 23253472, **Telefax:** 23267437, 43551305
  E-mail: contactus@newagepublishers.com • Visit us at www.newagepublishers.com
- **London** **NEW AGE INTERNATIONAL (UK) LTD.**
  27 Old Gloucester Street, London, WC1N 3AX, UK
  E-mail: info@newacademicscience.co.uk • Visit us at www.newacademicscience.co.uk
- **Nairobi** **NEW AGE GOLDEN (EAST AFRICA) LTD.**
  Ground Floor, Westlands Arcade, Chiromo Road (Next to Naivas Supermarket)
  Westlands, Nairobi, KENYA, **Tel.:** 00-254-713848772, 00-254-725700286
  E-mail: kenya@newagepublishers.com

BRANCHES

- **Bangalore** 37/10, 8th Cross (Near Hanuman Temple), Azad Nagar, Chamarajpet, Bangalore- 560 018
  Tel.: (080) 26756823, **Telefax:** 26756820, **E-mail: bangalore@newagepublishers.com**
- **Chennai** 26, Damodaran Street, T. Nagar, Chennai-600 017, **Tel.:** (044) 24353401, **Telefax:** 24351463
  E-mail: chennai@newagepublishers.com
- **Cochin** CC-39/1016, Carrier Station Road, Ernakulam South, Cochin-682 016, **Tel.:** (0484) 2377303, **Telefax:** 4051304
  E-mail: cochin@newagepublishers.com
- **Guwahati** Hemsen Complex, Mohd. Shah Road, Paltan Bazar, Near Starline Hotel, Guwahati-781 008,
  Tel.: (0361) 2513881, **Telefax:** 2543669, **E-mail: guwahati@newagepublishers.com**
- **Hyderabad** 105, 1st Floor, Madhiray Kaveri Tower, 3-2-19, Azam Jahi Road, Near Kumar Theater, Nimboliadda Kachiguda, Hyderabad-500 027, **Tel.:** (040) 24652456, **Telefax:** 24652457
  E-mail: hyderabad@newagepublishers.com
- **Kolkata** RDB Chambers (Formerly Lotus Cinema) 106A, 1st Floor, S N Banerjee Road, Kolkata-700 014
  Tel.: (033) 22273773, **Telefax:** 22275247, **E-mail: kolkata@newagepublishers.com**
- **Lucknow** 16-A, Jopling Road, Lucknow-226 001, **Tel.:** (0522) 2209578, 4045297, **Telefax:** 2204098
  E-mail: lucknow@newagepublishers.com
- **Mumbai** 142C, Victor House, Ground Floor, N.M. Joshi Marg, Lower Parel, Mumbai-400 013
  Tel.: (022) 24927869, **Telefax:** 24915415, **E-mail: mumbai@newagepublishers.com**
- **New Delhi** 22, Golden House, Daryaganj, New Delhi-110 002, **Tel.:** (011) 23262368, 23262370, **Telefax:** 43551305
  E-mail: sales@newagepublishers.com

ISBN: 978-81-224-1866-8
C-18-07-11523

Printed in India at Ajit Printing Press, Delhi.
Typeset at Dalip Graphics, Delhi.

NEW AGE INTERNATIONAL (P) LIMITED, PUBLISHERS
7/30 A, Daryaganj, New Delhi-110002
Visit us at **www.newagepublishers.com**
(CIN: U74899DL1966PTC004618)

VLSI DESIGN

"The Greatest Teacher"
One who makes you want to learn

Dedicated to
my beloved husband Dr. S. Eswaran
one of the greatest teachers.

Shanthi, A.

Dedicated to
my mom Smt. Sellammal,
dad Shri. P. Anbalagan and
hubby Shri. K. Srinivasan

Kavitha, A.

Preface

VLSI (Very Large Scale Integrated Circuits) is now the most sought after in the area of developing hi-fi ICs used in various applications in the field of information technology and communication electronics. Though it entered the curriculum of the undergraduate engineering courses few years back, it gained momentum recently as its advantages are better experienced.

This book is devoted to the study of VLSI design specifically for the students pursuing studies at the undergraduate level. Current silicon based IC technology has made possible the design of complex digital circuits with saving of space, power requirements and cost and also with high potential speed. The complexity and speed are very vital factors in VLSI systems. So majority of VLSI systems in silicon utilize nMOS, cMOS and BicMOS technology.

The book is intended to serve as first hand reference for the undergraduate engineering students who are studying VLSI design. As a pre-requisite, the reader is expected to have a good knowledge on digital circuits and design.

In the first Chapter, Introduction to VLSI, MOS device operation and its various characteristics are discussed. The successive chapter elaborates on Basic IC production processes viz. nMOS, cMOS, etc. Chapters 3 & 4 tell about how various circuits for digital systems can be implemented using MOS and the process involved in MOS circuit Design. While the 5th Chapter has a detailed look on combinational, sequential and data path circuits, Chapter 6 discusses on programmable logic devices, giving more attention to PLAs and memories. Hardware description language which offers the most versatile design entry is introduced in Chapter 7. Though VHDL and verilog are popular languages used, detailed study of VHDL precedes and a comparison of VHDL with verilog is done for better understanding. Chapter 8 is dedicated to analog VLSI which is another key area in VLSI design. It also briefs on GaAs technology—a backbone to high speed VLSI. The concluding chapter gives information on field programmable gate arrays—another important programming device. Though several FPGA vendors and manufacturers are in market, this chapter concentrates on the most versatile XILINX series. Sufficient coding on VHDL are given as appendix for the benefit of the students.

While this book covers most of the core areas of VLSI design, it is hoped that this will serve as a single reference book for the students pursuing undergraduate studies in electronics & communication engineering, electronics and instrumentation engineering under the Anna University, Chennai.

Although every endeavor is made to present an error free text, errors might have crept in due to reasons beyond the authors' control. Corrections are most welcome with authentication. Constructive criticism is solicited along with suggestions which will be duly attended to improve the quality of the book.

Shanthi Eswaran
Kavitha Srinivasan

Acknowledgement

During the course of lecture sessions, the authors required references of too many books due to the vastness of the subject. When the search for availability of a comprehensive single book for this subject was on, it was suggested by many, why the authors themselves could not endeavor to bring out one such. That was the prompt suggestions and there the seed for this work was sown. All those who encouraged this idea deserve the authors' first hand thanks.

The authors had excellent encouragement and support from ***Shri. M. SUBRAMANIAM***, Chairman, Jayaram College of Engineering and Technology and ***Dr. R. PONRAJ***, Principal, who were generous enough to give permission to utilize the EDP facilities of the institution in the making of this book. The authors gratefully acknowledge the help.

New Age International Publishers, readily accepted for the publication of this material and provided the syllabus being adopted by various Universities in the country for the subject which had been a great source in stimulating the authors' interest in this effort. The authors profusely thank the publishers.

References have sufficiently been made and some subject matter and sketches have been adopted from various books. The authors express their sincere thanks to those authors and publishers for having permitted adoption of notes and figures from their publications.

M/s Abaya Computer Graphics, Thiruvanaikoil, Trichy, Tamil Nadu have taken pains-taking efforts in bringing out the first electronic copy of this work. The authors owe their thanks to them.

There had been instances during the course of this Herculean task, the first ever work of this type for the authors, when the authors even intended to give up the idea, owing to pressure of other works and personal commitments. At such times of drop, the authors had words of encouragement and advice from Mr.A.Venkata Ganesh, Junior Engineer, Southern Railway (brother of Mrs. A.Shanthi). Also the authors received tips from him during fine tuning operations of this work. His contribution to this work is catalytic. The authors gratefully thank him.

Needless to say, but for the unstinted co-operation and support of all the teaching and non-teaching staff and students of the electronics and communication engineering department, Jayaram Engineering College, this work would not have come to its present shape. The authors place on record their heartfelt thanks to their direct/indirect help.

Shanthi, A

Kavitha, A

Contents

1

MOS Device Characteristics

1.1. INTRODUCTION TO MOS TECHNOLOGY AND VLSI

Electronics is characterized by reliability, low power dissipation, extremely low weight and volume, low cost, high degree of sophistication and complexity. Vacuum tubes were used for Electronic circuits during the first half of the 20th Century. When electronic components started emerging a new dimension, Transistors replaced vaccum tubes in 1948 which were superceded by IC demonstrated by Er. Jalk. Kilby of Texas Instruments in 1958. VLSI the small GIANT, was initially formulated in 1975 which now occupies the shelves of manufacturers of all micro electronic equipments. The work of both Jalky and Robert Noyce of fair child developed the field of VLSI.

Integrated Circuit (IC). It is an electronic assembly built in such a way that all the components in the circuit are fabricated on a single container called Chip. (Metallization is used for inter connection). Fig. 1.1 shows types of major processes used in IC fabrication.

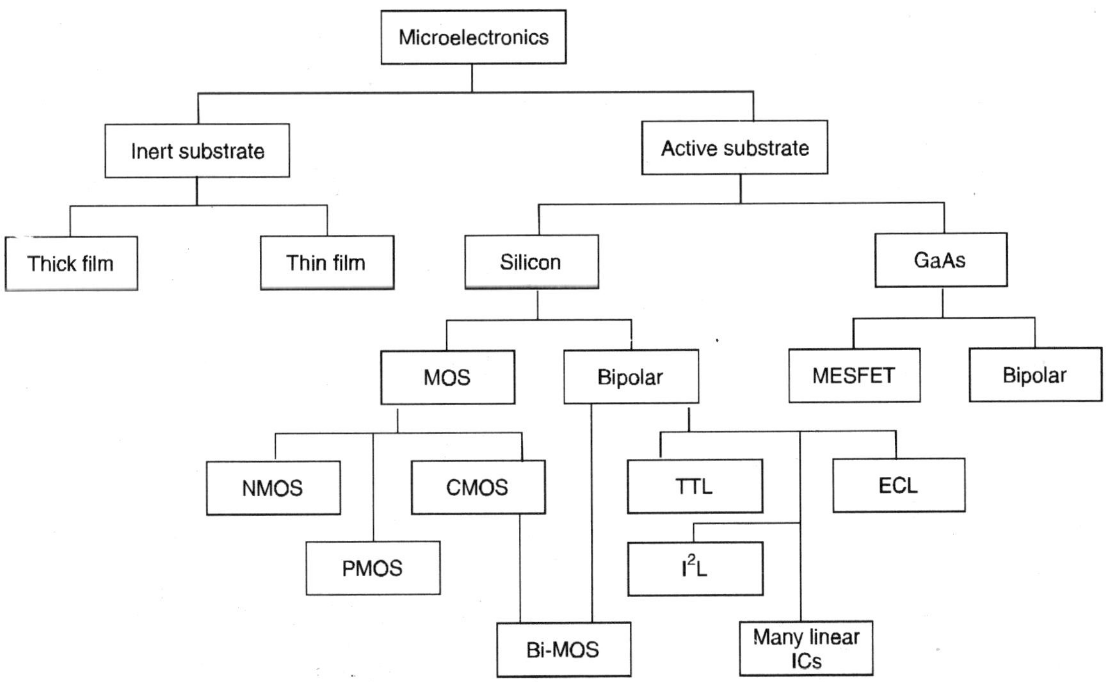

Fig. 1.1. Types of major processes used in IC fabrication.

CLASSIFICATION OF ICS

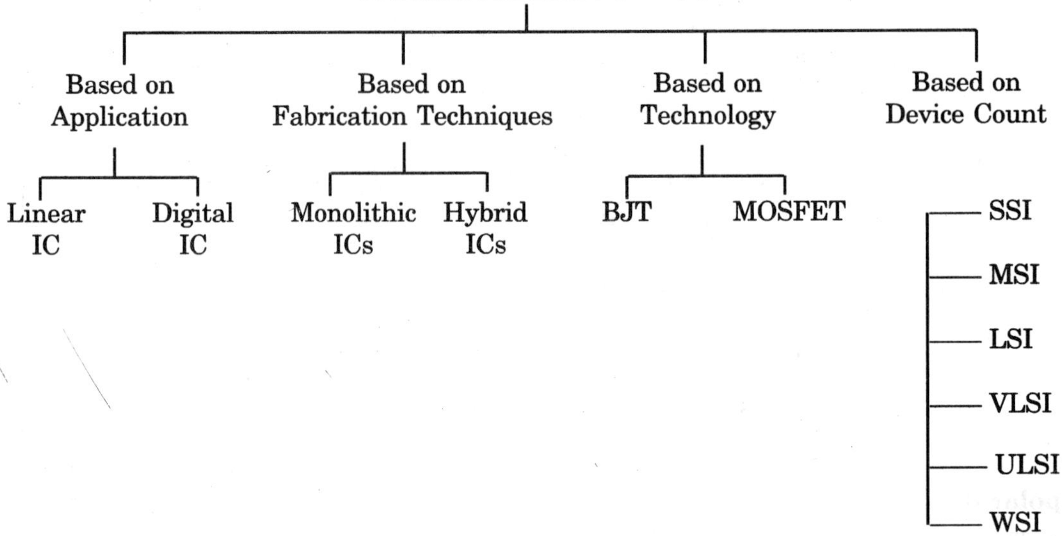

1. Based on application

Linear ICs : These are discrete circuits which need external components like resistor and capacitor for satisfactory performance apart from the input. They work on analog signals rather than digital signals. *e.g.* op-amp, ADC, DAC, Timer (555).

Digital IC: They require only digital input and not any other component for operation. *e.g.* all ICs like logic gates, adders, comparators, etc.

2. Based on fabrication techniques

Monolithic ICs : All the active and passive components of a circuit can be fabricated on a single piece of silicon. (Metallization is used for interconnection).

Hybrid ICs : Separate components parts are attached to a ceramic substrate and inter connected by means of either metallization pattern (or) wire bonds.

3. Based on device count

Device count is restricted to the number of active devices (FET and BJT)

Nomenclature	*Active Device*	*Typical function*
SSI	1-100	Gates, op amps......
MSI	100-1000	Registers, Filters, etc., ...
LSI	1000-100000	Microprocessors, A/D ....
VLSI	10^5 to 10^6	Memories, Computers, Signal Processors
ULSI	10^6	(Ultra Large Scale Integration Especially used in Japan)
WSI → Wafer scale integration under processing.		

4. Based on technology

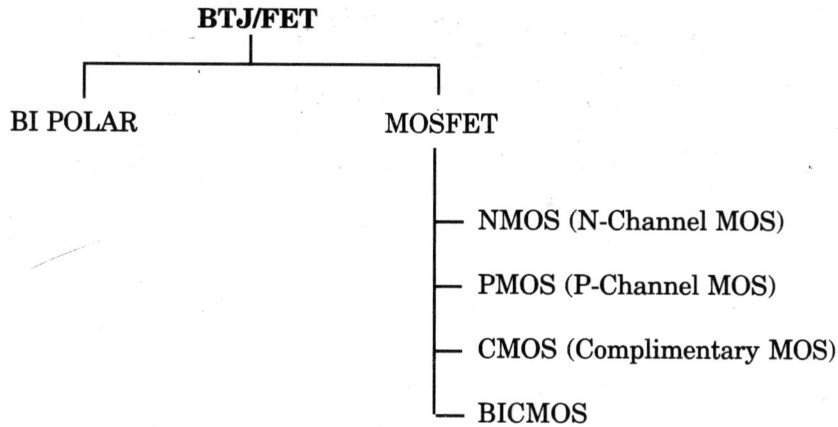

Bipolar devices

The basic active device used is the BJT. The speed is very high with bipolar process.

MOS Devices

- **NMOS** is primarily used in many applications, as mobility of p type material is poorer than N type. Also it has excellent density and very good performance.
- **PMOS :** It is dual of NMOS. The basic active device is the p-channel MOSFET.
- **CMOS :** Complementary MOS : It comprises both PMOS & NMOS. CMOS offers improvements in power dissipation and performance when compared with NMOS.
- **BiCMOS :** It is a combination of Bipolar CMOS transistors. Therefore in BICMOS technology we can have the advantages of both bipolar and CMOS technology. This seems to be effective in speeding up VLSI circuits.

Advantages of ICS over discrete components

- Interconnections between the components using wires are reduced (or) avoided.
- Power consumption is less.
- Fabrication process is simple.
- Production will be in bulk quantity and so the cost will be reduced.
- More number of components will be fabricated in a single chip of IC. So size will be reduced.
- Heat dissipation will be reduced.
- Reliability is more (can work for a long period than discrete type).
- Space is less.
- Weight is less.
- Cost of production of electronic circuits due to batch production is reduced.

For example in a 10 cm diameter wafer, 8000 rectangular chips can be manufactured. So from 10 wafers 8000 × 10 = 80000 ICs chips can be manufactured.

Disadvantages

- Cannot with stand rough handling and excessive heat. (If we increase the current, heat dissipation will be increased).
- Capacitors with values more than 30 pf cannot be fabricated on chip.
- Inductors and transformers cannot be fabricated.

Advantages of VLSI

- Low Power dissipation.
- Area Consumption is less.
- Power consumption is less.
- High speed of operation.

Some Basic Definitions

(*i*) **Wafer (Slice) :** It is a basic physical unit used in processing. It contains large number of identical ICs. Circular wafers with 4, 5 or 6-inch diameters are normally used.

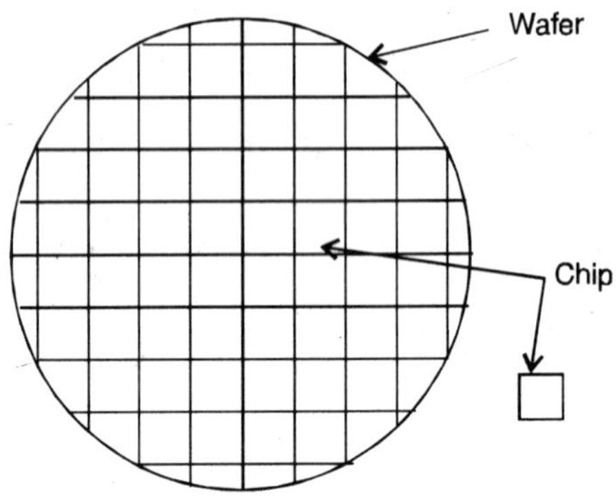

Fig. 1.1.0. Circular Wafer with repeated chips.

(*ii*) **Chip (die-bar) :** It is defined as one of the repeated IC on a wafer.

(*iii*) **Transistor Size:** Dimensions of transistors are measured in Micron (μ) [1 Micron = 10^{-6} m)

(*iv*) **Channel :** Space between source and drain, which allows the movement of majority carriers from source to drain.

(*v*) **Feature Size :** This refers to half of the channel length of the smallest transistors and is denoted by the symbol 'LAMDA'. (λ).

Now $\lambda = \frac{l}{2}$, where $\frac{l}{2}$ is half the channel length of the smallest transistor.

Development of Feature Size

- In the early to middle of 1970's – 7 μ to 10 μ
- In the late 1970 and in early1980 – 5 μ
- In the middle of 1980's – 2 μ (then 1 μ and 1¼ μ)

- In the early 1990 – 0.75 μ and 0.25 μ
- 1997 0.25 μ
- 1999 0.18 μ
- 2001 0.15 μ
- 2003 0.13 μ
- 2006 0.1 μ
- 2009 0.07 μ
- 2012 0.05 μ

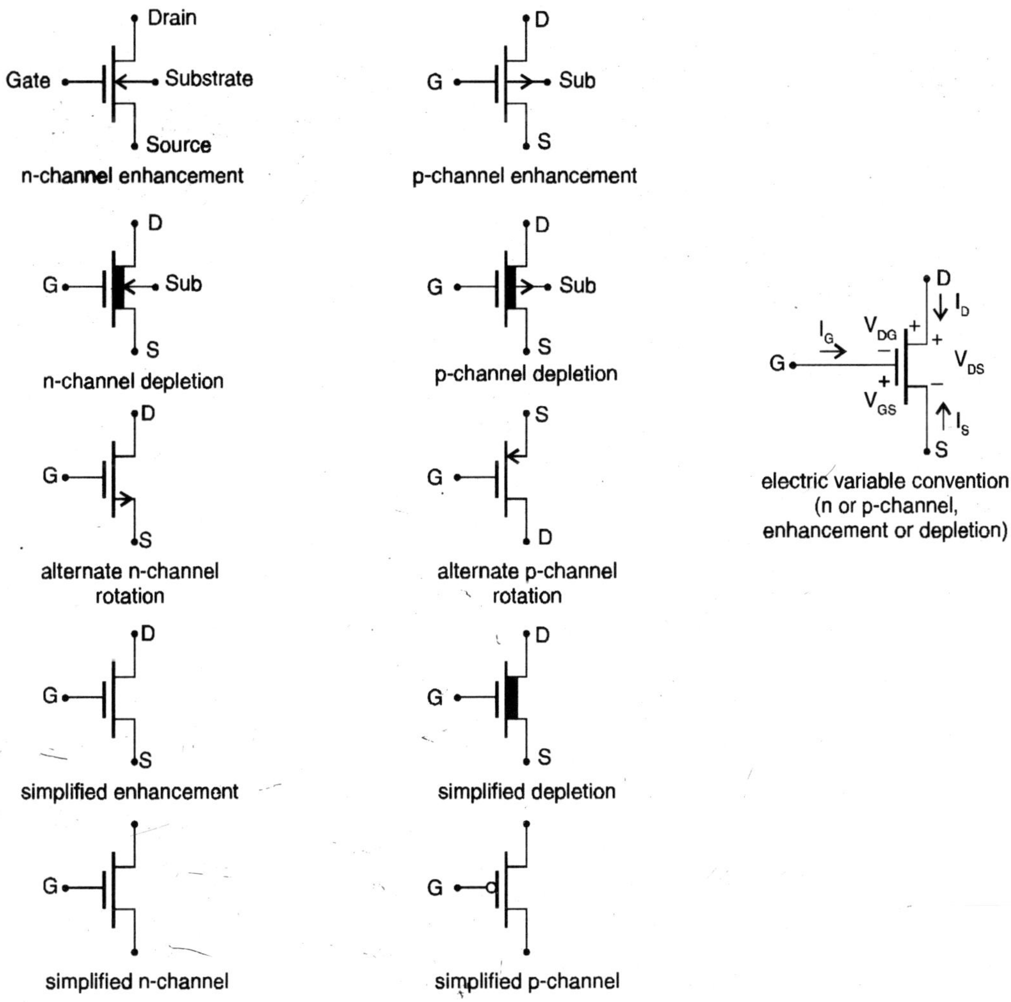

Fig. 1.1.1. Symbols for MOS transistors.

1.2. MOS TRANSISTOR THEORY

MOS Transistors. A MOS (Metal Oxide–Semiconductor) structure is created by super imposing several layers of conducting, insulating materials to create a sandwich like structure. These structures are created by a series of chemical processing etc. CMOS technology comprises of two types of transistors namely n-type transistor (nMos) and p-type transistor (p-MOS).

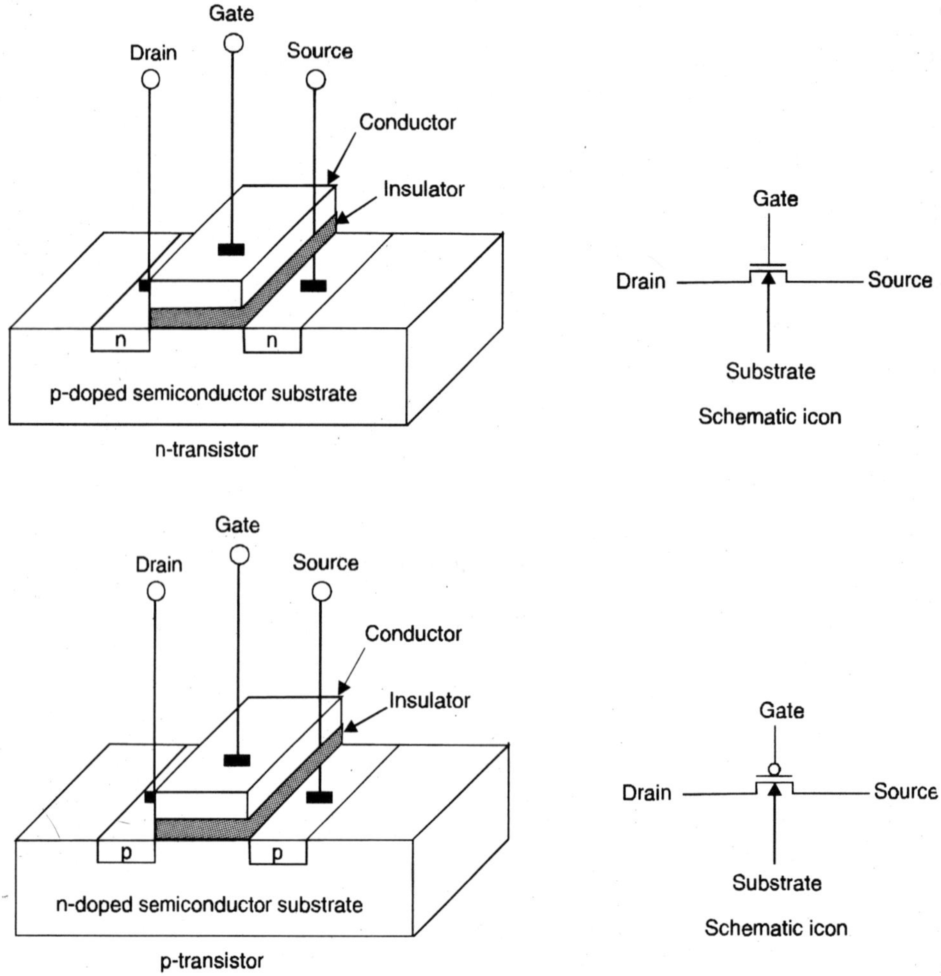

Fig. 1.2.1. Physical structure of MOS transistors and their schematic icons.

n-MOS devices are formed in a p-type substrate of moderate doping level. The source and drain regions are formed by diffusing n-type impurities through suitable masks. The source and drain are isolated from one another. The area separating the source and drain regions is capped with a sandwich consisting of SiO_2 (an insulator) and a conducting electrode (usually a poly crystalling silicon-poly) called the gate. Similarly for the p-type transistor (p-MOS) the structure consists of a section of n-type silicon separating two p-type areas. (source and drain). This has also a gate electrode. The physical structure of MOS transistors is shown in Fig. 1.2.1. The gate is the control input and it affects the flow of current between source and the drain.

Principle of MOS Transistor

A voltage applied to the gate controls the current in a conducting channel between the source and the drain. For example in an n-type Mos-transistor (nMOS), the majority carriers are electrons. A positive voltage applied on the gate with respect to the substrate enhances the number of electrons in the channel and thus increases the conductivity of the channel. For gate voltage less than a threshold value denoted by Vt, the channel is cut off, thus causing a very low drain-to-source current.

Threshold Voltage (Vt). It is defined as the voltage at which a MOS device begins to conduct (turn on).

Enhancement Mode. Devices that are normally cut-off (non-conducting) with zero gate-bias [*i.e.*, gate to source voltage (Vgs)] are called Enhancement–mode devices.

Depletion Mode. Devices that conduct with zero gate-bias are called depletion mode devices.

The threshold voltages for n-channel and p-channel devices are denoted by V_{tn} and V_{tp} respectively.

NMOS Enhancement Transistors. It consists of moderately doped p-type silicon substrate into which two heavily doped n^+ regions, the source and drain are diffused.

Between source and drain, there is a narrow region of p-type substrate called the channel, which is covered at the top with a thin insulating layer of silicon dioxide (SiO_2) called gate oxide. Over this oxide layer, polycrystalline silicon (polysilicon) electrode referred as gate, is formed. Polysilicon is a silicon that is not composed of a single crystal. Since the SiO_2 is an insulator, the DC-current from the gate to channel is essentially zero.

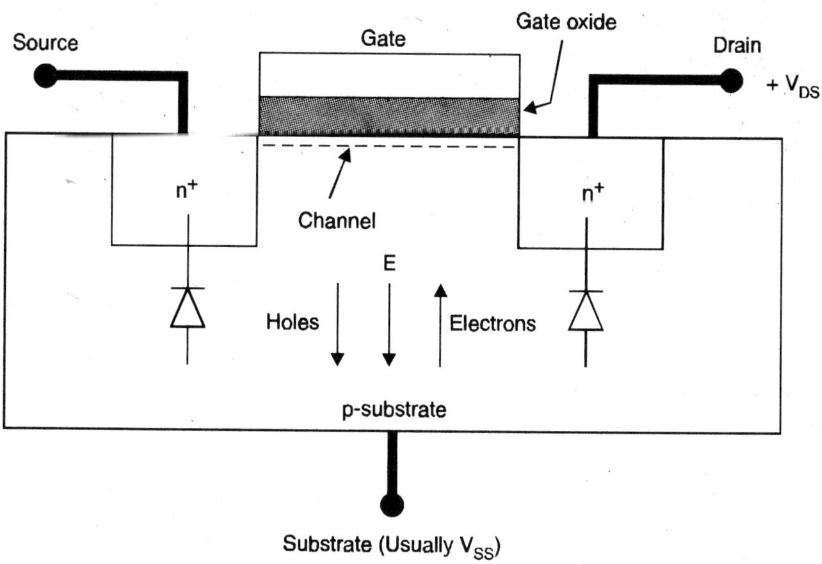

Fig. 1.2.2. Physical structure of a NMOS transistors.

Operation

- A positive voltage is applied between source and drain (V_{ds}). With zero gate bias ($V_{gs} = 0$), no current flows from source to drain because they are insulated from each other by the two reverse biased pn-junctions. However a voltage applied to the gate which is +ve with respect to source and substrate, produces an electric field 'E' across the substrate, which attracts electrons toward the gate and repels the holes. If the gate voltage is sufficiently large, the region under the gate changes from p-type to n-type (due to accumulation of attracted electrons) and provides a conduction path between source and drain. Under such condition, the surface of the underlying P-type silicon is said to be inverted.

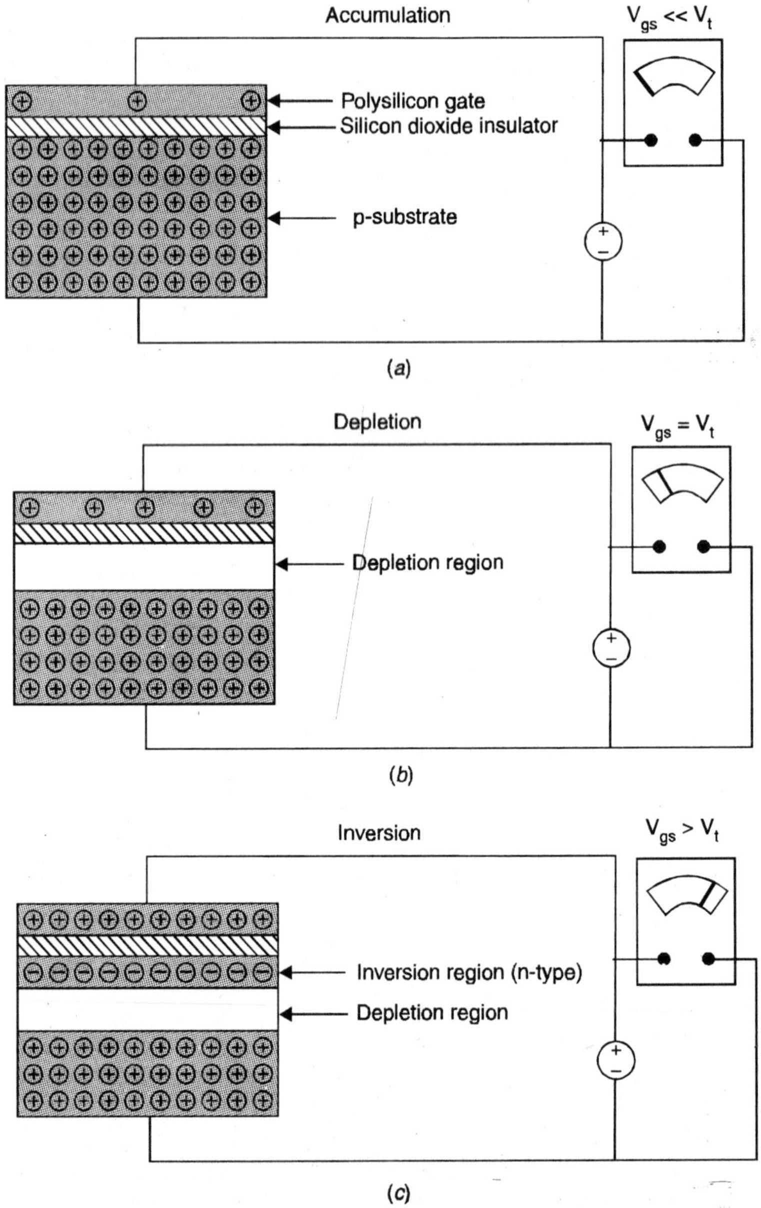

Fig. 1.2.3. Accumulation, depletion and inversion modes of a MOS transistor.

(*a*) **Accumulation Mode :** This shows the initial distribution of mobile positive holes in a p-type silicon substrate of a MOS structure for a Voltage, V_{gs} much less than threshold Voltage V_t.

(*b*) **Depletion Mode :** As V_{gs} is raised above V_t in potential, the holes are repelled causing a depletion region under the gate. Now the structure is in the depletion mode.

(*c*) **Inversion Mode :** Raising V_{gs} further above V_t, causes electrons in the substrate to be attracted to the region of the substrate under the gate. A conductive layer of electrons in the p-substrate give rise to the name inversion mode. As more +ve gate Voltage is applied, the gate attracts negative charge from the source and drain regions and the channel becomes n-region.

PMOS Enhancement Transistor

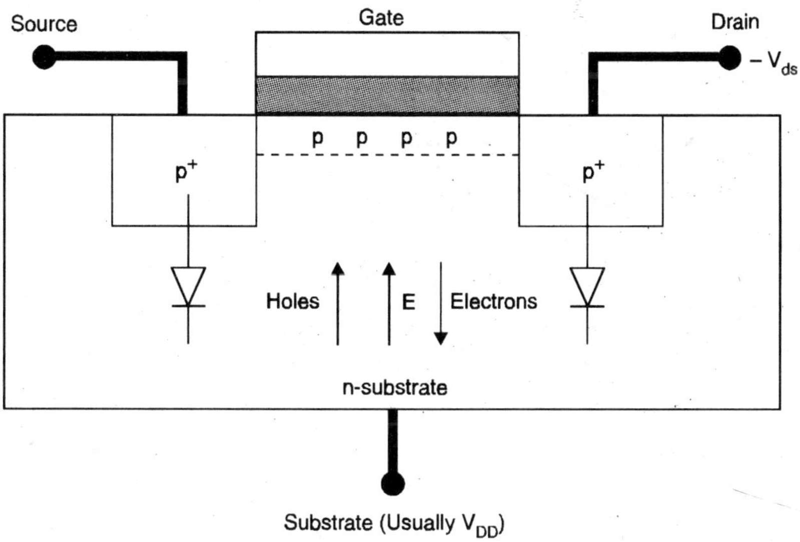

Fig. 1.2.4. Structure of PMOS transistor–Enhancement mode.

- Negative voltage applied to gate draws holes under the gate resulting in channel changing from n-type to p-type. Thus a conduction path is created between source and drain.
- The Conduction results from the movement of holes in the channel.

Depletion Mode Transistors

When Transistor is operated in depletion mode, even when $V_{gs} = 0$, it conducts. The channel may also be established under the gate under the condition of $V_{gs} = 0$ by implanting suitable impurities in the region between source and drain during manufacturing and prior to deposition of the insulation and gate. Even on applying negative voltage to the gate, the current flows in a n-type transistor and conduction takes place. Similarly, even on applying positive voltage to the gate, the conduction can be established on a p-type transistor. The Fig. 1.2.5 shows NMOS depletion mode transistor. PMOS depletion mode structure is also similar to this.

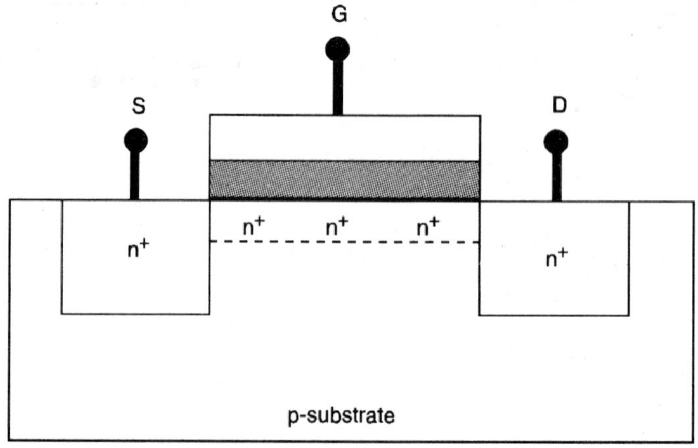

Fig. 1.2.5. Structure of NMOS depletion mode transistor.

Hybrid Technology : The hybrid approach to integrated circuit design involves attaching two (or) more integrated circuit dies (typically of difference types) along with some discrete components in some cases, in a single package to form what is called hybrid integrated circuit.

Normal conduction characteristics of a NMOS transistor

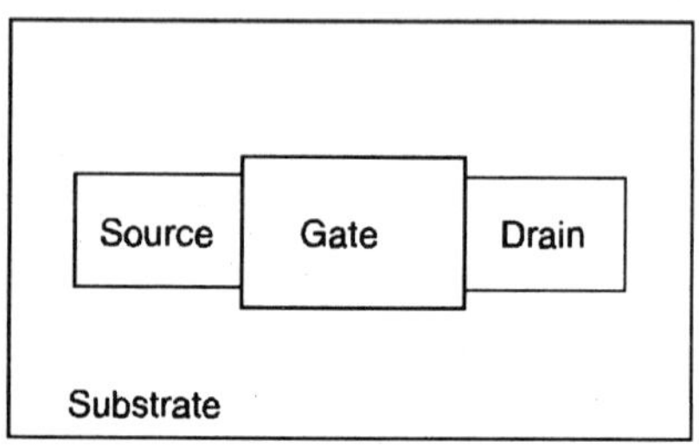

(a) Top view

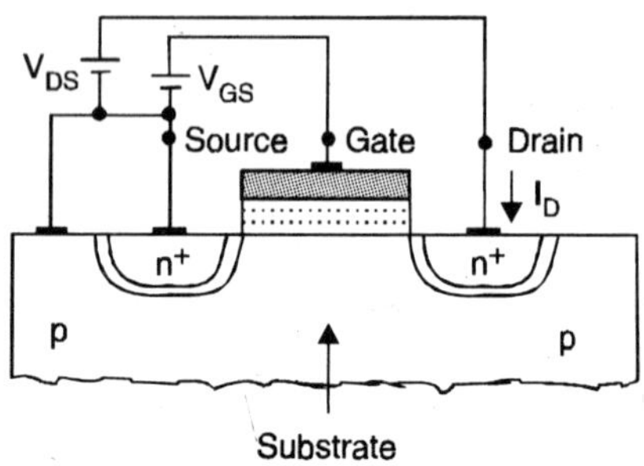

(b) $V_{GS} = 0$, $V_{DS} = 0$ (Cut off)

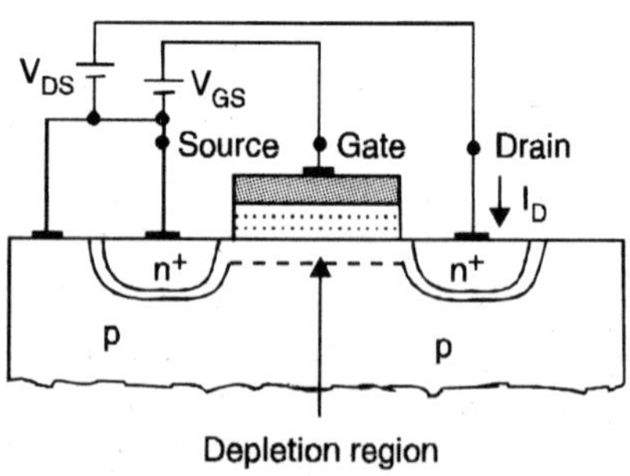

(c) $0 < V_{GS} < V_T$, $V_{DS} = 0$ (Cut off)

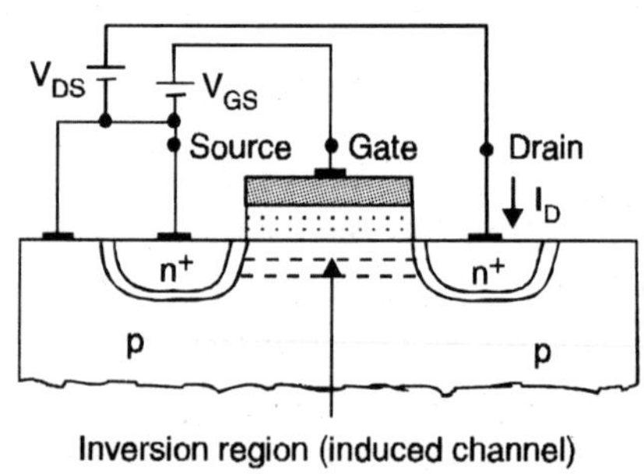

(d) $V_{GS} > V_T$, $V_{DS} = 0$ (Ohmic)

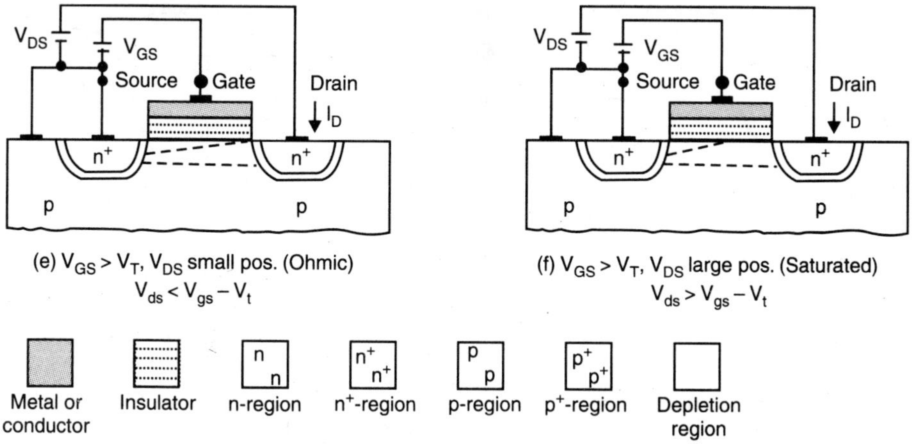

Fig. 1.2.6. Conduction characteristic of NMOS transistor.

(i) **Cut-off Region:** When $V_{ds} = 0$ and $V_{gs} < Vt$, fig. (c)

$I_D = 0$ *i.e.*, where the current flow is essentially zero (accumulation region)

(ii) **Linear Region:** (Resistive, non-saturated (or) unsaturated region), fig. (e)

$$V_{ds} < V_{gs} - V_t; \qquad I_D \text{ is increased}$$

Where $(V_{gs} - V_t)$ is the objective gate voltage, and the drain current is dependent on the gate and the drain voltage (inversion layer).

(iii) **Saturation Region:** $V_{ds} > V_{gs} - V_t$; $\qquad I_D = \text{constant}$

Under this condition the channel becomes pinched off, channel no longer reaches the drain. The Pinch-off voltage is the voltage at which the current becomes constant.

For fixed drain-to-source voltage and fixed gate voltage, the factors that influence the level of drain current, I_{ds}, flowing between source and drain are

- the distance between width
- the channel width
- the threshold voltage V_t
- the thickness of the gate insulation oxide layer
- the dielectric constant of the gate insulator
- the carrier (electron or hole) mobility, μ

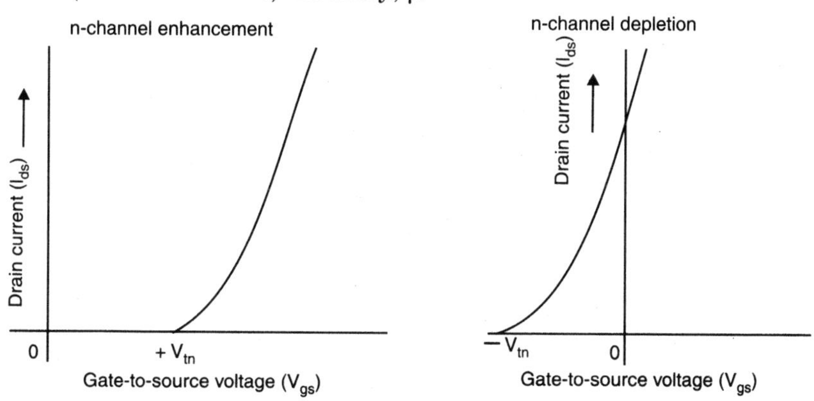

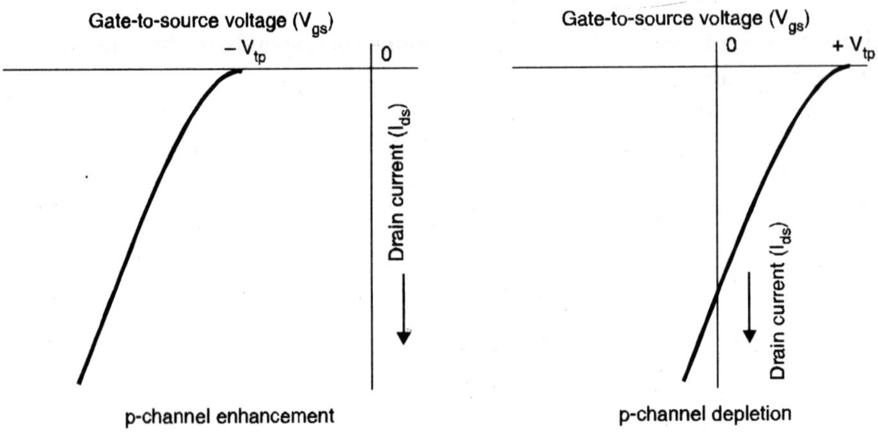

Fig. 1.2.7. VI Characteristics [I/P Characteristics] for enhancement and depletion mode MOS Transistors.

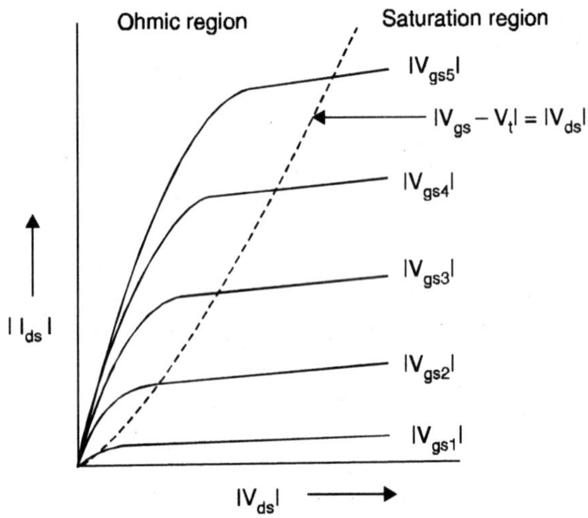

Fig. 1.2.8. Output characteristics of NMOS.

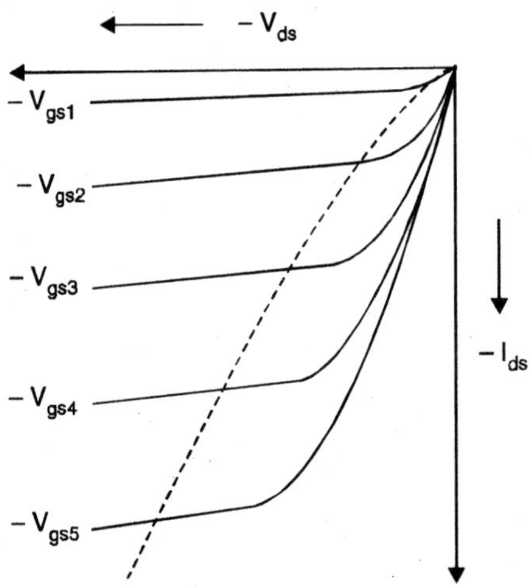

Fig. 1.2.9. Output characteristics of PMOS.

The boundary between linear and saturation regions corresponds to the condition $|V_{ds}| = |V_{gs} - V_t|$ and appears as dashed line.

1.3. THRESHOLD VOLTAGE (V_t)

The gate source voltage for which concentration of electrons under the gate is equal to the concentration of holes in the p-substrate is referred as threshold voltage or (The voltage at which the MOS device begins to conduct).

Threshold Voltage parameters : In general, the threshold voltage is a function of a number of parameters as following.

- Gate conductor material
- Gate insulation material
- Gate insulator thickness channel doping
- Impurities at the silicon insulator interface (impurities present at interface between insulator and substrate)
- Voltage between the source and the substrate (V_{sb}) (Substrate = bulk)

Threshold Voltage Equation

Threshold voltage, V_t, may be expressed as,

$$V_t = V_{t-mos} + V_{fb}$$

Where, V_{t-mos} is the ideal threshold voltage of an ideal MOS transistor. It is the threshold where there is no work function difference between the gate and substrate materials.

Work Function : When anything impinges on the photoelectric material, electrons will be emitted. To come out, energy is needed and some work should be done *i.e.*, called work function.

V_{fb} – flat-band voltage

(A flat band voltage V_{fb} that represents the built-in voltage offset across the MOS **structure**)

$$V_{t-mos} = 2\phi_b + \frac{Q_b}{C_{ox}}$$

ϕ_b → bulk potential
Q_b → bulk charge
C_{ox} → gate to substrate capacitance

$$\phi_b = \frac{kT}{q} \ln\left(\frac{N_A}{N_i}\right)$$

k → boltzmann's constant
T → absolute temperature
q → charge of electron
N_A → density of carriers in doped substrate
N_i → density of carriers in undoped substrate

$$Q_b = \sqrt{2qN_A . 2\phi_b \varepsilon_{si}} \quad \text{(Bulk charge potential)}$$

$\varepsilon_{si} \rightarrow$ permitivity of silicon subtrate

$$V_{fb} = \phi_{ms} - \frac{Q_{fc}}{C_{ox}}$$

$\phi_{ms} \rightarrow$ difference of work function between gate and substrate.

(Permitivity $\rightarrow$ a quantity measuring the ability of a substance to store electrical energy in an electric field)

$$\phi_{ms} = \phi_{gate} - \phi_{substrate}$$

ϕ_{fc} = fixed surface charge between insulating material and substrate.

C_{ox} = gate to substrate capacitance.

$$V_t = 2\phi_b + \frac{Q_b}{c_{ox}} + \phi_{ms} - \frac{Q_{fc}}{C_{ox}}$$

Controlling of V_T : Threshold voltage can be varied by the following factors.

1. Varying the impurities present at the interface between insulator and substrate, it will affect Q_{fc}.

2. By changing the insulating material. Instead of SiO_2, we can use SiN_4. It increases the dielectric constant (for SiN_4 = 6 is the dielectric constant) which in turn will affect C_{ox}.

The changes in Q_{fc} and C_{ox} will change the value of V_{fb}. If V_{fb} changes then V_t changes.

1.4. BODY EFFECT

All devices comprising MOS device are made on a common substrate. As a result, the substrate voltage of all devices is normally equal. In arranging the devices to form gating functions it might be necessary to connect several devices in series. The Fig. 1.4.1 shows the effect of substrate bias on series connected *n*-transistors.

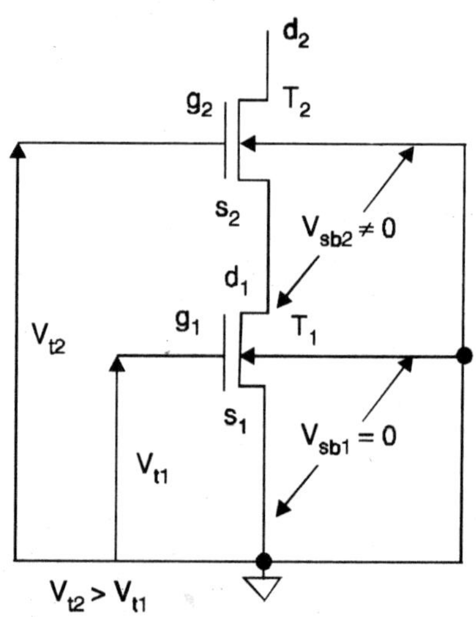

Fig. 1.4.1. The effect of substrate bias on series connected n-transistors.

This may result in an increase in source to substrate voltage as we proceed vertically along the series chain ($V_{sb1} = 0$, $V_{sb2} \neq 0$) under normal conditions, *i.e.*, when $V_{gs} > V_t$ the depletion layer width remains constant and charge carriers are pulled into the channel from the source. However, as the substrate bias V_{sb} (Vsource-Vsub) is increased the width of the channel-substrate depletion layer also increases resulting in an increase in the density of the trapped carriers in the depletion layer. For charge neutrality to hold, the channel charge must decrease. The resultant effect is that the substrate voltage, V_{sb}, adds to the channel substrate in potential. This increases the gate channel voltage drop. The overall effect is an increase in the threshold voltage, V_t ($V_{t2} > V_{t1}$). This effect is called body effect.

1.5. MOS DEVICE DESIGN EQUATIONS

1.5.1. First Order Effects

MOS transistors have 3 regions of operations. (1) Cut off (or) Sub threshold region. (2) Non-saturation (or) Linear region (or) Ohmic region and (3) Saturation region. Behaviour of an NMOS-Device in three regions are as follows :

1. Cut-off Region :

$$V_{gs} < V_t; \qquad I_{ds} = 0$$

2. Linear (or) Unsaturated Region :

$$0 < V_{ds} < V_{gs} - V_t$$

$$I_{ds} = \beta \left[(V_{gs} - V_t) V_{ds} - \frac{V^2_{ds}}{2} \right]$$

$\beta \rightarrow$ gain factor of MOSFET

If $\quad V_{ds} << (V_{gs} - V_t)$, $\frac{V^2_{ds}}{2}$ will be very small,

then, $\quad I_{ds} = \beta\,[(V_{gs} - V_t)\,V_{ds}]$, I_{ds} is linearly varied with V_{gs} and V_{ds}

3. Saturated Region :

$$V_{ds} > V_{gs} - V_t > 0$$

$$I_{ds} = \frac{\beta(V_{gs} - V_t)^2}{2}$$

$$\beta = \frac{\mu \in_{ox}}{t_{ox}} \left(\frac{W}{L}\right)$$

$$= \mu\, C_{ox} \left(\frac{W}{L}\right) \left[\text{because } \frac{\in_{ox}}{t_{ox}} = C_{ox}\right]$$

$\mu \rightarrow$ mobility of the carrier

$W \rightarrow$ width of the channel

$L \rightarrow$ length of the channel
$t_{ox} \rightarrow$ thickness of the oxide layer (insulator)
$\varepsilon_{ox} \rightarrow$ permittivity of the oxide layer.

1.5.2. Second Order Effects

1. Threshold Voltage–Body Effect: The threshold voltage V_t is not constant with respect to the voltage difference between the substrate and the source of the MOS transistor. This is known as the substrate-bias effect (or) Body Effect. The expression for the threshold voltage will be modified when we incorporate V_{sb} (the difference between the source and substrate voltage).

We know $$V_t = V_{tMOS} + V_{fb}$$

$$V_t = V_{fb} + 2\phi_b + \frac{Q_b}{C_{ox}}$$

$$V_t = V_{fb} + 2\phi_b + \frac{\sqrt{2\varepsilon_{si}\, qN_A(2\phi_b + |V_{sb}|)}}{C_{ox}}$$

$$V_t = V_{to} + \gamma\left(\sqrt{2\phi_b + |V_{sb}|} - \sqrt{2\phi_b}\right)$$

where V_{to} = Threshold voltage for $V_{sb} = 0$
γ = Constant that describes the substrate bias effect.

$$\gamma = \frac{\sqrt{2\varepsilon_{si}\, qN_A}}{C_{ox}}$$

$$C_{ox} = \frac{\varepsilon_{ox}}{t_{ox}} \text{ keeping } \phi_b = \frac{KT}{q}\ln\left(\frac{N_A}{N_i}\right)$$

$N_A \rightarrow$ density of carriers in doped substrate
$N_i \rightarrow$ density of carriers in undoped substrate
$q \rightarrow$ charge of an electron
$\varepsilon_{ox} \rightarrow$ dielectric constant of silicon dioxide
$t_{ox} \rightarrow$ thickness of the oxide layer

2. Subthreshold Region: Cut-Off Region:

$$V_{gs} \le V_t; \quad I_{ds} = 0 \qquad \text{...(A)}$$

The cut-off region described by equation (A) is also referred to as the sub-threshold region, where I_{ds} increases exponentially with V_{ds} and V_{gs} i.e., under this 2nd order design, $I_{ds} \neq 0$ and increases exponentially with V_{ds} and V_{gs}. Although I_{ds} is very small, ($I_{ds} \simeq 0$); it may be used to construct low power circuits.

3. Channel Length Modulation : Simplified equations that describe MOS devices assumes that carrier mobility is constant and do not take into account the variations of channel length due to change in V_{ds}. If V_{ds} changes, channel length will be varied linearly. But as devices are scaled down, this variation should be taken into account. When a MOS device is in saturation the effective channel length actually is decreased such that

$$L_{eff} = L - L_{short}$$

where
$$L_{short} = \sqrt{\frac{2 \in_{si} [V_{ds} - (V_{gs} - V_t)]}{qN_A}}$$

The reduction in channel length increases the (W/L) ratio, there by increasing β as the drain voltage increases. Thus rather than appearing as a constant current source with infinite output impedance, the MOS device has an finite output impedance. Considering this behaviour, I_{ds} is represented as:

$$I_{ds} = \beta \frac{(V_{gs} - V_t)^2}{2} (1 + \lambda V_{ds})$$

λ → Channel length modulation factor

N_A → Carrier density

β → gain factor

$$\beta = \frac{\mu\varepsilon}{t_{ox}} \ln\left(\frac{W}{L}\right)$$

μ → effective surface mobility of the carriers in the channel

ε → permittivity of the gate insulator

t_{ox} → thickness of the gate insulator

W → width of the channel

L → length of the channel

4. Fowler-Nordheim Tunnelling : When the gate oxide is very thin, a current can flow from gate to source or drain. This current is proportional to the area of the gate of the transistor as follows.

$$I = C\, WLE^2_{ox}\, e^{-\frac{E_0}{E_{ox}}}$$

where ;
$$E_{ox} = \left[\frac{V_{gs}}{t_{ox}}\right]$$

E_{ox} is the electric field across the gate oxide. E_0 and C are constants.

5. Mobility Variations : The flow of carriers on the substrate depends on mobility factor (μ).

$$\mu = \frac{\text{Average carrier drift velocity (V)}}{\text{Electric field intensity (E)}}$$

Mobility varies according to the type of charge carrier. Electrons in silicon have a much higher mobility than holes, resulting in n-devices having higher current producing capability than the corresponding p-device. Mobility decreases with increasing doping concentration and increasing temperature.

6. Drain Punch Through Effect : When the drain is at a high voltage with respect to the source, the depletion region around the drain may extend to the source, thus causing current to flow irrespective of the gate voltage (*i.e.*, even if it is zero). This is known as punch through effect.

7. Impact Ionization : As the length of the gate of an MOS-transistor is reduced the electric field at the drain of a transistor in saturation increase. For submicron gate lengths (<1μ), the field can become so high. So that electrons in the channel gets energized more which are termed as "hot" electrons. These hot electrons strike out the (impact) drain and so the holes in the drain are swept away towards the negatively charged substrate (channel) and which appear as a substrate current (*i.e.*, movements of holes from drain to substrate). This effect is known as "impact ionization". Moreover, the electrons can penetrate the gate oxide, causing a gate current. This effect can lead to degradation of the MOS device parameters (threshold voltage, sub threshold voltage...) which in turn can lead to the failure of circuits.

1.5.3. Small Signal AC-Characteristics

The MOS transistor can be represented by the simplified small signal equivalent model as shown below.

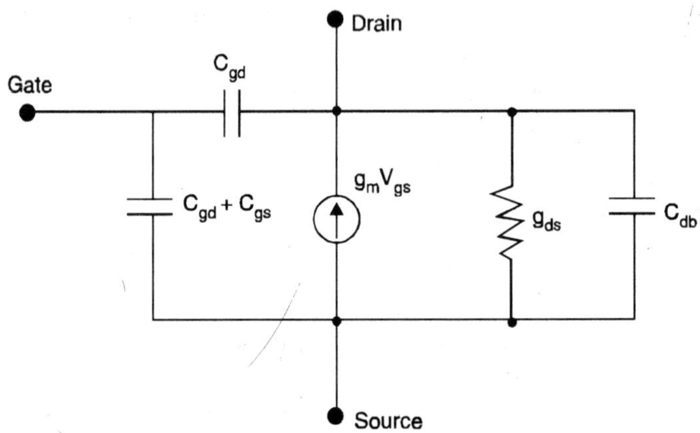

Fig. 1.5.3. Small signal model of a MOS transistor.

From Fig. 1.5.3

C_{gd} → Capacitance effect between gate and drain

C_{gb} → Capacitance between gate and bulk

g_m → trans conductance

g_{ds} → conductance

C_{db} → Capacitance between drain and bulk.

The above equivalent circuit is modeled with current source (g_m), an output conductance (g_{ds}) and interelectrode capacitances.

To find g_{ds} : The output conductance (g_{ds}) in the linear region can be obtained as follows.

$$r_{ds} = V_{ds} / I_{ds} \quad \text{and} \quad g = 1/r \Rightarrow g_{ds} = I_{ds} / V_{ds}$$

$$\therefore \quad g_{ds} = dI_{ds} / dV_{ds}$$

I_{ds} in linear region of 1st order relation is given by,

$$I_{ds} = \beta \left[(V_{gs} - V_t) V_{ds} - \frac{V_{ds}^2}{2} \right]$$

$$\frac{dI_{ds}}{dV_{ds}} = \beta \left[(V_{gs} - V_t) 1 - 2 V_{ds} /2\right]$$

$$= \beta \left[(V_{gs} - V_t) - V_{ds}\right]$$

In linear region, V_{ds} must be small compared to V_{gs} and so, V_{ds} is assumed as zero.

$$\frac{dI_{ds}}{dV_{ds}} \approx \beta \left[V_{gs} - V_t\right] \text{ (since } V_{ds} = 0)$$

I_{ds} in saturation region from 1st order relation is given by,

$$I_{ds} = \beta \left[V_{gs} - V_t\right]^2/2$$

$$g_{ds} = \frac{dI_{ds}}{dV_{ds}} = 0 \quad [\text{as } I_{ds} \text{ is independent of } V_{ds}]$$

3. The trans conductance g_m expresses the relationship between output current I_{ds} and input voltage V_{gs} and is defined by,

$$g_m = \frac{dI_{ds}}{dV_{gs}} \quad \text{keeping } V_{ds} = \text{constant}$$

g_m in Linear Region:

$$I_{ds} = \beta [(V_{gs} - V_t)V_{ds} - V_{ds}{}^2/2]$$

$$= \beta [V_{gs} g V_{ds} - V_t g V_{ds} - V_{ds}{}^2/2]$$

$$\frac{dI_{ds}}{dV_{gs}} = \beta [1 g V_{ds} - 0] = \beta g V_{ds}$$

$$\boxed{\therefore g_m = \beta [V_{ds}]}$$

g_m in Saturation Region:

$$I_{ds} = \beta (V_{gs} - V_t)^2/2 \quad g_m = \frac{dI_{ds}}{dV_{gs}} = \beta\, 2(V_{gs} - V_t)/2$$

$$\boxed{\therefore g_m = \beta (V_{gs} - V_t)}$$

1.6. ELECTRICAL PROPERTIES OF MOS

If the design is to be effectively carried out, proper understanding of MOS active devices is essential. Hence the basic characteristics of MOS transistor is explained.

1.6 (a) Drain-to-source Current (I_{ds}) versus Drain to Source Voltage (V_{ds}) relationship

Voltage on the gate induce a charge in the channel between source and drain. These charges will be caused to move from source to drain by voltage V_{ds} applied between drain and source. Since the charge is dependent on V_{gs}, I_{ds} is dependent on both V_{gs} and V_{ds}.

Consider a structure in which electrons will flow from source to drain

$I_{ds} = I_{sd}$ = Charge induced in channel (Qc) / Electron transit time (τ)

Transist time (τ_{sd}) = Length of channel (L) / Velocity (v)

But Velocity $v = \mu E_{ds}$

μ = electron or hole mobility (surface)

E_{ds} = electric field (drain to source)

$E_{ds} = V_{ds} / L$ so that $v = \mu V_{ds} / L$

Thus $\tau_{sd} = L^2/\mu\, V_{ds}$

1.6 (a) *(i) The non-saturated region [$V_{gs}-V_t>V_{ds}$]*

Charge induced in channel due to gate voltage is due to the voltage difference between the gate and the channel V_{gs} (Substrate Connected to Source). Voltage along the channel varies linearly with distance X from the source and assuming that the device is not saturated then the average value is $V_{ds}/2$. Further, the effective gate voltage $V_g = V_{gs} - V_t$ is the voltage needed to invert the charge under the gate and establish the channel.

Charge/unit area $= E_g\ \varepsilon_{ins}\ \varepsilon_0$.

Thus Induced charge $Q_c = E_g\ \varepsilon_{ins}\ \varepsilon_0\ WL$

where

E_g – average electric field gate to channel.

ε_{ins} – relative permittivity of insulation between gate and channel.

ε_0 – permittivity of free space.

Now $E_g = [(V_{gs} - V_t) - V_{ds}/2)/D]$

D = oxide thickness

Thus
$$Q_c = \frac{WL\varepsilon_{ins}\ \varepsilon_0}{D}\left[(V_{gs} - V_t) - \frac{V_{ds}}{2}\right]$$

$$I_{ds} = Q_c/\tau \qquad \tau_{sd} = L^2/\mu V_{ds}$$

Now
$$I_{ds} = \frac{\mu\varepsilon_{ins}\ \varepsilon_0}{D}\left[(V_{gs} - V_t) - \frac{V_{ds}}{2}\right].V_{ds}\frac{W}{L}$$

$$I_{ds} = \frac{KW}{L}\left[(V_{gs} - V_t) - \frac{V_{ds}}{2}\right]V_{ds}$$

where

$$K = \varepsilon_{ins}\, \varepsilon_0 \mu / D \qquad \text{and Let} \qquad \beta = \frac{KW}{L}$$

So that $$I_{ds} = \beta \left[(V_{gs} - V_t)\, V_{ds} - \frac{V_{ds}^2}{2} \right]$$

We know gate/channel capacitance

$$C_g = \text{(gate/channel capacitance)} = \varepsilon_{ins}\, \varepsilon_0 \frac{WL}{D} \qquad \text{also} \quad K = C_g\, \mu / WL$$

So that $$I_{ds} = \frac{C_g \mu}{L^2} \left[(V_{gs} - V_t)\, V_{ds} - \frac{V_{ds}^2}{2} \right]$$

But $C_g = C_o.WL$ [Sometimes it is convenient to use $C_o = C_{ox}$ (gate cap/unit area)

$$I_{ds} = \frac{C_o \mu W}{L} \left[(V_{gs} - V_t)\, V_{ds} - \frac{V_{ds}^2}{2} \right]$$

(ii) The saturated region : $V_{ds} > V_{gs} - V_t$

At $V_{ds} = V_{gs} - V_t$ IR drop in the channel equals the effective gate to channel Voltage at the drain and we may assume that the current remains fairly constant as V_{ds} increases, further

$$I_{ds} = \frac{KW}{L} \frac{(V_{gs} - V_t)^2}{2}$$

(or) $$I_{ds} = \frac{\beta}{2} (V_{gs} - V_t)^2 = \frac{Cg.\mu}{2L^2} (V_{gs} - V_t)^2 = \frac{C_o \mu W}{2L} (V_{gs} - V_t)^2$$

The expressions for I_{ds} hold for both enhancement and depletion mode devices, (But for the dep.mode devices V_t is –ve for nMOS).

1.6 (b) Threshold V_t

$$V_t = \phi_{ms} \frac{Q_B - Q_{ss}}{C_o} + 2\phi_{FN}$$

Q_B – Charge/unit area in the depletion layer beneath the oxide
Q_{ss} – Charge density at Si–SiO_2 interface } bulk charge
C_o – Cap./unit gate area
ϕ_{ms} – Work function difference between gate and silicon.
ϕ_{FN} – Fermi level potential between inverted surface and bulk Si

$$Q_B = \sqrt{2\varepsilon_o \varepsilon_{si}\, qN\, (2\phi_{FN} + V_{sb})}\ \frac{\text{coulomb}}{m^2}$$

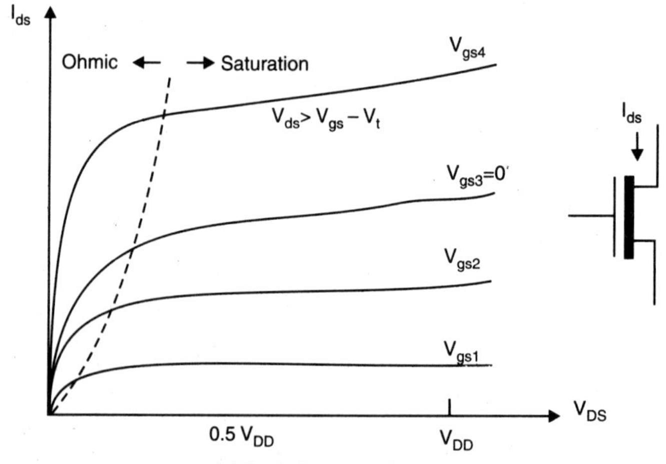

(a) Depletion mode device

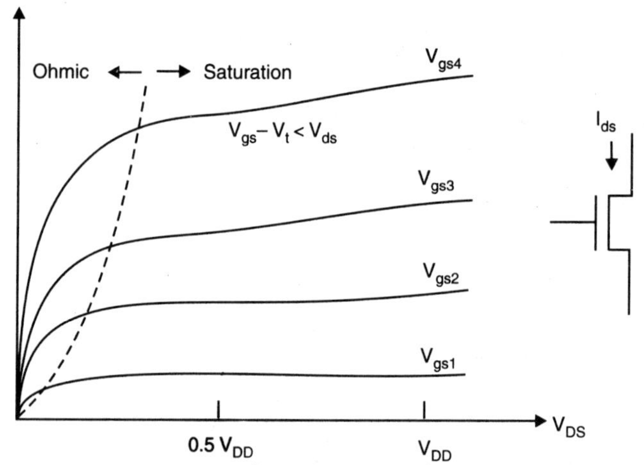

(b) Enhancement mode device

Fig. 1.6 (b) MOS transistor characteristics

$$\phi_{FN} = \frac{KT}{q} \ln\left(\frac{N}{N_i}\right) \text{volts}$$

V_{SB} = Substrate bias voltage (–ve w.r.t. source for n Mos)

N = Impurity concentration in the substrate

ε_{Si} = Relative permittivity of S_i = 11.7

n_i = Intrinsic electron concentration ($1.6 \times 10^{10}/cm^3$ @ 300 K)

K = Boltzmann's constant = 1.4×10^{-23} Joule/K

The body effects may also be taken into account. Increasing V_{SB} causes the channel to be depleted of charge carriers and thus V_t is raised.

Change in V_t is given by $\Delta vt = \delta (V_{SB})^{1/2}$ where δ (constant which depends on substrate **doping**) when substrate is lightly doped, body effect will be smaller.

$$V_t = V_t(0) + \frac{D}{\varepsilon_{ins}\varepsilon_0}\sqrt{2\varepsilon_0\varepsilon_{si}QN}\ (V_{SB})^{1/2}$$

Where $V_t(0)$ is the threshold voltage for $V_{SB} = 0$

1.6. (*c*) Transconductance (g_m) and O/P Conductance (g_{ds})

$g_m = (\delta I_{ds} / \delta V_{gs})$ keeping V_{ds} a constant

To find gm in terms of circuit and transistor parameter :

$Qc/I_{ds} = \tau$ (τ – Transit time)

Thus change incurrent

$\delta I_{ds} = \delta Qc/\tau_{ds}$ where $\tau_{ds} = L^2/\mu V_{ds}$

$$\delta I_{ds} = \frac{\delta Qc\, V_{ds}\mu}{L^2}$$

but change in charge $\delta Qc = C_g\, \delta V_{gs}$

So that $\delta I_{ds} = \dfrac{C_g \delta V_{gs}\, \mu V_{ds}}{L^2}$ and $g_m = \dfrac{C_g \mu V_{ds}}{L^2}$

In saturation $V_{ds} = V_{gs} - V_t$

$$g_m = \frac{C_g\mu}{L^2}(V_{gs} - V_t) \qquad C_g = \frac{\varepsilon_{ins}\varepsilon_0 WL}{D}$$

$$g_m = \frac{\mu\varepsilon_{ins}\varepsilon_0}{D}\frac{W}{L}(V_{gs} - V_t) \quad \text{or} \quad g_m = \beta(V_{gs} - V_t)$$

g_m can be increased by increasing **width** of MOS device. **But this will** also increase i/p capacitance and area occupied. If L decreases, W_o increases owing to higher g_m. But gain of Mos device decreases due to degradation of o/p resistance ($1/g_{ds}$)

$$g_{ds} = \frac{\delta I_{ds}}{\delta V_{gs}} - \lambda I_{ds}\ \alpha \left(\frac{1}{L}\right)^2$$

$\lambda \propto 1/L$ and $I_{ds} \propto 1/L$ ($\propto$ – denotes proportionality)

1.6 (*d*) Figure of Merit – ω_o

Frequency response may be obtained from the parameter ω_o

$$\omega_o = \frac{g_m}{c_g} = \frac{\mu}{L^2}(V_{gs} - V_t) = 1/\tau_{sd}$$

From the above, we understand that switching speed depends on gate voltage above threshold and on carrier mobility and inversely as the square of channel length. A fast circuit requires g_m be as high as possible.

1.6 (*e*) The Pass Transistors

Mos transistors can be used as switches in series with lines carrying logic levels in a way that is similar to the use of relay contacts. This application of the MOS transistor is called pass transistor. Pass transistor is one form of logic that is **popular in NMOS rich circuits.**

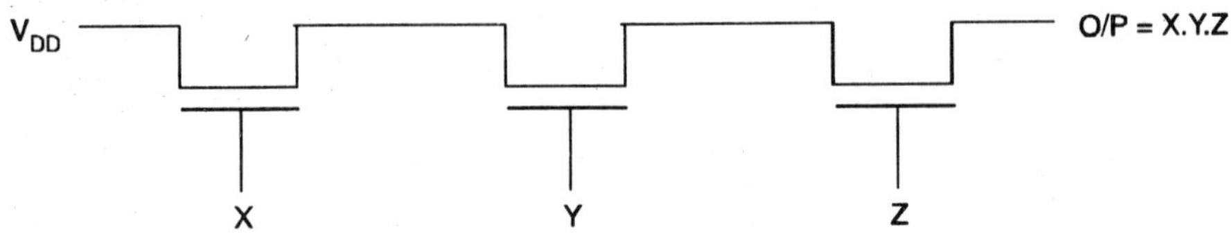

Fig. 1.6 (*e*) Pass transistor and gate.

1.6. (*f*) NMOS Inverter

The basic inverter circuit requires a transistor with source connected to ground and a load resistor of some sort connected from the drain to the positive supply rail V_{DD}. The O/P is taken from drain and the i/p is applied between gate and ground. But resistors are not conveniently produced on silicon substrate, therefore, depletion mode transistor used as the load as in Fig. 1.6 (*f*) (*i*). With no current drawn from the O/P, the currents I_{ds} for both transistors must be equal.

For depletion mode transistors the gate is connected to source and so it is always on. (V_{gs}= 0) In this configuration for Upper one PU (Pull Up) depletion mode devices are used and for Lower one PD (Pull Down) enhancement mode devices are used.

To obtain the inverter transfer characteristics we superimpose the V_{gs}= 0 depletion mode characteristic curve on the family of curves for the enhancement mode device, noting that maximum voltage across the enhancement mode device corresponds to minimum Voltage across the depletion mode transistors.

The point of intersection of the curves as in Fig. 1.6(*f*) (ii) below give points on the transfer characteristics which is of the form shown in Fig. 1.6(*f*) (*iii*).

As V_{gs} exceeds the V_t of P.D transistor current begins to flow. Thus $V_{out}\downarrow$ and as $V_{gs}\uparrow$. This will cause P.D to come out of saturation and becomes resistive. If V_{in} = '0', V_{out} = max. if V_{in} increases, V_{out} decreases, i.e. V_{ds} decreases. So P.D will come out of saturation and become resistive. Note that P.U transistors is initially resistive as the P.D turns on. During transition, the slope of the transfer characteristic determines the gain.

$$\text{Gain} = \delta V_{out}/\delta V_{in}$$

The point at which $V_{out} = V_{in}$ is denoted as V_{inv}. The transfer characteristic and V_{inv} can be shifted by variation of the ratio of Pu to Pd resistance and (Z_{pu}/Z_{pd}). Z is determined by the length to width ratio of the transistor.

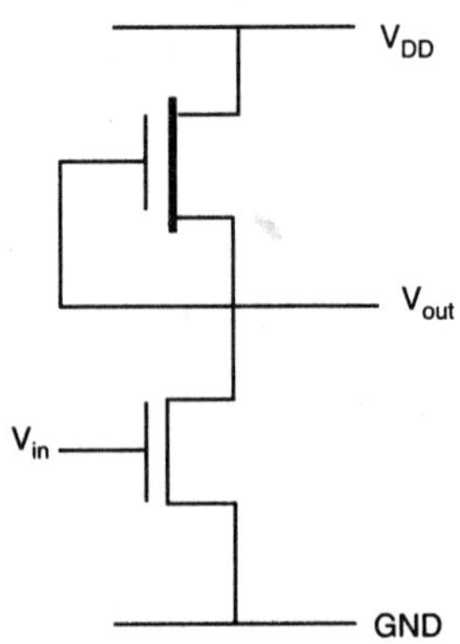

Fig. 1.6 (*f*) (*i*) NMOS inverter.

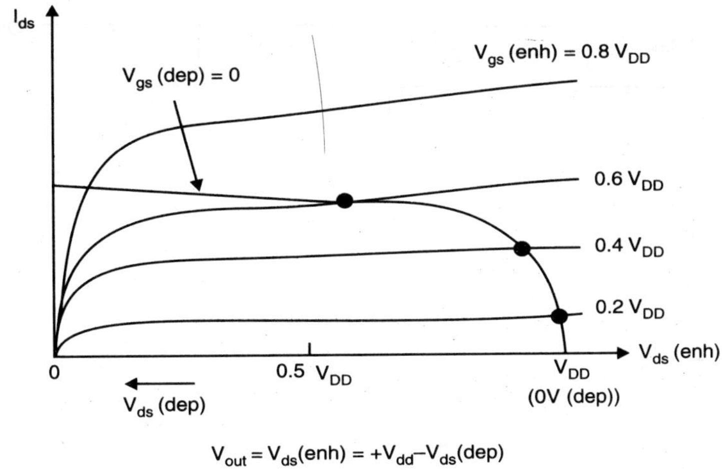

Fig. 1.6 (*f*) (*ii*) Intersection of depletion and enhancement mode characteristics.

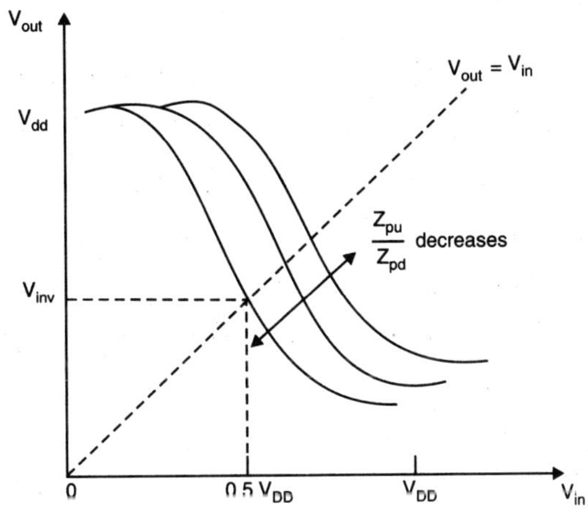

Fig. 1.6 (*f*) (*iii*) NMOS inverter transistor characteristics

1.6. (*g*) Determination of (Z_{PU}/Z_{PD}) for an NMOS Inverter Driven by Another NMOS inverter

Consider the depletion mode transistor for which $V_{gs} = 0$ under all conditions and further in order to cascade inverters without degradation of levels. We are aiming to meet the requirement.

$$V_{in} = V_{out} = V_{inv}$$

We get $\qquad V_{inv} = 0.5V_{DD}$.

At this point both transistors are in saturation and

$$I_{ds} = K\frac{W}{L}\frac{(V_{gs} - V_t)^2}{2}$$

In depletion mode $\quad I_{ds} = K\frac{W_{pu}}{L_{pu}}\frac{(-V_{td})^2}{2}$; since $V_{gs} = 0$

In enhancement mode $\quad I_{ds} = K\frac{W_{pd}}{L_{pd}}\frac{(V_{gs} - V_t)^2}{2}$; since $V_{gs} = V_{inv}$

Equating $\quad \frac{W_{pd}}{L_{pd}}(V_{inv} - V_t)^2 = \frac{W_{pu}}{L_{pu}}(-V_{td})^2$

Let $\quad Z_{pd} = \frac{L_{pd}}{W_{pd}}; \qquad Z_{p.u} = L_{p.u}/W_{p.u}$

We have $\quad \frac{1}{Z_{pd}}(V_{inv} - V_t)^2 = (-V_{td})^2\frac{1}{Z_{pu}}$

Hence $\quad V_{inv} = V_t \frac{V_{td}}{\sqrt{Zp.u/Zp.d}}$

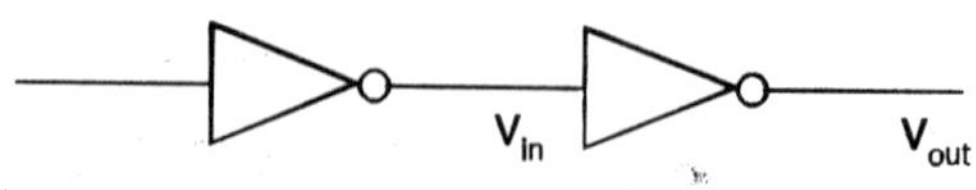

Fig. 1.6 (*g*) NMOS inverter driven directly by another inverter.

1.6. (*h*) Alternative Forms of Pull-up

So far we have assumed depletion mode pull-up transistor as load for inverter circuit. The other forms are :

1. Load resistance R_L : This is not often used because of the large space requirements of resistors.

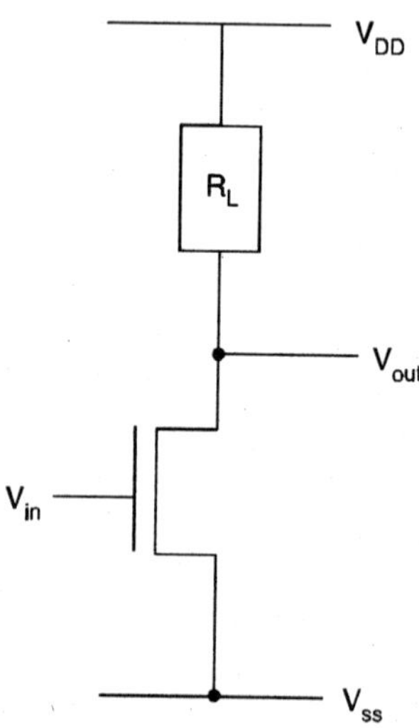

Fig. 1.6 (*h*) (*i*) Resistor pull up.

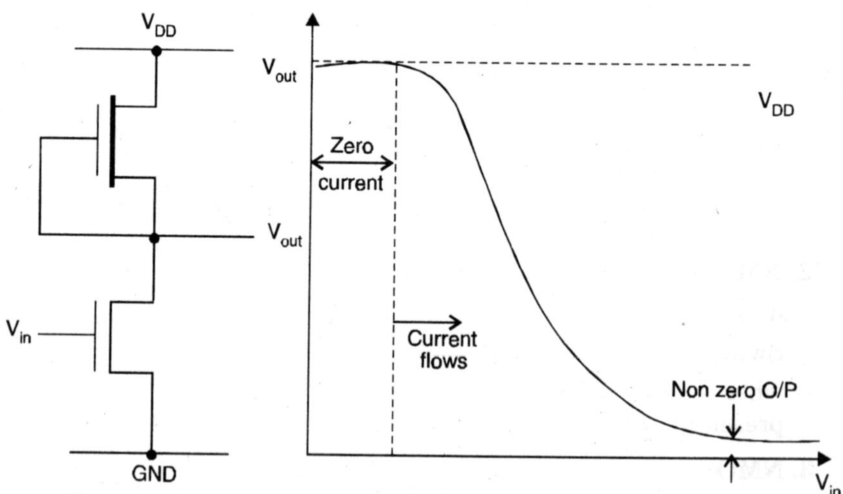

Fig. 1.6 (*h*) (*ii*) NMOS depletion mode transistor pull up transfer characteristic.

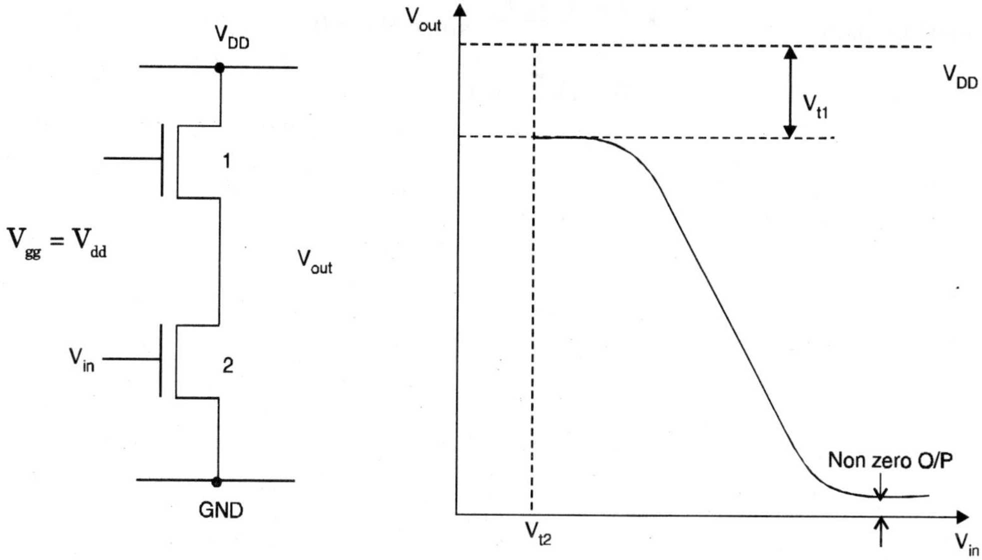

Fig. 1.6 (*h*) (*iii*) NMOS enhancement mode pull up.

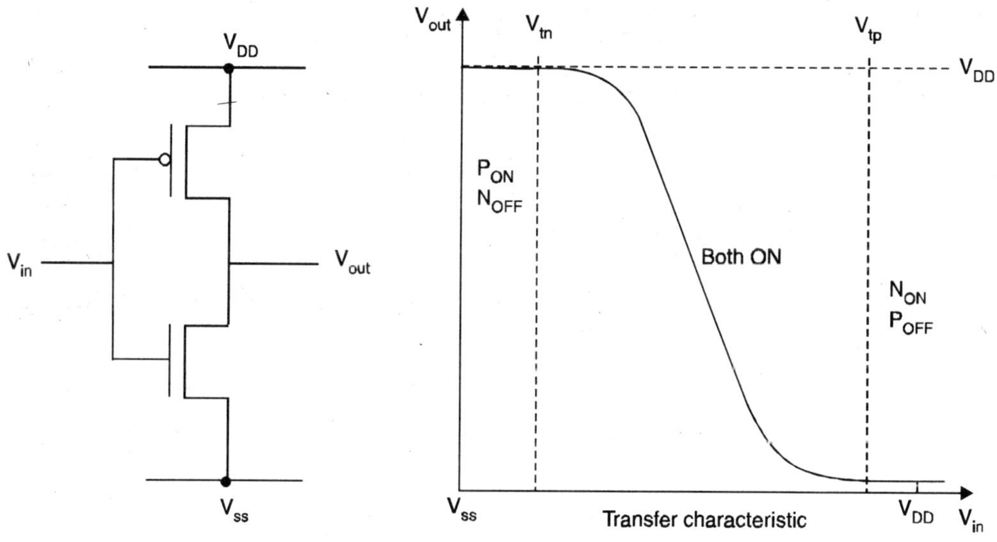

Fig. 1.6 (*h*) (*iv*) Complementary transistor pull up CMOS.

2. NMOS depletion mode transistors pull-up [Fig. 1.6 (*h*) (*ii*)]

- dissipation is high since rail to rail current flows when V_{in} = logical 1.
- Switching of O/P from 1 to 0 begins when V_{in} exceeds V_t of p.d. device.
- when switching the O/P from 1 to 0, the P.U device is non saturated initially and this presents lower resistance thro' which to charge capacitive loads.

3. NMOS enhancement pull up [Fig. 1.6 (*h*) (*iii*)]

- dissipation is high since current flows when V_{inv} = logical 1. (V_{GG} is returned to V_{DD})

- V_{out} can never reach V_{DD} (logical 1) if $V_{GG} = V_{DD}$ as is normally the case.
- V_{GG} may be derived from a switching source, for *e.g.*, one phase of a clock, so that dissipation can be greatly reduced.
- if V_{GG} is higher than V_{DD} then an extra supply rail is required.

4. Complementary Transistor Pull up (CMOS) [Fig. 1.6 (*h*) (*iv*)]

- No current flow either for logical 0 or for logical 1 inputs
- Full logical 1 and 0 levels are presented at the output
- For devices of similar dimensions the p-channel is slower than the n-channel device.

1.7. CMOS INVERTER – DC CHARACTERISTICS

A complementary CMOS inverter is realized by the series connection of *p* and *n* devices as shown in Fig. 1.7.1. The objective is to find the variation in output voltage (V_{out}) for changes in the input voltage (V_{in}).

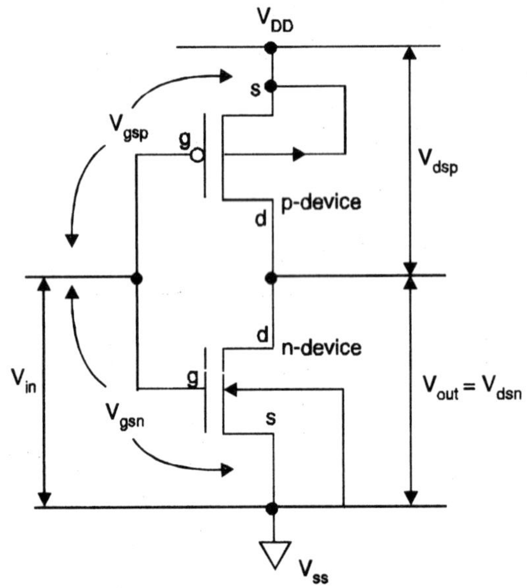

Fig. 1.7.1. A CMOS inverter.

Transfer Characteristics of CMOS Inverter: This is shown in Fig. 1.7.2(*a*). The absolute value of the p-transistor drain current I_{ds} inverts this characteristics. This allows VI-characteristics for the p-device to be reflected about x-axis. This shown in Fig. 1.7.2 (*b*). The above step is followed by taking the absolute value of p-device, V_{ds} and superimposing the two characteristics yielding the resultant curves as shown in Fig. 1.7.2 (*c*). A '•' shows equal current points. By joining equal current points, the inverter characteristics is obtained. The input/output transfer curve may be determined by the points of common V_{gs} intersection. Thus solving for $V_{in\text{-}n} = V_{in\text{-}p}$ and $I_{ds\text{-}n} = I_{ds\text{-}p}$ gives the desired transfer characteristics of a CMOS-inverter. The switching point is designed to be 50% of the magnitude of the supply voltage = $V_{dd}/2$. Then both transistors are momentarily 'ON' resulting in a short pulse of current drawn from the supply. This is shown in dotted line in Fig. 1.7.3.

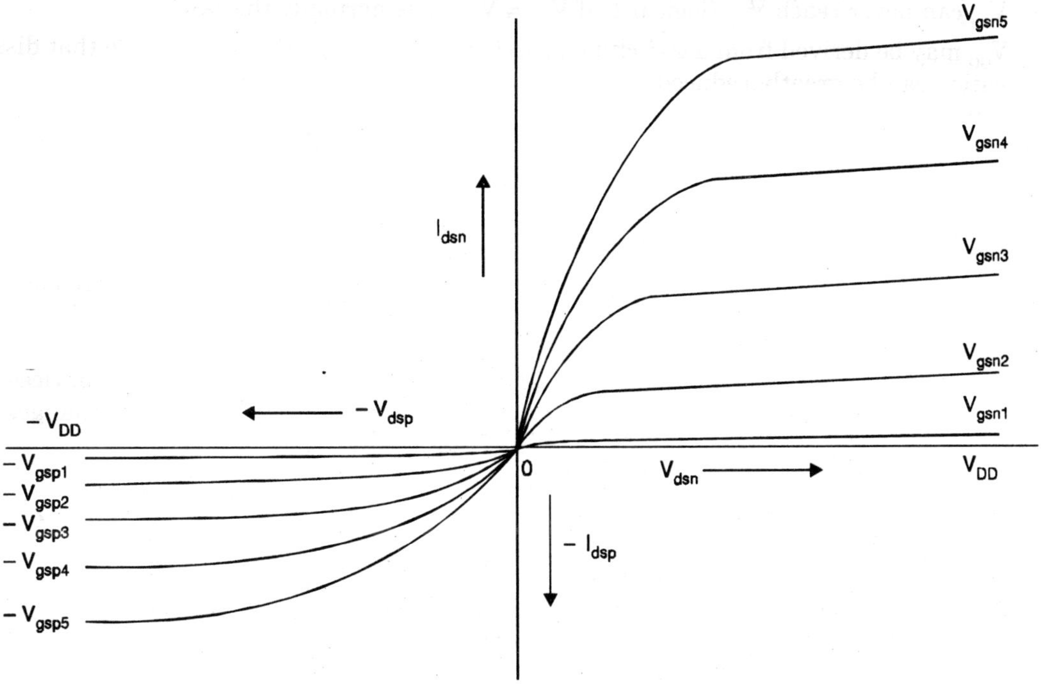

Fig. 1.7.2 (*a*) Output characteristics of CMOS.

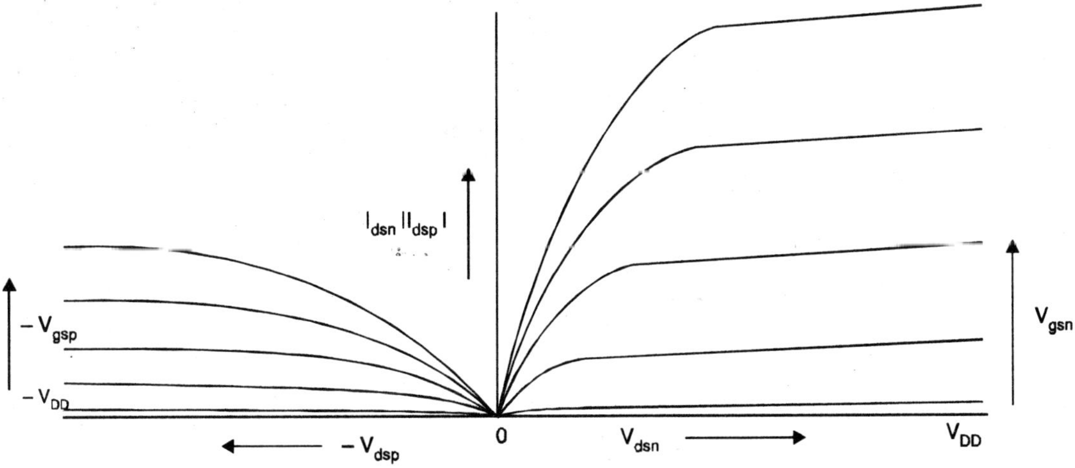

Fig. 1.7.2 (*b*) Output characteristics of p-device reflected about x-axis.

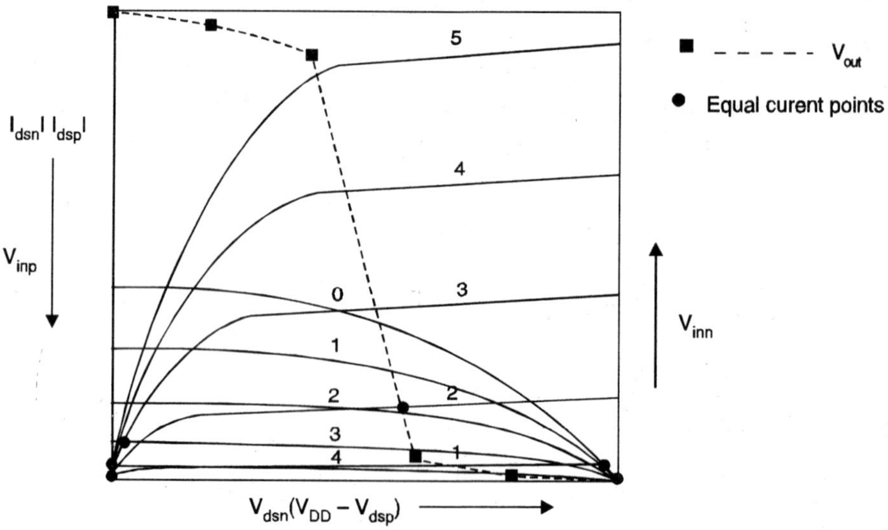

Fig. 1.7.2 (*c*) Super imposition of *p* and *n* device characteristics.

DC –Transfer Characteristics and Operating Region

The Fig. 1.7.3 shows various operating regions. The operation of CMOS-inverter can be divided into five regions.

Region	V_{in}	*P-device*	*n-device*
A	$V_{in} < V_{tn}$	Linear region (unsaturated region)	Cut-off
B	$V_{tn} < V_{in} < V_{DD}/2$	Linear region (unsaturated region)	Saturated
C	$V_{in} = V_{DD}/2$	Saturated	Saturated
D	$V_{DD}/2 < V_{in} < V_{DD} - \lvert V_{tp} \rvert$	Saturated	Linear
E	$V_{in} > V_{DD} - \lvert V_{tp} \rvert$	Cut-off	Linear (unsaturated)

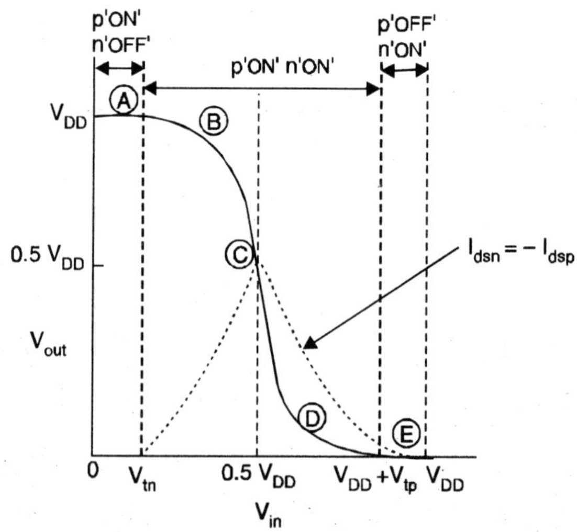

Fig. 1.7.3. CMOS inverter DC transfer characteristic and operating region.

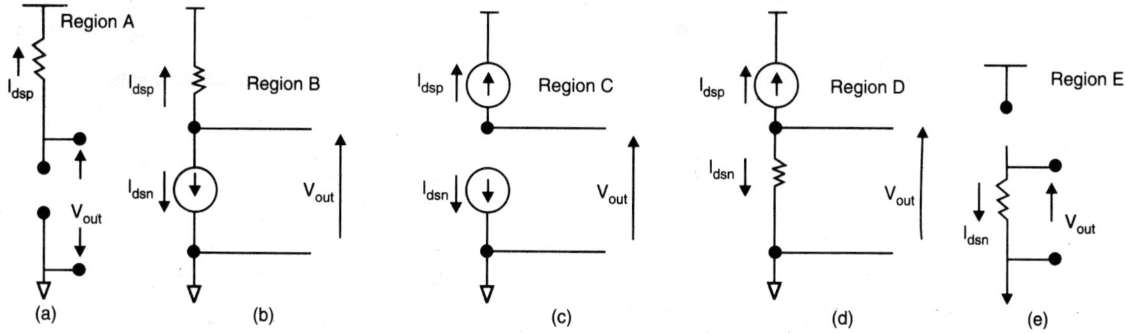

Fig. 1.7.4. Equivalent operating regions of CMOS inverter

In Region-A: $V_{in} < V_{tn}$ (n-device is cut-off and p-device is in linear region)

$I_{dsn} = 0$ (as it is in cut-off) But $I_{dsp} = I_{dsn}$

$I_{dsp} = 0$ (as $I_{dsn} = 0$, I_{dsp} is also zero)

The O/P voltage from inverter Fig. of 1.7.4 is given by

$$V_{out} = V_{DD} - V_{dsp}$$

$V_{dsp} = 0$ (as $I_{dsp} = 0$) and $V_{out} = V_{DD} - 0 = V_{DD}$

$$\therefore \quad V_{out} = V_{DD}$$

In Region B: $V_{tn} \leq V_{in} < V_{DD}/2$ (In this region, P-device is in Linear region and n-device is in saturation region)

$$I_{dsp} = -\beta_p\left[(V_{gsp} - V_{tp})V_{dsp} - V_{dsp}^{\ 2}/2\right] \quad \ldots(1)$$

(For p-device, I_{ds} is –ve so, –ve sign is introduced)

$$I_{dsn} = \beta_n \frac{(V_{gsn} - V_{tn})^2}{2} \quad \ldots(2)$$

We know that $I_{dsp} = -I_{dsn}$ ...(3)

Substituting (1) and (2) in equation (3) we get,

$$-\beta_p\left[(V_{gsp} - V_{tp})V_{dsp} - \frac{V_{dsp}^{\ 2}}{2}\right] = -\beta_n \frac{(V_{gsn} - V_{tn})^2}{2} \quad \ldots(4)$$

From the Fig. 1.7.4.

$$-V_{dsp} = V_{DD} - V_{out} \qquad V_{dsp} = -V_{DD} + V_{out}$$

$$V_{dsp} = V_{out} - V_{DD} \quad \therefore \quad V_{gsn} = V_{in}$$

$$-V_{gsp} = V_{DD} - V_{gsn} \quad \therefore \quad V_{gsp} = V_{gsn} - V_{DD}$$

Substituting above values in equation (4)

$$\beta_p\left[(-V_{DD} + V_{gsn} - V_{tp})\, V_{out} - V_{DD}) - \frac{(V_{out} - V_{DD})^2}{2}\right] = \beta_n \frac{(V_{in} - V_{tn})^2}{2}$$

$$V_{out} = (V_{in} - V_{tp}) + \sqrt{(V_{in} - V_{tp})^2 - 2\left(V_{in} - \frac{V_{DD}}{2} - V_{tp}\right)V_{DD} - \frac{\beta_n}{\beta_p}(V_{in} - V_{tn})^2}$$

The output voltage V_{out} can be expressed as

$$V_{out} = (V_{in} - V_{tp}) + \sqrt{(V_{in} - V_{tp})^2 - 2\left(V_{in} - \frac{V_{DD}}{2} - V_{tp}\right)V_{DD} - \frac{\beta_n}{\beta_p}(V_{in} - V_{tn})^2}$$

where,

$$\beta_p \rightarrow \frac{\mu_p \varepsilon}{t_{ox}} \frac{W_p}{L_p}$$

$V_{tp} \rightarrow$ threshold voltage of p-device

$\mu_p \rightarrow$ mobility of holes

$W_p \rightarrow$ channel width of p-device

$L_p \rightarrow$ channel length of p-device

Region-C:

$$V_{in} = \frac{V_{DD}}{2}; I_{dsp} = -I_{dsn}$$ (In this region, both the n and p-devices are in saturation)

In saturated condition,

$$I_{dsn} = \frac{\beta_n}{2}(V_{gsn} - V_{tn})^2$$

$$I_{dsp} = \frac{\beta_p}{2}(V_{gsp} - V_{tp})^2$$

But $\quad I_{dsp} = -I_{dsn}$

$$\frac{\beta_n}{2}(V_{gsn} - V_{tn})^2 = \frac{\beta_p}{2}(V_{gsp} - V_{tp})^2$$

Sabstituting $V_{gsn} = V_{in}$ and $V_{gsp} = V_{gsn} - V_{DD}$

We get $\quad \beta_n(V_{in} - V_{tn})^2 = \beta_p[V_{gsn} - V_{DD}) - V_{tp}]^2$

Taking roots on both sides, (Take negative roots on RHS)

$$\sqrt{\frac{\beta_n}{\beta_p}}(V_{in} - V_{tn}) = -(V_{in} - V_{DD} - V_{tp})$$

$$\sqrt{\frac{\beta_n}{\beta_p}}V_{in} - \sqrt{\frac{\beta_n}{\beta_p}}V_{tn} = -V_{in} + V_{DD} + V_{tp}$$

$$\sqrt{\frac{\beta_n}{\beta_p}}V_{in} + V_{in} = \sqrt{\frac{\beta_n}{\beta_p}}V_{tn} + V_{DD} + V_{tp}$$

$$V_{in}\left(\sqrt{\frac{\beta_n}{\beta_p}} + 1\right) = \sqrt{\frac{\beta_n}{\beta_p}}V_{tn} + V_{DD} + V_{tp}$$

$$V_{in} = \frac{\sqrt{\frac{\beta_n}{\beta_p}} \cdot V_{tn} + V_{DD} + V_{tp}}{\left(\sqrt{\frac{\beta_n}{\beta_p}} + 1\right)}$$

By setting, $\beta_n = \beta_p$ and $V_{tp} = -V_{tn}$

$$\therefore \quad V_{in} = \frac{1.V_{tn} + V_{DD} - V_{tn}}{1+1} = \frac{V_{DD}}{2}$$

Region-D:

$\frac{V_{DD}}{2} < V_{in} < V_{DD} - V_{tp}$ (In this region, p-device is in saturation and n-device is in Linear region.)

$$I_{dsn} = \beta_n \left[(V_{gsn} - V_{tn})V_{dsn} - \frac{V^2{}_{dsn}}{2} \right]$$

$$I_{dsp} = -\frac{\beta_p}{2}(V_{gsp} - V_{tp})^2$$

But $\quad I_{dsp} = -I_{dsn}$

$$\beta_n \left[V_{gsn} - V_{tn})V_{dsn} - \frac{V^2{}_{dsn}}{2} \right] = \frac{\beta_p}{2}[V_{gsp} - V_{tp}]^2$$

$\therefore \quad V_{gsn} = V_{in}, \; V_{gsp} = V_{gsp} - V_{DD}, \; V_{dsn} = V_{out}, \; V_{dsp} = V_{out} - V_{DD}$

Substituting in the above, we get,

$$\beta_n \left[(V_{in} - V_{tn})V_{out} - \frac{V^2{}_{out}}{2} \right] = \frac{\beta_p}{2}[-V_{DD} + V_{gsn} - V_{tp}]^2$$

$$\beta_n \left[(V_{in} - V_{tn})V_{out} - \frac{V^2{}_{out}}{2} \right] = \frac{\beta_p}{2}[-V_{DD} + V_{in} - V_{tp}]^2$$

By simplifying this, the output voltage in the region-D becomes,

$$V_{out} = (V_{in} - V_{tn}) - \sqrt{(V_{in} - V_{tn})^2 - \frac{\beta_p}{\beta_n}(V_{in} - V_{DD} - V_{tp})^2}$$

Region-E:

$V_{in} > V_{DD} - |V_{tp}|$ In this region, p-device is in cut-off and n-device is in Linear mode.

$$I_{dsp} = -I_{dsn}$$

Since p is in cut-off mode, $I_{dsp} = 0$

As $I_{dsp} = 0$, I_{dsn} is also 0

Now $V_{dsn} = V_{out} = 0$ (Since $I_{dsn} = 0$, $V_{dsn} = 0$)

At this point when β_n

$$\frac{\mu_p W_p}{L_p} = \frac{\mu_n W_n}{L_n}$$

But the mobilities are inherently unequal.

∴ W/L of p. device is to be 2 or 3 times that of n-device.

$$W_p / L_p \approx 2.5\ W_n / L_n$$

However, mobility μ is affected by the transverse electric field in the channel and is thus dependent on V_{gs}. So the actual mobility is

$$\mu = \mu_z (1 - \phi (V_{gs} - V_t))^{-1}$$

Where ϕ is a constant = 0.05. V_t – includes body effect and μ_z – Mobility with zero transverse field.

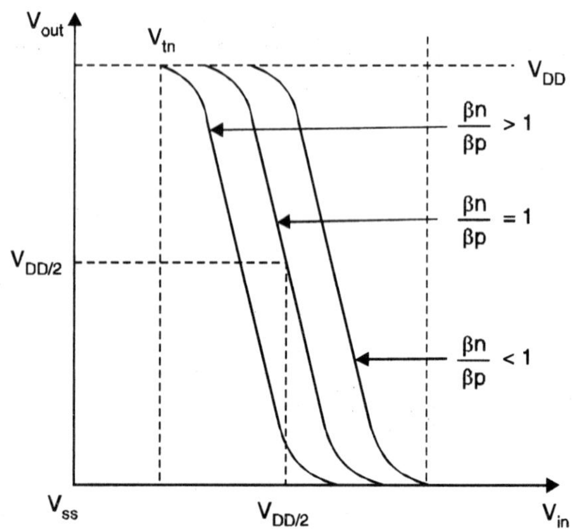

Fig. 1.7.5. Influence of $\frac{\beta_n}{\beta_p}$ on inverter characteristics.

β_n/β_p ratio :

The Fig. 1.7.5 shows transfer characteristics [V_{in} V_s V_{out}) for several values of β_n / β_p. Here the gate threshold Voltage, V_{inv}, where $V_{in} = V_{out}$ is dependent on β_n/β_p. For the CMOS inverter a ratio of $\beta_n / \beta_p = 1$ may be desirable. β_n / β_p can be changed by changing the channel dimensions. As the ratio of β_n / β_p decreases the transition region shifts from left to right.

As temperature increases, μ decreases and β also decreases.

$$\beta \propto T^{-1.5} \quad \therefore I_{ds} \propto T^{-1.5}$$

2

Fabrication Processes ana Device Modelling

INTRODUCTION

To provide optimized integrated circuit design, the designer should have the knowledge of processing and fabrication technology. For improving device performance, several innovative techniques are available, but all are process dependent. The various processes used for fabrication of monolithic ICs containing active devices are (1) NMOS Process, (2) CMOS Process, (3) Bipolar Process. The NMOS and CMOS processes are very similar in that both have FET as the basic active device. In NMOS process only *n*-channel transistors are used where as in CMOS process (complementary MOS) both *n*-channel and *p*-channel devices are used. The PMOS process is also available and it is simply the dual of the NMOS process. *i.e.*, in PMOS *p*-channel FETS are used. Comparing NMOS and PMOS processes, NMOS is preferred because of the superiority of *n*-channel FET over its counter part. But CMOS is still better than NMOS in its performance. A comparison between NMOS and CMOS is given in Table 2.1.1.

Table 2.1.1. Comparison between NMOS and CMOS Technology

NMOS	*CMOS*
1. Design is Complex	Design is simple
2. Requires less processing steps	Requires more processing steps
3. Power dissipation is high	Power dissipation is less
4. Inferior to CMOS in performance	Performance is better

In bipolar process the active device used is the Bipolar junction transistor. Bipolar processes with higher speed than MOS processes are currently available. A comparison between CMOS and Bipolar processes is given in Table 2.1.2.

Table 2.1.2. Comparison between CMOS and Bipolar Technologies

CMOS technology	*Bipolar technology*
Low static power dissipation	High power dissipation
High input impedance (low drive current)	Low input impedance (high drive current)
Scalable threshold voltage	
High noise margin	Low voltage swing logic
High packing density	Low packing density
High delay sensitivity to load (fan-out limitations)	Low delay sensitivity to load
Low output drive current	High output drive current
Low gm (gm $\propto V_{in}$)	High gm (gm $\propto (e^{V_{in}})$), High f_1 at low currents
Bidirectional capability	Essentially uni directional
A near ideal switching device	

2.1. IC PRODUCTION PROCESS

Processing Steps

1. Crystal preparation. The substrate of bipolar and MOS ICs is generally a single crystal of silicon that is lightly doped with either *n*- or *p*-type impurities. This substrate serves both as the physical medium upon and within which the IC is built and as part of the electrical circuit itself. The basic raw material used is a wafer a disc of silicon. Wafers are cut from ingots of single crystal silicon got from polycrystalline silicon by CZ method.

Construction of transistors and other structure of interest depends on ability to control where, how many and what type of impurities are introduced into the silicon surface.

- What type of impurities is controlled by dopant source. Boron, arsenic and phosphorus are commonly used dopants.
- How much used is determined by
  (*a*) Energy and time of the ion implantation step or
  (*b*) Time and temperature of the deposition and diffusion step.
- Where it is used is determined by using special materials such as masks. The masks act as a barrier against doping impurities to provide selective diffusion.

2. Masking (Pattern generation). Masks are made of glass covered with thin film of opaque metal. The masks are produced through photolithographic process. In other methods laser beam, E-beam is also used to generate the actual patterns directly onto the final masks. In photolithographic process, photoresists is applied to entire surface of wafer. The physical characteristics of photoresist can be changed by exposure to light. And there by unwanted areas can be selectively removed. Commonly used mask materials 1. Photo resist; 2. Polysilicon; 3. SiO_2 & SiN.

Negative and Positive Photoresists

Negative: By passing uv rays exposed areas are hardened. Unexposed areas are removed during development.

Positive: Exposed areas are removed.

3. Deposition. Films of various materials must be applied to the wafer during processing e.g., Insulators, resistive films, conductive films, dielectrics, *n* and *p*-type semiconductor materials, dopants. Some of the available deposition techniques are Physical vapour deposition (evaporation and sputtering), Screen printing and Chemical vapour deposition. Deposition involves evaporating dopant material on to the silicon surface followed by a thermal cycle to drive the impurities from the surface into the bulk.

4. Etching. Etching refers to selectively removing unwanted material from the surface of the substrate. Photoresist and masks are used to selectively pattern the surface of the substrate.

5. Diffusion. It refers to controlled forced migration of impurities into the substrate or adjacent material. The commonly available diffusion techniques are deposition and ion implantation. Diffussion will occur between any silicon that has different densities of impurities with impurities tending to diffuse from areas of high concentration to areas of low concentration. Ion implantation involves subjecting the Si substrate to highly energized donor or acceptor atoms. The atoms impinge on the silicon surface and they travel below the surface to form regions with varying doping concentrations.

6. Oxidation. It is a process where by oxide is grown on the surface of the substrate to produce SiO_2 layer. This serves as a very good insulator. The two common approaches are wet oxidation and dry oxidation. The oxidation process consumes silicon.

7. Epitaxy (carried out only in Bipolar processes). This is also a CVD. These are ideally single crystalline extensions of the substrate. *i.e.,* Epitaxy involves growing a single crystal film on the silicon surface.

8. Conductors and resistors. Aluminium or other metals are often used as conductors for interconnection of components on an IC.

9. Packaging and Testing. After processing, the IC's are tested and packaged.

2.2. NMOS FABRICATION PROCESS

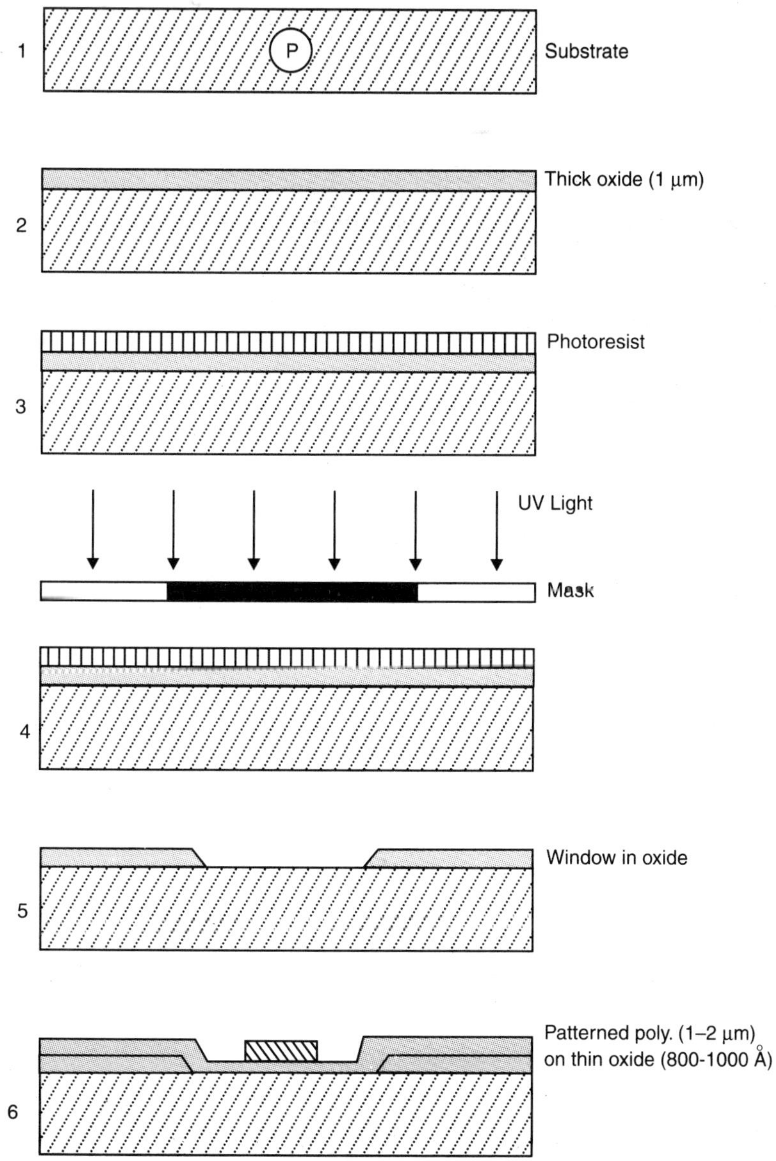

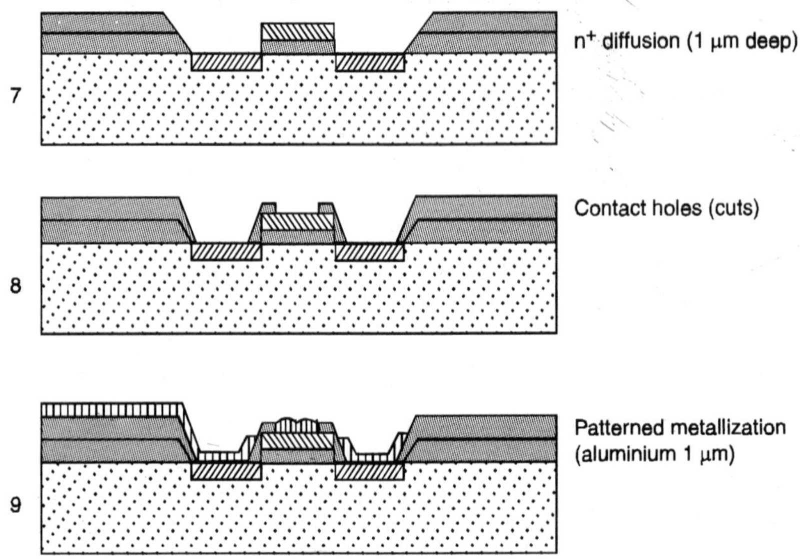

Fig. 2.2. NMOS fabrication process.

1. A thin wafer is cut from a single crystal of silicon of high purity. It is doped with *p*-type impurities to form substrate. Wafer diameter is of 75 to 150 mm. It is 0.4 mm thick and doped with boron.

2. A layer of SiO_2 (1 Um thick) is grown all over surface. (acts as barrier to dopants during processing).

3. Surface is now covered with photoresist.

4. It is exposed to UV light through a mask. The mask defines those regions into which diffusion is to take together with transistor channels. Note : Areas exposed to UV light are hardened. The areas required for diffusion are shielded by the mask and remain unaffected by UV rays. Unexposed areas are removed – negative photo resist.

5. The areas (which are unaffected by UV rays) are subsequently readily etched away together with the underlying SiO_2 so that the wafer surface is exposed in the window defined by the mask.

6. Remaining photoresist are removed, a thin layer of SiO_2 is grown over the entire chip surface. Polysilicon is deposited on top of this to form the gate structure (by CVD).

7. Further photoresist coating and masking allows the polysilicon to be patterned and then the thin oxide is removed to expose areas into which *n*-type impurities are to be diffused to form the source and drain as shown. Diffusion is achieved by heating the wafer to a high temperature and passing a gas containing the desired *n*-type impurity (Phosphorous) (Self aligned process).

8. Thick oxide is grown over all again and is then masked with photoresist and etched to expose selected areas of the polysilicon gate and the drain and source areas where connections (is contact cuts) are to be made.

9. Aluminium (metal) is deposited over entire surface. This metal layer is then masked and etched to form the required interconnection pattern.

Note : In order to form depletion mode devices it is only necessary to introduce a masked ion implantation step between steps 5 and 6.

2.3. CMOS FABRICATION PROCESS

Basic CMOS Technology : CMOS-(Complementary metal oxide silicon)

CMOS Provides low power static circuit technology that has the capability of providing a low power-delay product than Bi-polar NMOS (or) GaAs technology.

Four Main CMOS Technologies are :

- *n*-well process
- *p*-well process
- Twin tub process
- Silicon on insulator

2.3.1. N-well CMOS Process

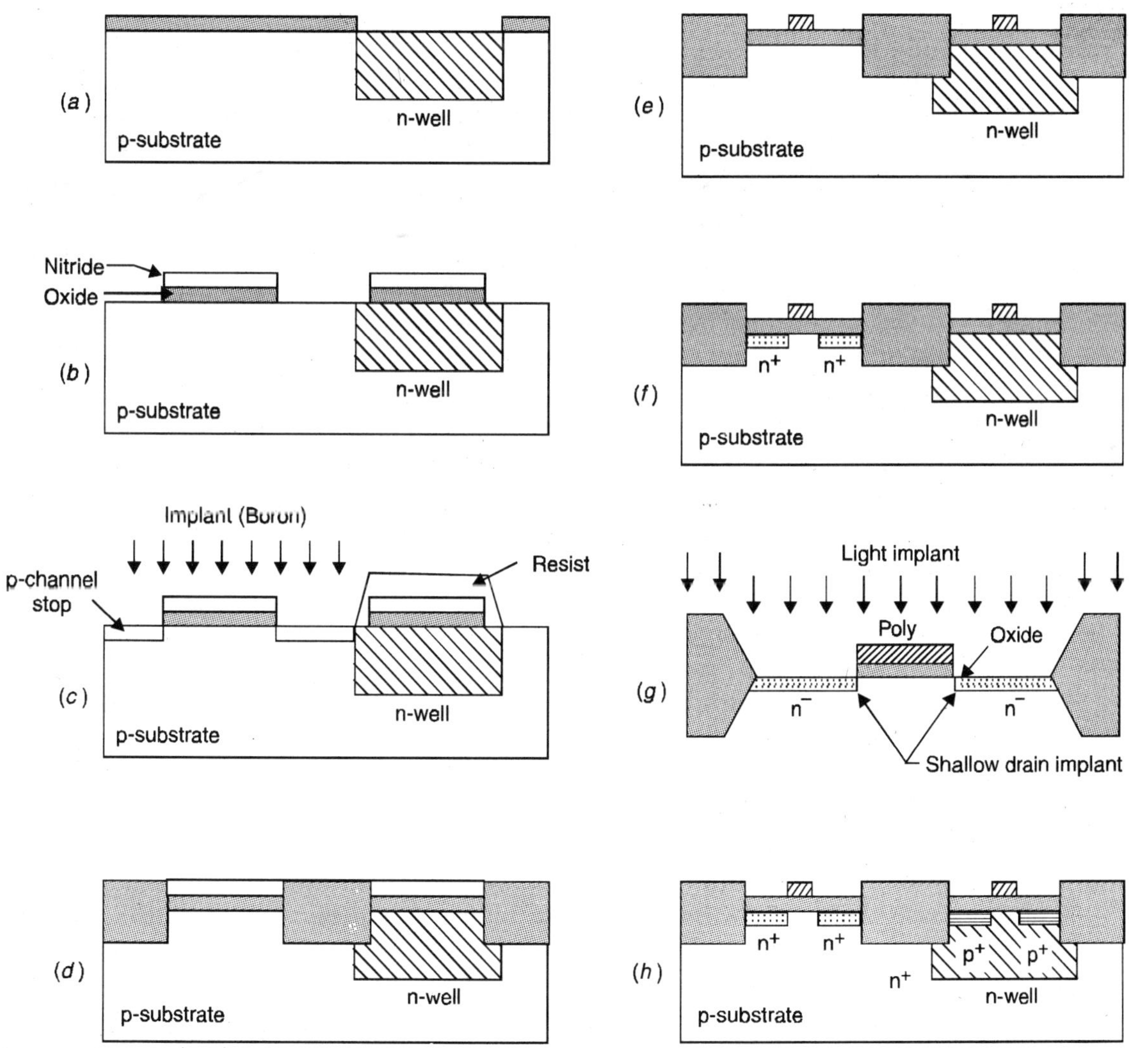

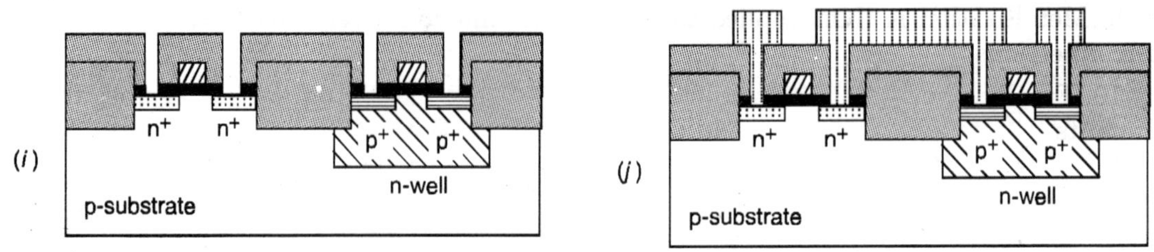

Fig. 2.3.1. (*a*) N-Well CMOS Process.

To start with a lightly doped *p*-type substrate (wafer) is taken, *n*-type well is created for *p*-channel devices and the *n*-channel transistor is built in the native *p*-substrate itself.

1. The first mask defines *n*-well. *p*-channel transistor will be fabricated in this well. Ion implementation (or) deposition or diffusion is used to produce *n*-well. The former one tends produce shallower wells which are compatible with fine dimension processes.

2. The next mask is called "active mask" because it defines where areas of thin oxide are needed to implement transistor gates and allow implantation to form *p*- (or) *n*-type diffusions for transistor source/drain region. A thin layer of SiO_2 is grown and covered with SiN (Silicon Nitride). This is used as masking layer for the following two steps.

3. The channel stop implant is usually then completed. It dopes the *p*-substrate in areas where there are no *n*-transistors with p^+ impurity using a photoresist mask. This is conjunction with, thick field oxide which will cover these areas, helps in preventing conduction between unrelated transistor source/Drain. After the doping of boron, the photoresist is removed.

Negative resist. Unexposed photoresist is dissolved by solvent. So, they can be removed (*i.e.*, area **unexposed** to UV-rays can be removed).

Positive resist. UV light is passed through a mask containing desired pattern. The coating can be polymerized where the pattern is to appear. The polymerized areas may be removed with an organic solvent. (*i.e.*, Area **exposed** to UV can be removed).

4. The thick field oxide is then grown. This grown in areas where the SiN-layer is absent. The oxide grows in both directions vertically and also laterally under the SiO_2/SiN sandwich. This lateral movement results in what is called a birds beak because of the shape of the oxide encroachment under the gate oxide mask. This is shown in Fig. (*d*). The general oxide construction technique is called LOCOS (local oxidation of silicon). This oxide encroachment results in an active area that is smaller than the patterned.

5. Silicon gate is then completed. It involves covering the surface with polysilicon and then etching the required pattern.

6. Mask is provided over the *n*-well and all other areas except the area where n^+ impurities are to be diffused (self aligned process). (In self aligned process Drain and Source regions are themselves formed without any separate aligment step). The n^+ areas acts as source and drain for the *n*-transistor.

7. Mask is provided over all the areas except the area where the p^+ is to be diffused. The p^+ acts as source and drain for *p*-transistor.

8. Contact cuts are then defined. It involves etching of SiO_2 down to the surface to be contacted. SiO_2 is deposited and then patterned.

9. Metallization is then applied to the surface and selectively etched to produce circuit interconnections.

10. Finally passivation is completed which protects the silicon surface against the contaminants that can modify the circuit behaviour.

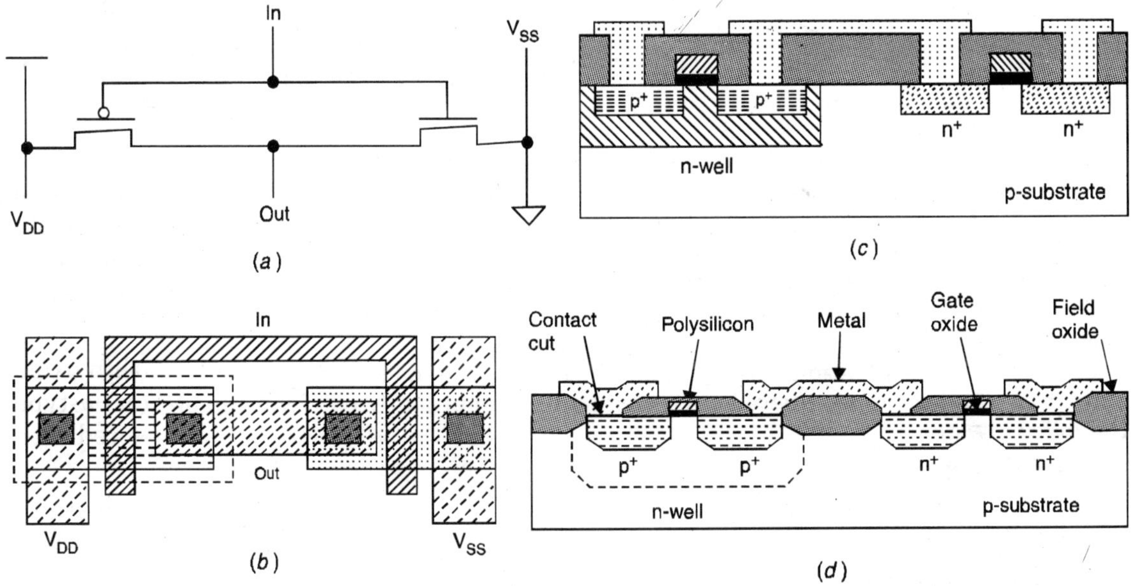

Fig. 2.3.1 (*b*) Cross section of a CMOS inverter in a *n*-well process.

The Fig. 2.3.1 (*b*) shows the circuit representation of CMOS inverter and also it shows how it can be formed in *n*-well process.

2.3.2. The *p*-well Process

p-well fabrication steps are similar to a *n*-well process, excepting that a *p*-well is implanted rather than an *n*-well. The first masking step defines the *p*-well regions. Then low dose boron is driven in for the formation of *p*-well.

Further steps such as other diffusions, growing field oxide, making contact cuts and metallisation are done similar to *n*-well process. But whenever we get '*n*' that is to be replaced with *p*. *p*-well processes are preferred in circumstances where the characteristic of the *n*- and *p*-transistors are required to be more balanced than that achievable in an *n*-well process. The transistors that resides in the native substrate tends to have better characteristics, *i.e.*, *p*-well process has better *p*-devices than *n*-well process.

2.3.3. Twin Tub Process

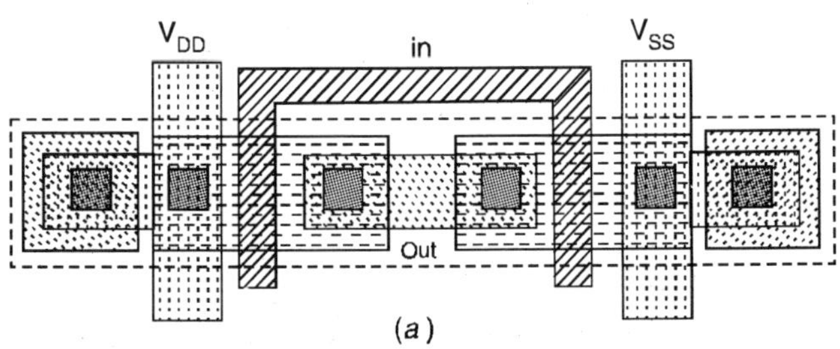

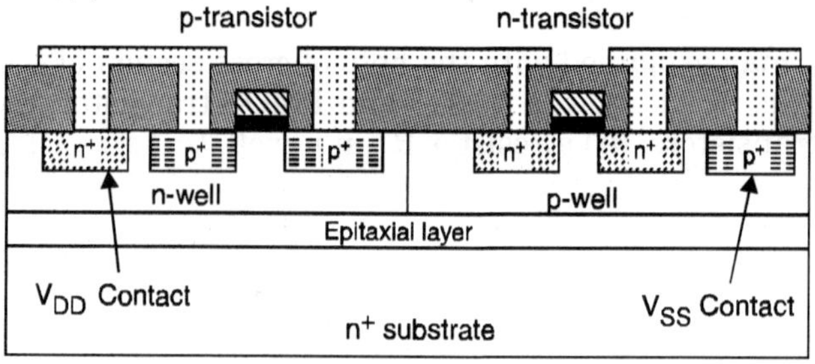

(b)

Fig. 2.3.3. Twin tub CMOS process-cross section.

In twin tub CMOS technology p-type and n-type transistors can be separately optimized. *i.e.*, threshold voltage, body effect and the gain of each device can be separately optimized. The starting material is either an n^+ (or) p^+ substrate with a lightly doped epitaxial (or) epi layer which is used for protection against latch up. The aim of epitaxy is to grow high-purity silicon of controlled thickness with accurately determined dopant concentrations distributed homogeneously throughout the layer. The electrical properties of this layer are determined by the dopant and its concentration in the silicon. The process sequence is similar to the n-well process apart from tub formation, where both p-well and n-well are utilized. Twin-tub process contains the following steps.

- Tub formation
- Thin oxide construction
- Gate formation
- Source and Drain implantations
- Contact cut definitions
- Metallisation

Since this process provides separately optimized wells, balanced performance n-transistors and p-transistors may be constructed.

2.3.4. Silicon on Insulator

Rather than using silicon as the substrate, insulating substrate is used to improve process characteristics such as latch up and speed.

Latch up. Because of the parasitic circuit effect the shorting of the V_{DD} and V_{SS} lines occur resulting in destruction of chip. This effect is called Latch-up.

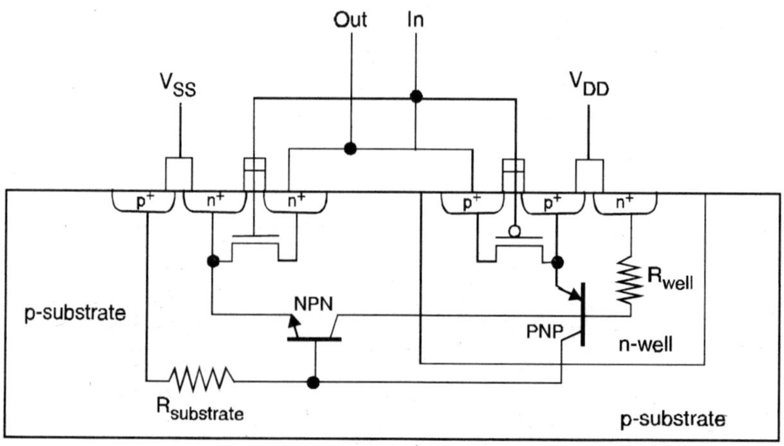

Fig. 2.3.4 (a) Cross section of CMOS inverter with parasitic devices.

The latch up effect may explained with the help of cross section of CMOS inverter shown in Fig. 2.3.4 (*a*). As in Fig. in addition to NMOS and PMOS transistor, the parasitive devices such as (*i*) NPN, PNP transistors, (*ii*) a resistor R_w in *n*-well, (*iii*) a resistor R_{sub} in *p*-substrate are also available which are unlikely. The effect of the parasitics is to cause a short circuit above a critical voltage by drawing heavy current which will result in complete destruction of chip.

The latch up may be prevented in two ways : 1. By using Latch up resistant CMOS process such as Silicon on insulator process, 2. By using proper layout technique. Also latchup can be eliminated by reducing the resistor values and reducing the gain of the parastic transistor. The gain of parasitic can be reduced through the use of guard rings. (*i.e.*, A p^+ guard ring for an n^+ source/drain and a n^+ guard ring for p^+ source/drain). The guard rings act as dummy collectors for parasitic transistors and collects the minority carriers and prevents them from being injected into the respective bases and hence reduces the gain of parasitic transistors.

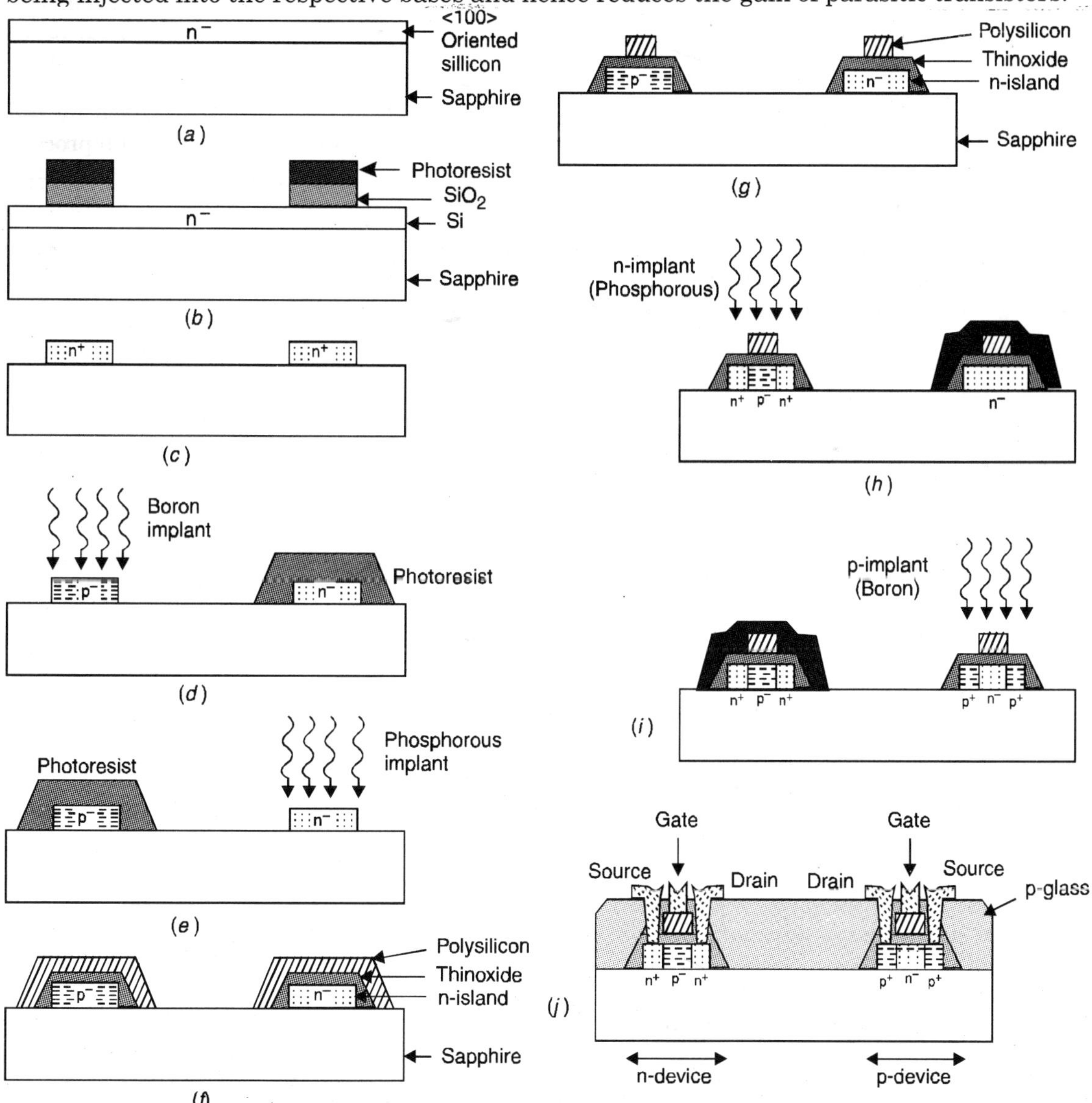

Fig. 2.3.4. (*b*) SOI process.

The Fig. 2.3.4 (*b*) illustrates the process.

1. A thin film of (7–8 m) of very lightly doped *n*-type Si is grown over an insulator such as magnesium aluminate (or) SiO_2.

2. Etching technique is used to etch away the Si except where a diffusion area will be needed.

3. P islands are formed next by masking the *n*-islands with a photoresist. A *p*-type dopant, boron for example is then implanted. The *p*-islands will serve as the *n*-channel devices.

4. The *p*-island is then covered with a photoresist and a *n*-type dopant phosphorus is implanted to form the *n*-islands. The *n*-island will serve as the *p*-channel devices.

5. A thin gate oxide is grown over all of the Si-substrate structure. This is normally done by thermal oxidation. Then polysilicon film is deposited over the oxide.

6. Polysilicon is patterned by photo masking and is etched. This defines the polysilicon layer in the structure.

7. The next step is to form the *n*-doped source and drain of n-channel devices in the *p*-islands. The n^+ is covered with a photoresist and a *n*-type dopant phosphorus is implanted.

8. The *p*-channel device is formed next by masking *p*-islands and implanting a *p*-type dopant such as boron.

9. A layer of phosphorus glass (or) some other insulator such as silicon dioxide is then deposited over the entire structure. The glass is etched at contact cut location. Metallization layer is formed using aluminium. Finally passivation is done.

SOI CMOS Process Advantage

→ Closer packing of *p* and *n*-transistors

→ Absence of latch up problems

→ Lower parasitic substrate capacitances.

2.4. BIPOLAR PROCESS

The basic active devices in the bipolar process are the *npn* and *pnp* transistors. The components available in the bipolar process are

1. *npn* and *pnp* bipolar transistors.

2. Resistors, Capacitors, Diodes, Zener diodes.

3. JFETs (available only in some bipolar processes).

The basic characteristics of the devices available for bipolar IC's are determined by the process, and the range of practical values and parameter is severely limited. In addition, the methods of interconnection are strictly limited. The basic devices have characteristics that are quite temperature dependent, and the passive component values are typically somewhat dependent on the signal applied. The bipolar process is used for the TTL, ECL and I^2L digital logic families (741, 723,565). It requires 7 mask and the process steps are as follows. The construction of npn and pnp transistors only will be considered here. The steps are as follows :

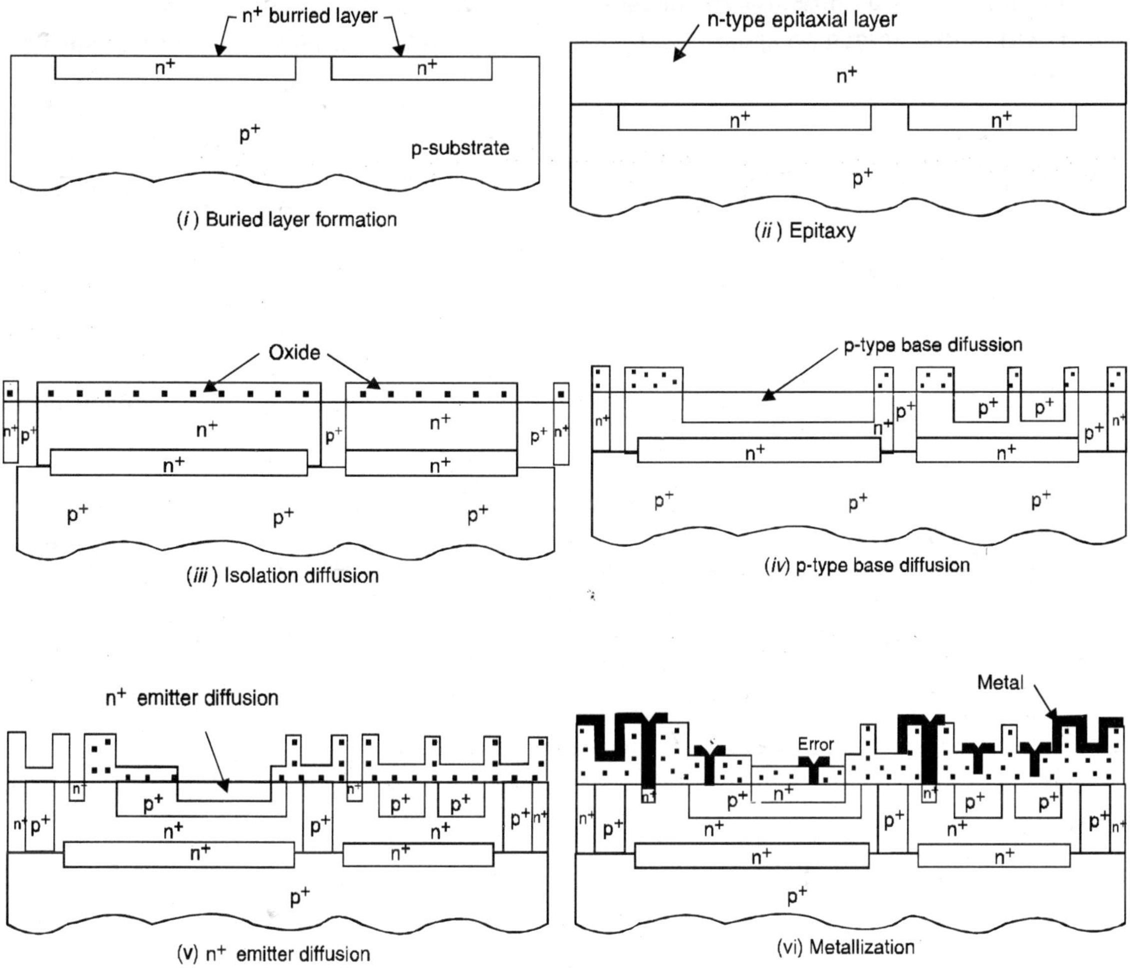

Fig. 2.4. (*a*) Process description for the bipolar process.

(*a*) A *p*-type silicon wafer is taken and a layer of SiO_2 is grown over that. Then a layer of photo resist is applied.

(*b*) Using M ≠ 1 *n* + buried layer is patterned.

- It decreases collector resistance
- Minimizes parasitic current flow from collector to substrate in npn transistors.
- Decreases base resistance in lateral *pnp* transistors.
- *n*-type impurities are introduced into the substrate through the openings provided by M ≠ 1.

(*c*) An *n*-type epitaxial layer of thickness 2μ –15μ is then grown over the entire wafer. This layer will be used for the collector region in npn transistors. A thick layer of SiO_2 is then grown over the entire surface.

(*d*) Mask ≠ 2 is used to pattern the SiO_2 layer for the P^+ isolation diffusion. This is used to electrically separate adjacent transistors. Following this another thick layer of SiO_2 is grown over the entire wafer.

(*e*) Mask ≠ 3 is used to pattern the SiO_2 layer and define the base regions for *npn* transistors as well as the collector and emitter regions for lateral *pnp* devices. A *p*-type deposition and a subsequent drive in diffusion create these regions in the unprotected areas. This diffusion is much shallower than isolation diffusion and must not penetrate the epitaxial layer.

(*f*) Following growth of another layer of SiO_2, M ≠ 4 is used to pattern the emitter regions. Openings are also made in the oxide above the collector region to provide for better electrical contact from the surface. The emitter diffusions must be shallow so as not to penetrate the relatively shallow *p*-type base regions already created.

The gain of transistors is influenced by

1. Impurities in n^+ emitter
2. Thickness of *p*-type base (between n^+ emitter and n^+ collector)

(*g*) M ≠ 5 is used to pattern contact openings to allow for top contact of the circuit with the metallisation. The entire circuit is then covered with a thin layer of metal. M ≠ 6 is used to pattern the metal followed by the addition of a passivation layer. M ≠ 7 patterns pad openings to allow for electrical contact to the bonding pads.

Using additional mask a deep collector diffusion can be done. This will reduce collector resistance in high current applications.

By using additional mask step an additional *p*-diffusion in the *p*-channel stops can be made. This is used to avoid surface inversion in high voltage parts.

The *npn* and *pnp* transistors in this process are shown in Fig. 2.4 (*b*). *npn* is formed as a vertical *npn* device (as E, B and C are stacked vertically). *pnp* is formed as a lateral device since it is stacked horizontally. The base width cannot practically be made as narrow and the base area is not as accurately controllable as for the *npn* device. The characteristics of the lateral transistors are generally considered poorer than those of the vertical devices.

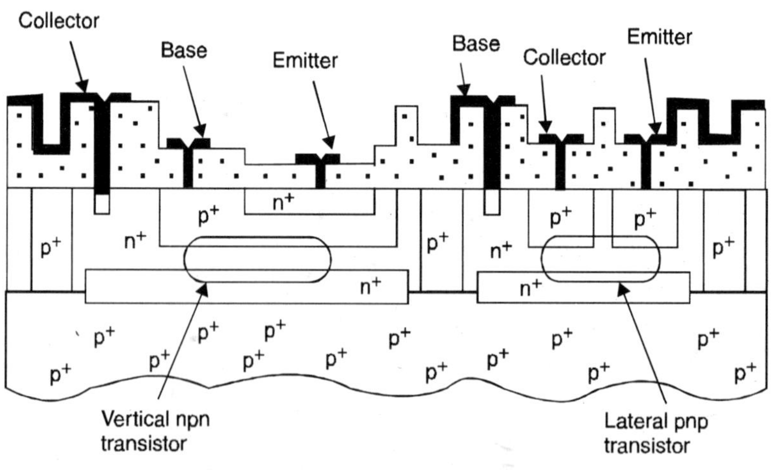

Fig. 2.4. (*b*) Vertical and lateral transistors in a bipolar process.

Although minimum feature sizes are comparable for the bipolar and MOS processes, standard bipolar processes require more area/device than do the NMOS processes. The reason for this increased area is the deep and wide p^+ channel stops that are required for device isolation in standard Bipolar processes. To have increased component density bipolar process using trench isolation can be done.

2.5. PROCESS PARAMETER

The following parameters influence the device characteristics and operation of the devices.

1. Threshold voltage.
2. Body effect.
3. Channel length modulation.
4. Surface potential.
5. Gate of channel capacitance.
6. Gate oxide thickness-t_{ox}.
7. Channel Mobility.
8. Conduction factor.
9. Breakdown voltage.
10. Leakage currents.
11. Feature size λ.
12. Lateral diffusion.
13. Width reduction.

Hence, enough care should be taken while dealing with these parameters. Enough description of the above said parameters have been already given.

2.6. PROCESS CONSIDERATIONS

The equipment needed for the CMOS process is basically the same as is needed for the NMOS process. With this equipment the minimum feature size for the CMOS process is comparable to that for the NMOS process. Comparing with NMOS process, more processing steps and more number of masks are needed for the CMOS process. Also more area is required for same number of devices in CMOS process. This is because of the requirement of large *p* tubs and the *n* and *p*-type guard rings. But this increase in area is often offset by superior performance that is attainable with the CMOS process. Several physical and processing dependent material characteristics cause the MOSFET to differ from the ideal.

1. Width and Length reduction. Where W – Width of the active region and L – Distance between drain and source diffusions. During field oxide growth, encroachment into the active region effectively reduces the width of the transistor. This is termed as bird's beaking. A second factor that affects the effective width is the accuracy with which the protective Si_3N_4 layer used to pattern the field oxide can be controlled. The effective width of this layer is affected by both the patterning of the photoresist and the problems associated with etching. The major source of length reduction is associated with the lateral diffusion of the drain and source diffusions, which are difficult to precisely control. This is very significant and problematic in short channel transistors. To summarize, both length and width reduction are inherent with existing process technologies. They can be partially compensated for by considering these reductions during design, *i.e.*, by adjusting mask size.

2. Lateral well diffusion. This will not limit the device performance, but consumes considerable surface area. This can be minimized by grouping large number of n-channel transistors into a single p-well.

3. Latch-up. Several parasitic diodes, transistors (bipolar) and SCR are formed in CMOS process. The Fig. 2.6 shows how parasitic transistors are formed in a p-well CMOS Process. These parasitic diodes and bipolar transistors have got some practical applications. However SCR is very undesirable. As, if it is caused to fire, excessive current will usually flow, causing destructive failure of the IC. The firing of the SCR is termed latch-up in CMOS circuits. The designer must make certain that latch-up cannot occur in any design. One way to obtain latch-up protection is to use guard rings.

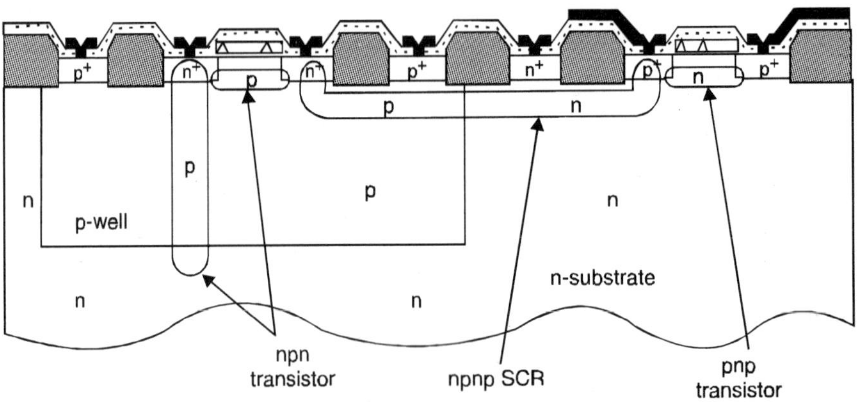

Fig. 2.6. Parasitic transistors in a p-well CMOS process.

Input Protection

The protection of inputs must be provided to prevent destructive breakdown. When the inputs have a direct connection to a region separated from the rest of the circuit only by a thin oxide, with no direct connection to any diffused region, breakdown may occur when electric fields exceeds the oxide breakdown voltage. One common protection scheme involves connecting the inputs pads through a small polysilicon resistor to a reverse biased diode that nondestructively breaks down at voltages below the critical gate oxide breakdown voltage.

2.7. MOS DEVICE CAPACITANCES

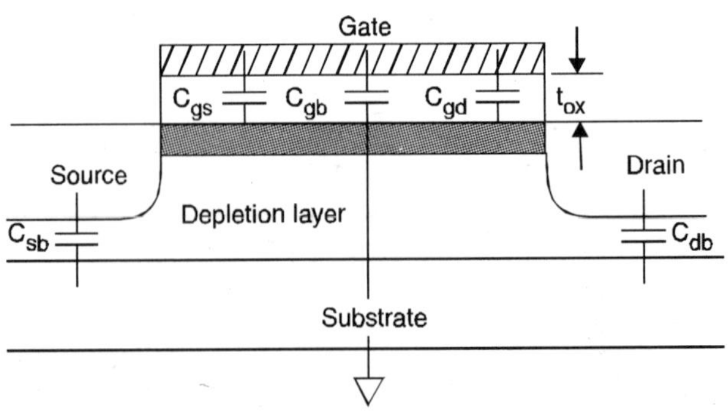

Fig. 2.7.1. Parasitic capacitance in a MOS transistor.

The Fig. 2.7.1 shows the diagrammatic representation of the parasitic capacitances present in a MOS transistor.

From the Fig. 2.7.1 the following capacitive components have been identified.

$C_{gs} \rightarrow$ gate to source capacitances

$C_{gd} \rightarrow$ gate to drain capacitances

$C_{sb} \rightarrow$ source to bulk capacitances

$C_{db} \rightarrow$ drain to bulk capacitances

$C_{gb} \rightarrow$ gate to bulk capacitances

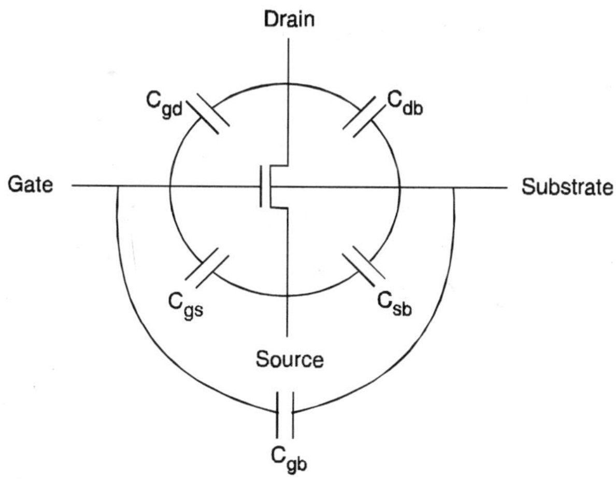

Fig. 2.7.2. Circuit symbols for parasitic capacitance.

The total gate capacitance 'C_g' of an MOS transistor is given by,

$$C_g = C_{gb} + C_{gs} + C_{gd} \rightarrow (1)$$

The behaviour of the gate capacitance of an MOS device can be explained in terms of the following simple models in 3 regions of operations.

1. Off–Region ($V_{gs} < V_t$). When the MOS device is 'off', there is no channel and hence $C_{gs} = C_{gd} = 0$, C_{gb} can be modeled as the series combination of the two capacitors, C_o and C_{dep}

$C_{dep} \rightarrow$ depletion capacitance

$C_o \rightarrow$ gate oxide capacitance

2. Non-Saturated Region ($V_{ds} < V_{gs} - V_t$). Now, as the result of formation of the channel, the gate to source and drain capacitances becomes significant. These capacitances are dependent on gate voltage. Their values can be estimated as,

$$C_{gd} = C_{gs} = 1/2 \left(\frac{\varepsilon_o \varepsilon_{SiO_2}}{t_{ox}} \right) A$$

C_{gb} = effectively falls to zero.

3. Saturated Region ($V_{ds} > (V_{gs} - V_t)$). In this mode, the channel is heavily inverted. The drain region of the channel is pinched off causing C_{gd} to be zero. C_{gs} increases to approximately,

$$C_{gs} = \frac{2}{3}\left(\frac{\varepsilon_o \varepsilon_{SiO_2}}{t_{ox}}\right)A$$

MOS capacitor Characteristics. When $V_{gs} < 0$, the –ve charge on the gate attracts holes towards the silicon surface. *i.e.*, accumulation layer is formed. Now, the MOS structure behaves like a parallel plate capacitor, *i.e.*, the gate forms one plate of the capacitor and the high concentration of holes in a
p-substrate forms the second plate.

Since the accumulation layer is directly connected to the substrate, the gate capacitance may be approximated by,

$$C_o = \left(\frac{\varepsilon_{SiO_2}\varepsilon_o}{t_{ox}}\right)A$$

where, A $\rightarrow$ area of gate

ε_{SiO_2} $\rightarrow$ dielectric constant (or) relative permittivity of SiO_2

ε_o $\rightarrow$ permittivity of free space.

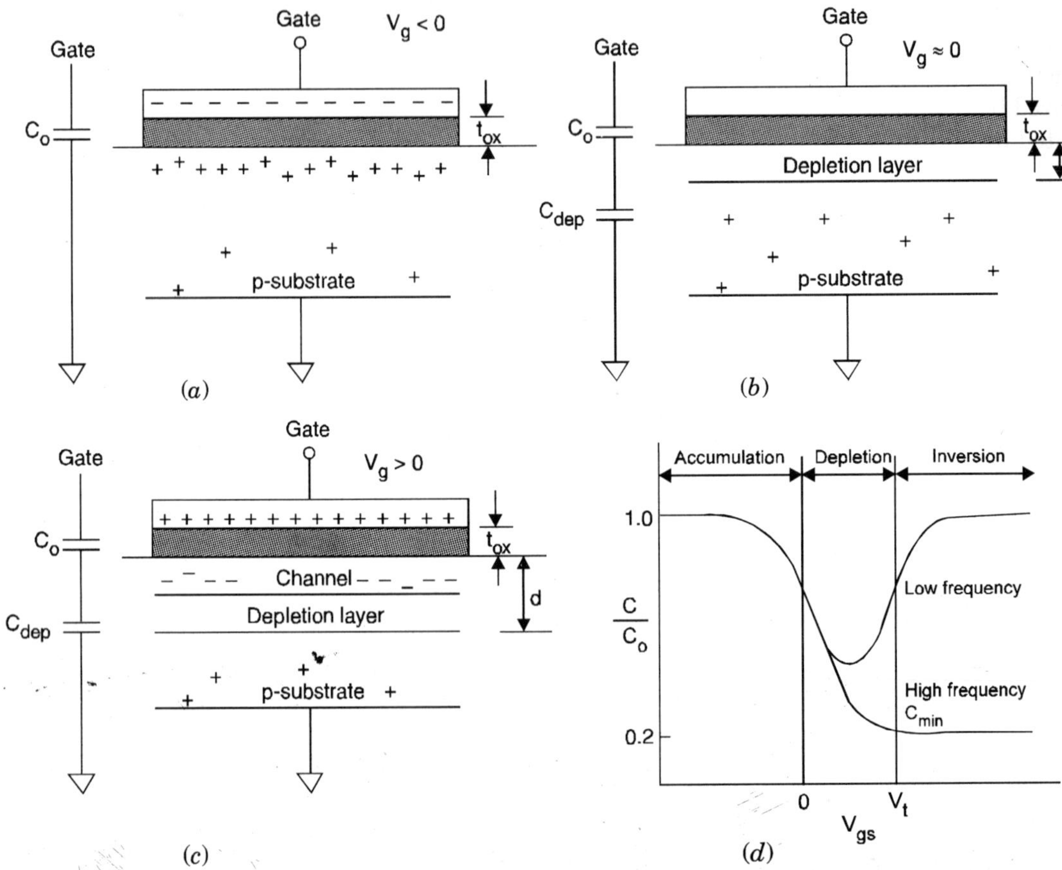

Fig. 2.7.3. MOS capacitance in accumulation, depletion and inversion region.

When a small +ve voltage is applied to the gate with respect to the substrate, a depletion layer is formed in the *p*-substrate under the gate. The +ve gate voltage repels holes, leaving a negatively charged region depleted of carriers. The depletion capacitance C_{dep} is given by,

$$C_{dep} = \left(\frac{\varepsilon_o \varepsilon_{Si}}{d}\right) A$$

where, d = depletion layer depth

ε_{Si} = dielectric constant of silicon

Thus as the depth of the depletion region increases, the capacitance from gate to substrate will decrease. The total capacitance from gate to substrate due to the gate-oxide capacitance, Co in series with C_{dep}.

$$C_{gb} = \frac{C_o C_{dep}}{C_o + C_{dep}}$$

As gate voltage is further increased channel is inverted. But the capacitance remains the same as for the maximum depletion situation.

Diffusion Capacitance. The n^+ and p^+ diffusions form the source and drain terminals of n and p channel devices. All diffusion regions have a capacitance to substrate that depends on the voltage between the diffusion regions and substrate as well as on the effective area. The diffusion capacitance 'C_d' is proportional to the total diffusion to substrate junction area. 'C_d' is function of 'base area' and also of the 'sidewall' periphery. Sidewall capacitance can be characterized by a periphery capacitance per unit length. Side wall total capacitance C_d can be represented by

$$C_d = C_{ja} * (ab) + C_{jp} * (2a + 2b)$$

where C_{ja} → junction capacitance per μ^2.

C_{jp} → Periphery capacitance per μ.

a → width of diffusion region (μ).

b → length of diffusion region (μ).

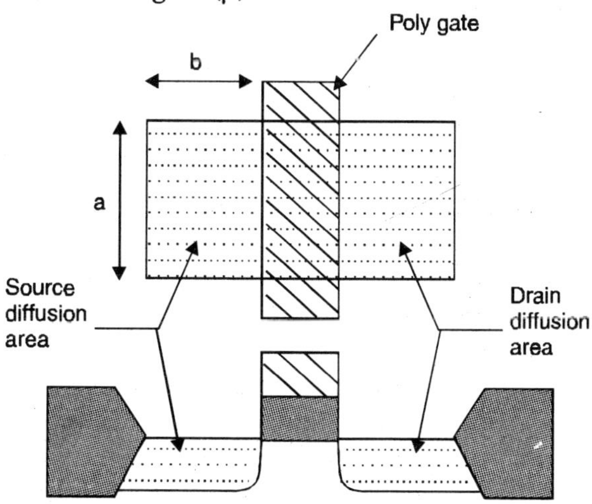

Fig. 2.7.4. Diffusion area of MOS.

Capacitance contributed by the sidewall facing the channel will be reduced by the presence of the channel depletion region. As the diffusion area is reduced, the contribution of the peripheral capacitance becomes more important. Since the thickness of the depletion layer depends on the voltage across the junction, both C_{ja} and C_{jp} are functions of junction voltage V_j. A general expression that describes the junction capacitance is,

$$C_j = C_{jo}\left(1-\frac{V_j}{V_b}\right)^{-m}$$

where V_j = junction voltage

C_{jo} = zerobias capacitance

V_b = built-in-junction potential

m = constant, which depends on the distribution of impurities near the junction and whether the junction is due to the bottom (or) side of the diffusion (m → has an effective value from 0.3 to 0.5).

2.8. DEVICE MODELING

Device modeling is performed to obtain the functional relationship among the terminal electrical variables of the device that is to be modeled. These electrical characteristics depend upon a set of parameters including both geometric variables and variables dependent upon the device physics. This finds its application in computer simulation of circuits employing these devices. DC Modelling is otherwise called as large signal analysis where DC bias voltage V and current I are taken into consideration. Similarly in AC Modeling or small signal analysis, instantaneous voltage v and current i are considered.

DC Models. It relates the actual terminal voltages and currents of the devices at dc and low frequency. For MOS, DC model is not linear and DC analysis is tedious.

2.8.1. Small Signal Modelling

MOS circuits perform the task for which they were intended only over a limited excitation range. This range is specified in terms of a maximum input signal excursion about some nominal point. Internal to the circuit these input variations typically cause excursions around some non-zero DC operating point. Often these inputs are sinusoids of small amplitude compared to the supply voltages providing power to the circuit. An analysis of how these small sinusoids propagate thro' the ckt is termed **small signal analysis or ac analysis**. The points (nodal voltage and branch currents) about which the circuit operates are termed the bias points or Q-points. Although it might be desirable to have all quiescent points @ either zero volts or zero amps, it is not practical and generally not possible to do this.

V or V_c – Total instantaneous value

υ – Small signal value (instantaneous value)

$V_{u.c}$ – Quiescent value

$V_{u.c} = \upsilon + V_c$

For small signal analysis V_c is assumed to be periodic with period T and quiescent value is defined by

$$V_c = 1/T \int_0^T V_{u.c}(t)\, dt$$

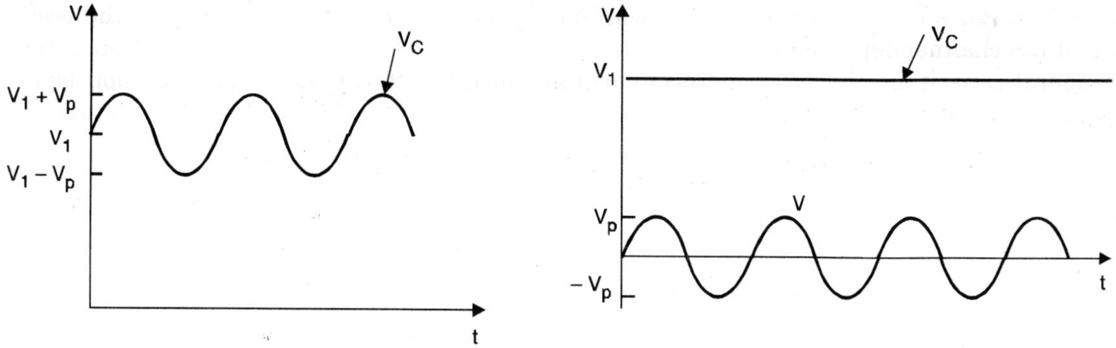

(*i*) Total instantaneous value. (*ii*) Quiescent and small signal values.

Fig. 2.8.1. (*a*) Relationship between total (V_C), small signal (υ) and quiescent (V_c) values.

The electrical behaviour of linear multiple terminal networks can typically be modeled in terms of one or more established sets of transfer function parameters, such as the h, y or g. But 'y' model has been widely adopted for modeling these devices. Consider a linear small-signal four terminal network. One terminal is selected as reference (Fig. 2.8.1. (*b*) 'y' parameter relate the terminal voltages and currents by the expressions.

$$i_1 = y_{11}\upsilon_1 + y_{12}\upsilon_2 + y_{13}\upsilon_3$$
$$i_2 = y_{21}\upsilon_1 + y_{22}\upsilon_2 + y_{23}\upsilon_3$$
$$i_3 = y_{31}\upsilon_1 + y_{32}\upsilon_2 + y_{33}\upsilon_3$$

where $Y_{kj} = \dfrac{i_k}{\upsilon_j} \mid \upsilon_m = 0,\ m \in \{1, 2, 3\},\ m \neq j$

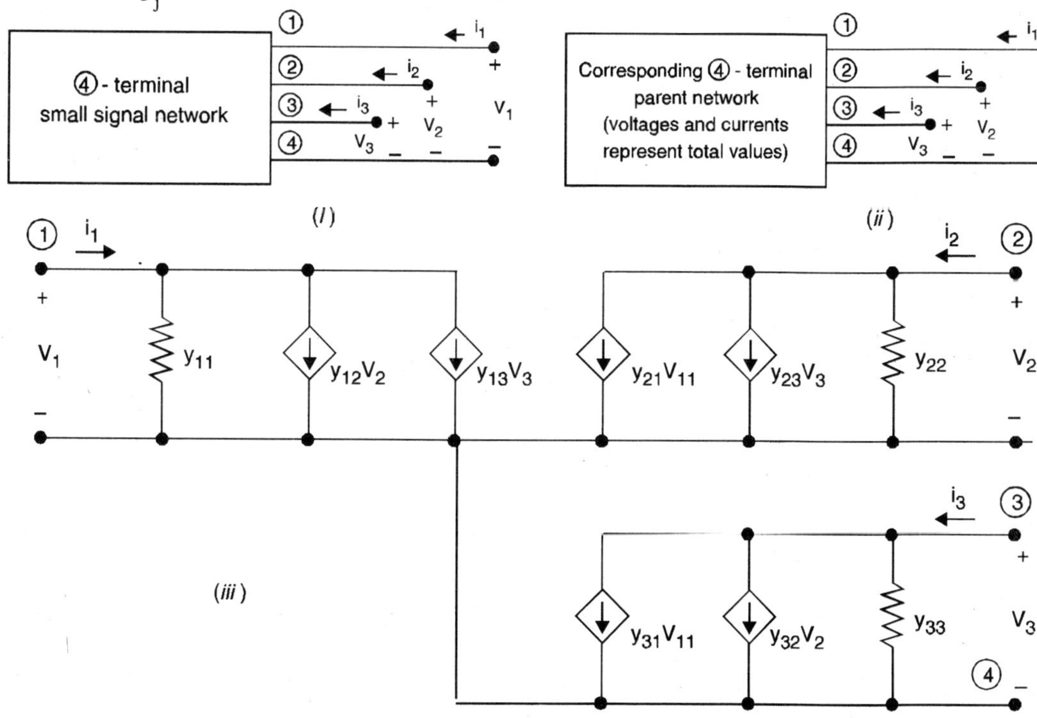

(*i*) Small signal, (*ii*) Parent network, (*iii*) Equivalent circuit for small signal network.

Fig. 2.8.1. (*b*) General four-terminal network.

Since the small signal voltage and current variables are the time-varying part of the corresponding total terminal voltage and current variables in the parent network relative to the Q-point (Fig. (*ii*)), then it follows that the y-parameters can be obtained from the large signal variables and thus from the dc model by the expression.

$$y_{kj} = \frac{\partial I_k}{\partial Vj} \;|V_m = \text{quiescent value, m} \in \{1, 2, 3\}$$

A small signal equivalent circuit of the multiport network is often found useful for circuit analysis and design. It is easy to verify that the circuit shown in Fig. (*iii*) has the same i-v characteristic as the four terminal network of Fig. (*i*) which is characterized by the above equations and hence it is electrically indistinguishable.

2.8.2. Large Signal Modelling

The triode region equation for a MOS transistor relates the drain current to the gate source and drain source voltages. It can be shown that this relationship is given by

$$I_D = \mu_n C_{ox} \left(\frac{W}{L}\right) \left[(V_{ss} - V_t)\, V_{ds} \frac{V_{ds}^2}{2}\right] \qquad \text{... (1)}$$

As V_{ds} increases, I_D increases until the drain end of the channel becomes pinched off, and then I_D no longer increases. This pinch off occurs for $V_{ds} = -V_{tn}$ (or) approximately.

$$V_{ds} = V_{gs} - V_{tn} = V_{eff} \qquad \text{... (2)}$$

At the edge of pinch off, the drain current resulting from equation (1) and the drain current in the active region must have the same value.

Therefore, the active region equation can be found by substituting equation (2) into equation (1)

$$I_D = \mu_n C_{ox} \left(\frac{W}{L}\right) \left[(V_{gs} - V_{tn})\,(V_{gs} - V_{tn}) - \frac{(V_{gs} - V_{tn})^2}{2} (V_{gs} - V_{tn})^2\right]$$

$$= \mu_n C_{ox} \left(\frac{W}{L}\right) \left[(V_{gs} - V_{tn})^2 - \frac{(V_{gs} - V_{tn})^2}{2}\right]$$

$$\therefore \qquad I_D = \frac{\mu_n C_{ox}}{2} \left(\frac{W}{L}\right) (V_{gs} - V_{tn})^2 \qquad \text{... (3)}$$

For $V_{ds} > V_{eff}$, the current stays constant at the value given by equation (3). This equation is most important one that described the large signal operation of a MOS transistor. It should be noted that equation (3) represents a squared current voltage relationship for a MOS transistor in the active region. Equation (3) implies that the drain current, I_D is independent of the drain source voltage (V_{ds}). This is true only to a first order approximation. The major source of error is due to the channel length shrinking as V_{ds} increases. To see this effect, consider the following diagram, which shows a cross section of a transistor in the active region.

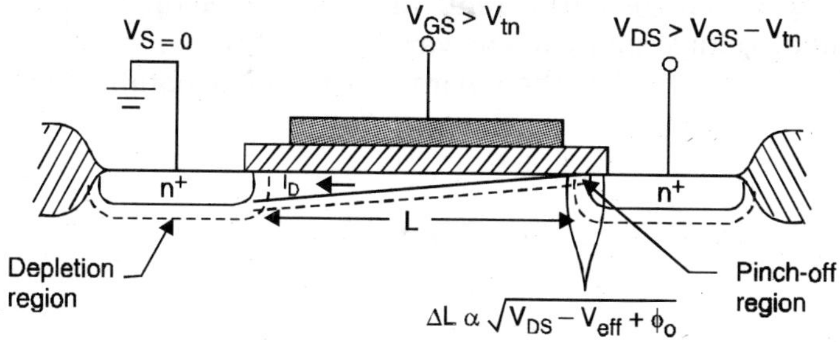

Fig. 2.8.2. (*a*) Cross-section of NMOS in active region.

A pinched off region with very little charge exists between the drain the channel. The voltage at the end of the channel closest to the drain is fixed at ($V_{gs} - V_{tn} = V_{eff}$). The voltage difference between the drain and the near end of the channel lies across a short depletion region often called the pinch off region. As V_{DS} becomes larger than V_{eff}, this depletion region surrounding the drain junction increases its width in a square root relationship with respect to V_{DS}. This increase in the width of the depletion region surrounding the drain junction decreases the effective channel length. In turn, this decrease in effective channel increases, the drain current, resulting in what is commonly referred to as channel length modulation.

To derive an equation to account for channel length modulation.

The width of the depletion region by X_d is

$$X_d = K_{ds}\sqrt{V_{D-ch} + \phi_0}$$

$$= K_{ds}\sqrt{V_{dg} + V_{tn} + \phi_0}$$

where, $$K_{ds} = \frac{\sqrt{2K_s\varepsilon_0}}{qN_A}\ \frac{m}{\sqrt{V}}$$

N_A is used here since the *n*-type drain region is more heavily doped than the *p*-type channel.

ϕ_0 = The built in voltage of the open circuit *pn* Junction

$\phi_0 = V_T\, l_n\, (N_A N_D / n_1^2)$

where $V_T = KT/q$

T → Temperature in degree kelvin (≃300°K)

K → Boltzmann's constant (1.38×10^{-23} JK^{-1})

q → Charge of an electron (1.602×10^{-19} C)

At room temperature, $V_T \simeq 0.26$ mv. Writing a Taylor approximation I_D around its operating value of

$$V_{DS} = V_{GS} - V_{tn} = V_{eff},$$

we find I_D to be given by

$$I_D = I_D \text{ sat} + (\partial I_D/\partial L)(\partial L/\partial V_{DS})\, \Delta V_{DS}$$

$$\simeq I_D \text{ sat } \frac{[1 + K_{ds}(V_{DS} - V_{eff})]}{2L\sqrt{V_{DS} + V_{tn} + \phi_n}} \qquad \text{... (A)}$$

Where, I_D sat is the drain current when $V_{DS} = V_{eff}$. In deriving the final equation (A), the relationship $\partial L/\partial V_{DS} = -\partial X_d/\partial V_{DS}$ is used. Usually equation (A) is written as

$$I_D = \frac{\mu_n C_{ox}}{2}\left(\frac{W}{L}\right)(V_{gs} - V_t)^2 [1 + \lambda (V_{ds} - V_{eff})]$$

where, λ = is the O/P impedance constant and is given by,

$$\lambda = \frac{K_{ds}}{2\lambda\sqrt{V_{dg} + V_{tn} + \phi_o}} = \frac{K_{ds}}{2L\sqrt{V_{ds} - V_{eff} + \phi_o}}$$

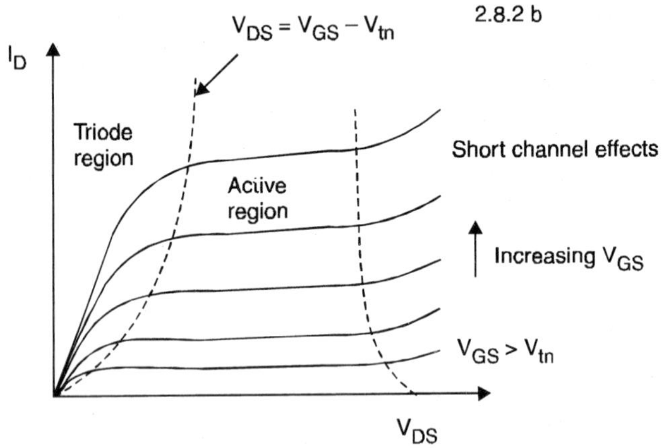

Fig. 2.8.2. (*b*) Plot of I_D versus V_{DS} for different value of V_{gs}.

The triode region is also referred to as linear region. Note that in the active region, the small slope indicates the small dependence of I_D on V_{DS}.

2.8.3. Use of Device Models in Circuit Analysis

Electrical circuits are composed of one or more active devices along with some passive components. The designer is generally interested in controlling the relationship between the electrical port variable. A typical two port network is shown in Fig. 2.8.3 (*a*).

It is often required to find

Large signal Characteristic → V_O Versus V_I or V_I versus I_I

Small signal Characteristic → Voltage gain $A\upsilon = \upsilon_o/\upsilon_i$ or (input impedence $Zin = \upsilon_i / i_i$)

Both are frequency dependent

The performance specification that the analog designer must meet are typically expressed in terms of the small signal characteristics. The knowledge of large signal characteristics is necessary for biasing. (Setting the Q point).

Use of device Models

1. The dc transfer **characteristic of a circuit** are obtained by using the **dc device models** in the network.

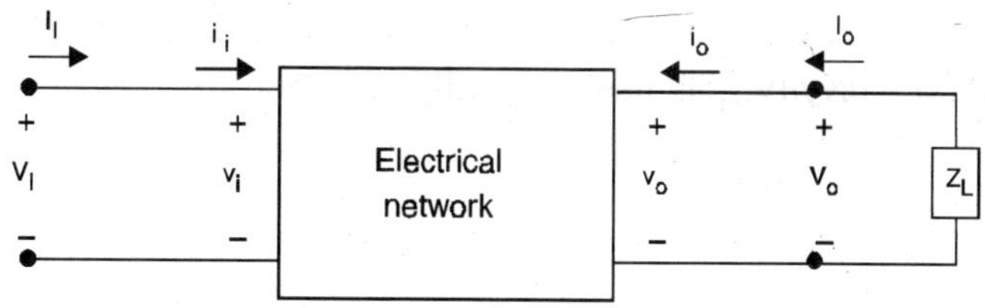

Fig. 2.8.3. (*a*) Typical electrical network structure.

2. To develop insight into the dc performance of even simple networks, assumptions on the device models are necessary.

The low frequency small signal characteristics of a circuit are obtained in two ways.

1. Replacing all devices and elements with a linear small signal equivalent circuit.
2. Based on dc transfer characteristics.

If x_i and x_0 are the small signal input and output variable of interest and X_i and X_0 are the corresponding large signal variables, then these variables are related by

$$\frac{X_0}{X_i} = \frac{\partial X_0}{\partial X_i}\Bigg|_{\text{Q-point}}$$

The results obtained in both cases will be identical provided that the small signal models of the devices are obtained from the same dc model used to obtain the dc transfer characteristics.

But there are some advantages in obtaining small signal transfer characteristic directly from a small signal analysis. 1. Small signal equivalent circuit is linear, 2. Accuracy is high. To find frequency response of the network add parasitic capacitors and devices to the small signal equivalent circuit and the small signal device models. (Additions of capacitors does not affect linearity). Note that it can't be obtained directly from the large signal transfer characteristic.

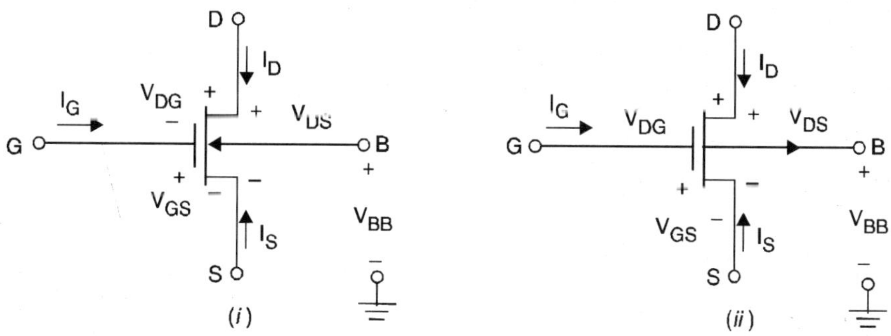

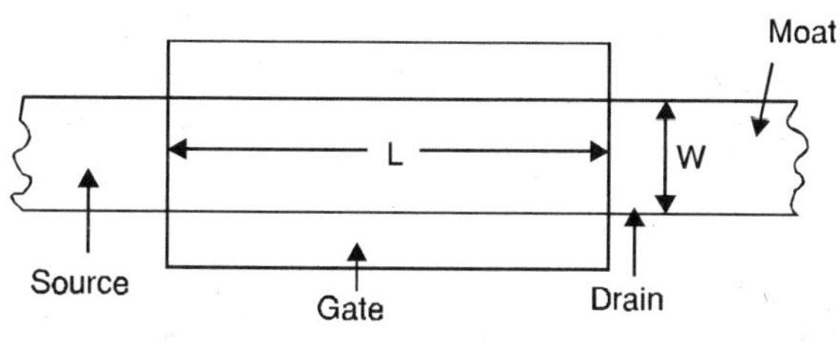

(*iii*)

Fig. 2.8.3. (*b*) (*i*) *n*-Channel MOS, (*ii*) *p*-Channel MOS, (*iii*) Top view of the device.

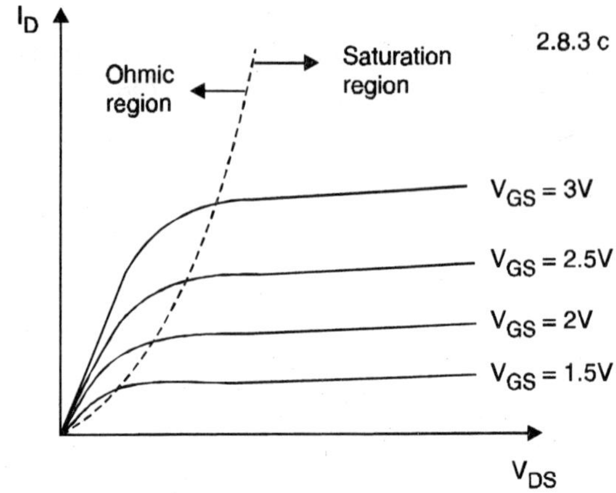

Fig. 2.8.3. (*c*) Output characteristics of *n*-channel MOS.

2.8.4. DC MOSFET Model

For a typical MOSFET, I_{D}-V_{DS} characteristics as a function of V_{GS} is known. To obtain a mathematical model of the MOSFET that will accurately predict the experiment characteristics over a wide range of process parameters and operating conditions we start with sah's model.

Sah's Model

$$I_D = \frac{K'W}{L}\left[(V_{GS} - V_T) - \frac{V_{DS}}{2}\right]V_{DS} \qquad [V_{GS} > V_T \text{ and } V_{DS} \text{ is small (ohmic region)}] \qquad \ldots (1)$$

$I_G = 0 \quad ; \qquad V_{GS} < V_T$

K′ – Transconductance parameter

L, W – design parameters

K′ , V_T – process parameters

If K′ and V_T are constant, the device is electrically symmetric with drain and source. For the Sah's equation a plot of I_D V_S V_{DS} is shown below.

Sah's model is good only in ohmic region

1. When $V_{DS} = V_{GS} - V_T$

   $\delta I_D / \delta V_{DS} = 0$ from equation (1). This causes channel to pinch off.

2. For $V_{DS} > V_{GS} - V_T$, current becomes constant independent of V_{DS}

3. At $V_{DS} = V_{GS} - V_T$

$$I_D = \left(\frac{K'W}{2L}\right)(V_{GS} - V_T)^2$$

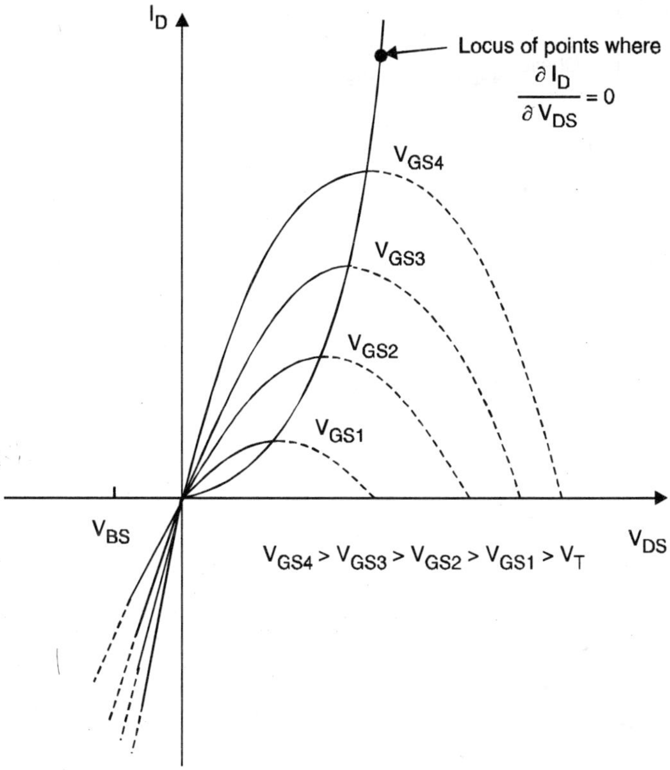

Fig. 2.8.4. (*a*) $I_D - V_{DS}$ Characteristics of Sah's model.

The above equation is good for $V_{DS} > V_{GS} - V_T$ and $V_{GS} > V_T$. This region of operation is saturation region. In this I_D is independent of V_{DS}. The MOSFET is inherently a Transconductance type device. MOSFET is operated in the saturation region in most linear applications. But it can be shown both theoretically and experimentally that the drain current in the saturation region increases slightly in a approximately linear manner with V_{DS}. This is physically due to slight shortening of the channel as V_{DS} is increased in the Saturation region.

Let λ be the Coefficient that represents the linear dependence of I_D on V_{DS}. Now the accurate expression for the drain current in region is given by

$$I_D = \left(\frac{K'W}{L}\right)(V_{GS} - V_T)^2 (1 + \lambda V_{DS}) \qquad V_{DS} > V_{GS} - V_T,\ V_{GS} > V_T$$

Where λ is channel length modulation (λ is very small for long devices but increases considerably for very short transistors.)

The above one is termed as Shichman – Hodges model. Since λV_{DS} is so small in the ohmic region it has little effect on I_D. But in order to incorporate the effects of λ slight discontinuity is created in the model at $V_{DS} = V_{GS} - V_T$. The discontinuity due to λ effects is shown in Fig. 2.8.4. (*b*). Here the device is operated at the saturation ohmic transition region.

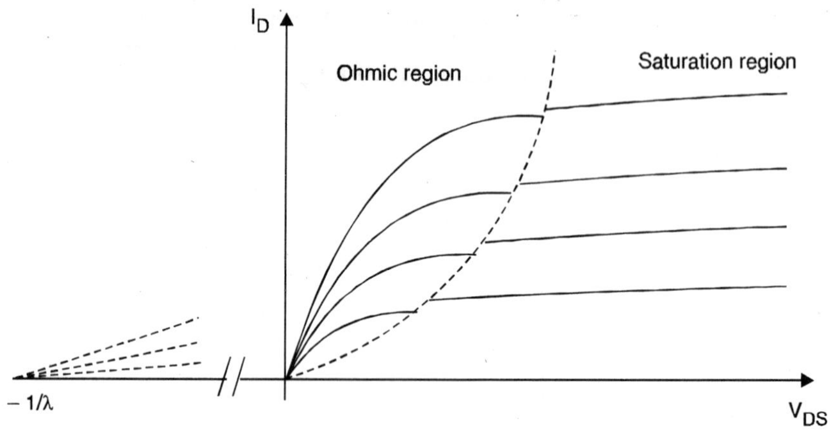

Fig. 2.8.4. (*b*) Discontinuity in model due to λ effects.

V_T – is some what dependent upon the bulk source voltage.

$$V_T = V_{TO} + \gamma\left(\sqrt{\phi - V_{BS}} - \sqrt{\phi}\right)$$

V_{BS} – bulk source Voltage .

V_{TO} – Threshold voltage for $V_{BS} = 0$

γ – Bulk threshold parameter

ϕ – Strong inversion surface potential

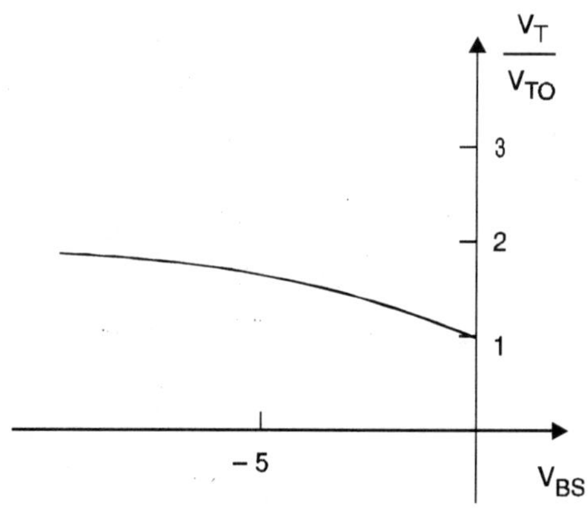

Fig. 2.8.4. (*c*) Effects of V_{BS} on threshold voltage.

The transconductance $K' = \mu C_{OX}$

C_{OX} = Capacitance density of gate channel capacitor.

μ = Channel mobility.

Table. 2.8.4. Low Frequency MOSFET Model

***n*-channel MOSFET**

$$I_G = 0 \tag{1}$$

$$I_D \begin{cases} 0 & V_{GS} < V_T \text{ (cutoff)}, V_{DS} \geq 0 \quad (2) \\ \dfrac{K'W}{L}\left(V_{GS} - V_T - \dfrac{V_{DS}}{2}\right)V_{DS} & V_{GS} > V_T, 0 < V_{DS} < V_{GS} - V_T \text{ (ohmic)} \quad (3) \\ \dfrac{K'W}{2L}(V_{GS} - V_T)^2(1 + \lambda V_{DS}) & V_{GS} > V_T, V_{DS} > V_{GS} - V_T \text{ (saturation)} \quad (4) \end{cases}$$

where $$V_T = V_{T0} + \gamma(\sqrt{\phi - V_{BS}} - \sqrt{\phi}) \tag{5}$$

***p*-channel MOSFET**

$$I_G = 0 \tag{6}$$

$$I_D \begin{cases} 0 & V_{GS} < V_T \text{ (cutoff)}, V_{DS} \leq 0 \quad (7) \\ \dfrac{K'W}{L}\left(V_{GS} - V_T - \dfrac{V_{DS}}{2}\right)V_{DS} & V_{GS} < V_T, 0 > V_{DS} > V_{GS} - V_T \text{ (ohmic)} \quad (8) \\ \dfrac{K'W}{2L}(V_{GS} - V_T)^2(1 + \lambda V_{DS}) & V_{GS} < V_T, V_{DS} < V_{GS} - V_T \text{ (saturation)} \quad (9) \end{cases}$$

where $$V_T = V_{T0} + \gamma(\sqrt{\phi + V_{BS}} - \sqrt{\phi}) \tag{10}$$

Design parameters : W = channel width
L = channel length

Process parameters : K' = transconductance parameter
V_{T0} = threshold voltage for $V_{BS} = 0$
γ = bulk threshold parameter
ϕ = strong inversion surface potential
λ = channel lengh modulationparameter

K', γ, ϕ and λ are positive for both n-channel and p-channel transistors, n-channel transistors are termed enhancement if $V_{T0} > 0$ and depletion if $V_{T0} < 0$, p channel transistors are termed enhancement if $V_{T0} < 0$ and depletion if $V_{T0} > 0$.

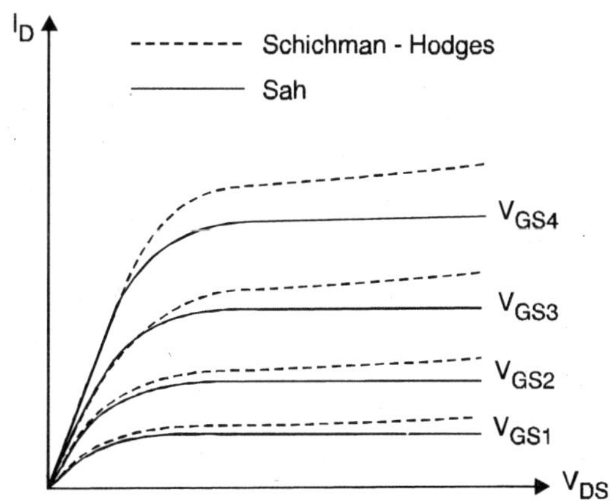

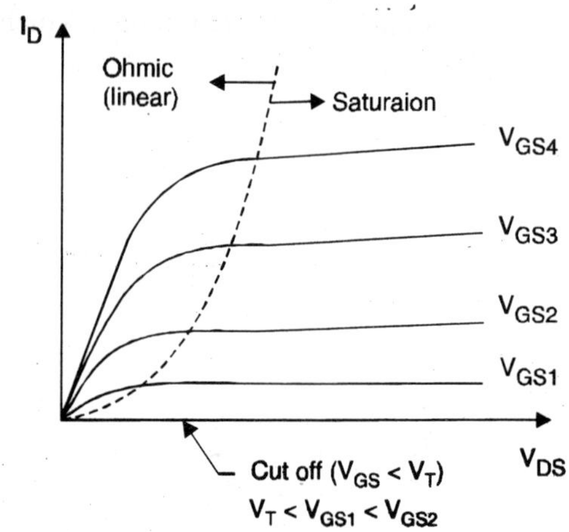

Fig. 2.8.4. (*e*) Regions of operation of MOSFET in $I_D - V_{DS}$ plane.

The operating regions of the MOSFET in the $I_D - V_{DS}$ plane are shown in Fig. 2.8.4 (*e*). The near linear relationship between V_{DS} and I_D can be seen in the ohmic (linear region). The dependence of I_D on V_{GS}, and essential independence from V_{DS}, should be apparent in the saturation region.

Equivalent Current for the dc operation of the MOSFET

We know in ohmic region

$$I_D = \left(\frac{K'W}{L}\right)\left[(V_{GS} - V_T) - \frac{V_{DS}}{2}\right] V_{DS} \text{ for } V_{gs} > V_T$$

If $V_{ds}/2$ is negligible compared to $V_{GS} - V_T$. $R_{FET} = \dfrac{V_{DS}}{I_D} = \dfrac{L}{WK'}(V_{GS} - V_T)$

As V_{DS} increases towards $(V_{GS} - V_T)$, the relationship between I_D and V_{DS} becomes nonlinear.

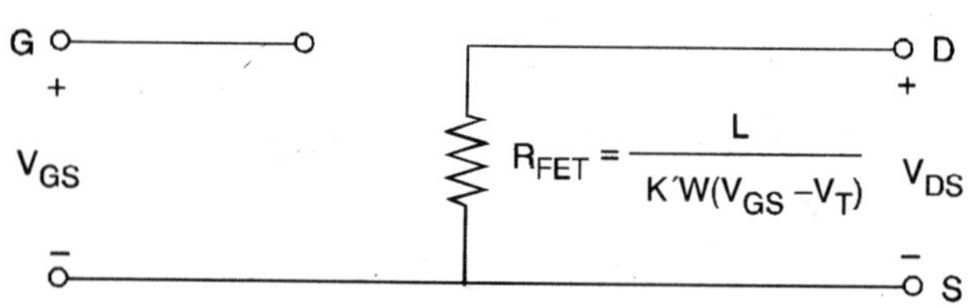

Fig. 2.8.4. (*f*) Equivalent circuit of MOS in ohmic region.

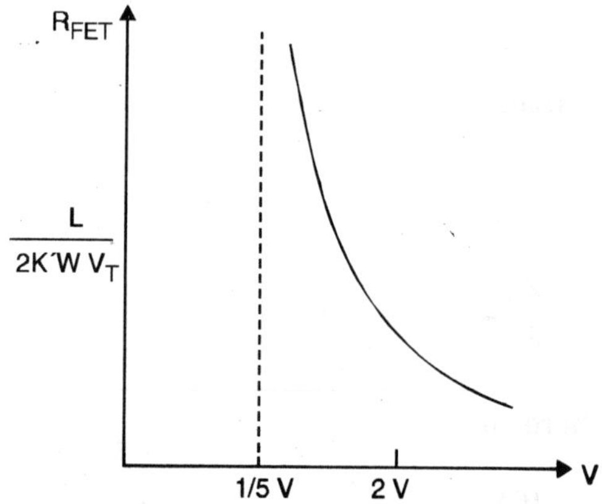

Fig. 2.8.4. (*g*) MOS resistance versus V_{GS} in ohmic region.

In ohmic region it is used as a VVR (voltage variable resistor) with V_{GS} as controlling Voltage.

R_{FET} is very sensitive to small changes in V_{GS}. The model which we have used is not very accurate for small $(V_{GS} - V_T)$. For the above said reasons this device is not used as a very high impedance in this region.

2.8.5. Small Signal MOSFET Model

Small signals model can be obtained directly from the dc model. There are three regions of operation, namely cut off, sat, ohmic and correspondingly we have small signal models for each region.

1. In cut off $I_D = 0$ resulting in a trivial small signal model.

2. Because of performance limitations MOSFET is not biased for small signal operation in the ohmic region.

3. Most small signal applications empty the MOSFET biased in sat region.

In saturation region

$$I_D = \frac{K'W}{L}(V_{GS} - V_T)^2 (1 + \lambda V_{DS})$$

When, $I_G = I_B = 0$

Let $g \to 1, d \to 2 \quad \text{and } b \to 3$

$$Y_{dg} = g_m = \partial I_d / \partial V_{GS}$$

$$V_{DS} = V_{DSQ}$$

$$V_{BS} = V_{BSQ}$$

$$V_{GS} = V_{GSQ}$$

$$= \frac{K'W}{L}(V_{GS} - V_T)(1 + \lambda V_{DS})$$

upon relating $V_{GS} - V_T$ to the quiescent drain current and noting that the λ effects are typically negligible in this model, it follows that

$$g_m \approx \sqrt{\frac{2K'W}{L}} \sqrt{|I_{DQ}|}$$

Nonzero small signal model parameters for MOSFET

1. $$g_m = \sqrt{\frac{2K'W}{L}} \sqrt{|I_{DQ}|}$$

2. $$g_{mb} = \eta g_m$$

3. $$g_{ds} = \lambda |I_{dq}| (r_{ds} = 1/g_{ds})$$

where $\eta = \frac{\delta}{2} 2\sqrt{Q - V_{BSQ}}$

If $V_{bs} \neq 0$, any ac signals that appear on the bulk will also modulate the drain current. Although g_{mb} is less than gm, g_{mb} limits the performance of many circuits. Since the current $g_m V_{gs}$ typically dominates the drain current, it should be apparent that the small signal MOSFET is inherently a good Tran conductance amplifier. But its performance as a current amplifier, voltage amplifier or tranresistance amplifier would be poor. (Drain is connected to source and also to substrate) so we have two transconductance.

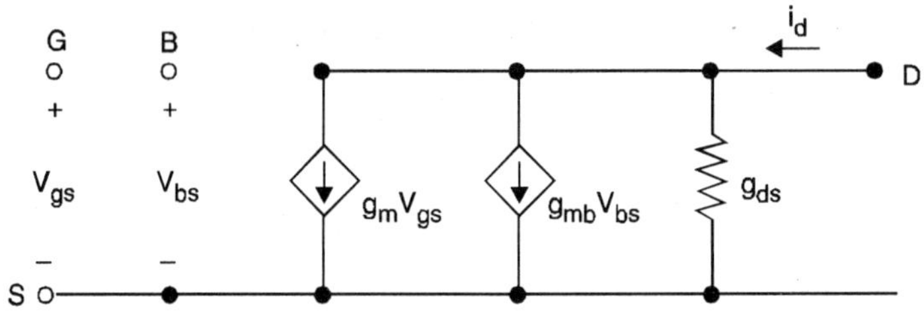

Fig. 2.8.5. Small signal model for MOS.

2.8.6. High Frequency MOSFET Model

MOS Device Capacitances. In this model we have to include the effects of parasitic capacitance. These parasitic capacitance can be divided into two groups. The first group is composed of those parasitic capacitors formed by sandwiching an insulating dielectric of fixed geometric dimensions between two conductive regions. (These capacities are not voltage dependent)

$$C = \xi A/\delta \quad ; \quad \xi/d = C_d$$

$$C = C_d$$

where C_d – process parameter

A – design parameter.

Second group is composed of the capacitors formed by the separation of charge associated with a PN junction. The depletion region associated with the semiconductor junction serves as the dielectric. These junction capacitances are voltage dependent.

$$C = C_{jo} A / (1 - V_F / \phi_B)^n$$

where C_{jo} – Process parameter junction capacitance density at zero volt bias.

A – junction area

V_F – dc forward bias voltage of *pn* junction.

ϕ_B – Barrier potential ≈ (0.7V)

n – constant (1/2 for step graded and 1/3 for linearly graded junctions.)

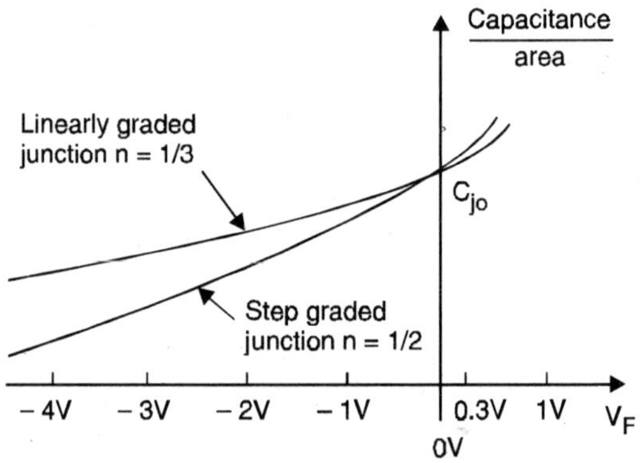

Fig. 2.8.6. (*a*) Voltage-dependent capacitance of *pn* junctions.

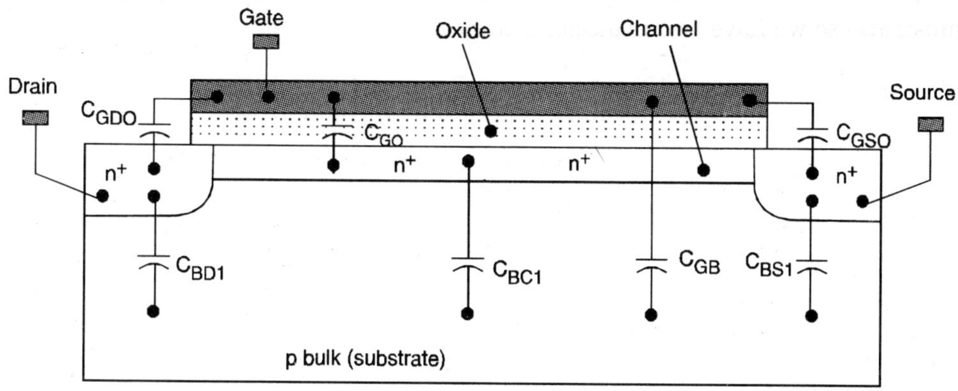

Fig. 2.8.6. (*b*) Parasitic capacitors in MOS transistors, shown for *n*-channel device.

A plot of junction capacitance density versus forward bias is shown in Fig. 2.8.6 (*a*) and the Fig. 2.8.6 (*b*) shows the parasitic capacitors that dominate the high frequency behaviour of MOS transistors as C_{BDI}, C_{BCI}, and C_{BSI}. These three are voltage dependent junction capacitors. And the remaining are parallel plate capacitors. C_{OX} – capacitance density of the gate/oxide/channel capacitor.

$$C_{OX} = \xi_0 S_iO_2/t_{ox}$$

ξ_0 – permittivity of free space.

C_{GDO} and C_{GSO} – Gate drain and Gate source overlap capacitors.

$C_{GDO} = C_{GSO} = C_{ox} WL_D$ in cut off region.

L_D = Distance of the lateral moat diffusion under the gate.

$C_{GC}0$ – Gate to channel capacitance.

In cut off region as the channel is not formed $C_{GC} = 0$. In ohmic region it extends uniformly from drain to source under the entire gate region. $C_{GC} = C_{ox}$ WL. In saturation region, the majority of the channel area acts as ohmic extension of the source region. $C_{GC} = \frac{2}{3}(C_{ox} WL)$. It is modeled as a lumped element between gate and source and the remaining $\frac{1}{3}(C_{ox} WL)$ is neglected.

In ohmic region $C_{GO} = C_{GS} = C_{OX} WL_D + \frac{1}{2} WL\, C_{OX}$. In saturation region channel is heavily inverted. The drain region is pinched off.

$$\therefore \quad C_{GO} = C_{OX} WL_D \text{ and } C_{GS} = C_{OX} WL_D + \frac{2}{3} C_{ox} WL$$

$$G_B = \text{gate bulk capacitance.}$$

In cut off region there is no depletion region in the channel and no inversion layer exist. So the bulk region extends to the bottom side of the gate oxide layer and $C_{GB} = C_{OX}$ WL. In both saturation and ohmic regions, the existence of the inversion layer makes C_{GB} zero.

$$C_{BCJ} = \text{Bulk to channel junction capacitance.}$$

In cut off no inversion layer and $C_{BCI} = 0$. In ohmic and saturation regions, a junction

In cut off no inversion layer and $C_{BCI} = 0$. In ohmic and saturation regions, a junction capacitance exists. It is distributed between drain and source with ½ : ½ in ohmic and 0 : 2/3 in satu-ration regions.

C_{BSJ} and C_{BDJ} are capacitors of bulk/source and bulk/drain junctions.

The lateral junction Area determines all the junction capacitances discussed so far. For short channel transistors (less than 3–5μ in length) the lateral capacitance associated with the edge of the junction becomes significant. These are termed as sidewall capacitances. The lumped model and small signal equivalent circuit are shown in Fig. 2.8.6 (c) below.

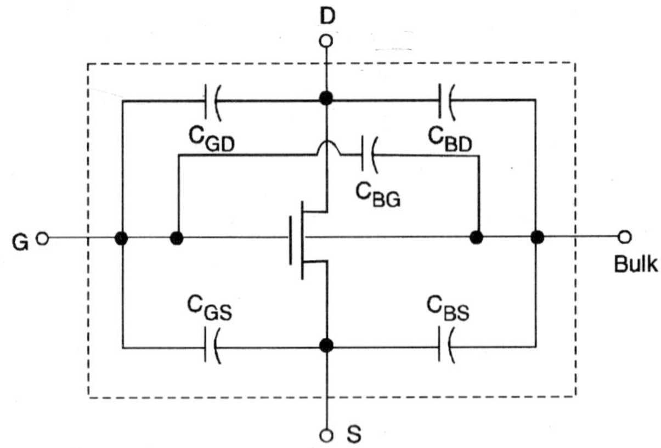

(*i*) Lumped model.

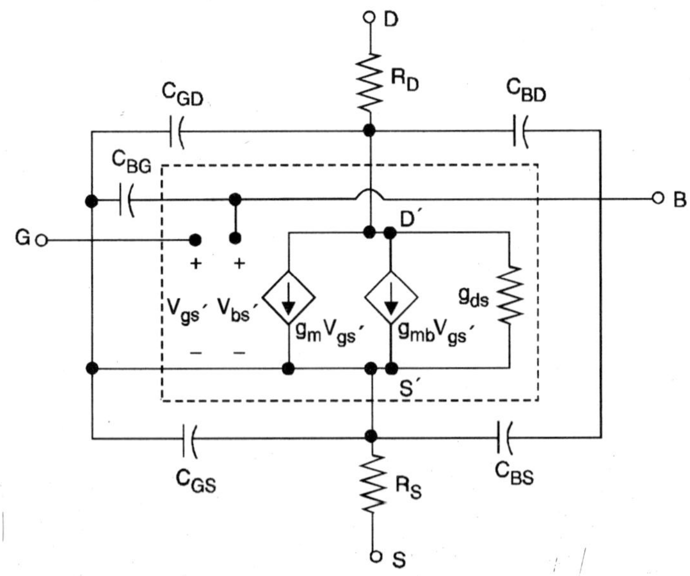

(*ii*) Small signal equivalent circuit.

Fig. 2.8.6. (*c*) Parasitic capacitors.

Table 2.8.6. Values of Parasitic Capacitors

	Region		
	Cut off	*Ohmic*	*Saturation*
C_{GD}	$C_{OX}\,WL_D$	$C_{OX}\,WL_D + \frac{1}{2}\,WLC_{OX}$	$C_{OX}\,WL_D$
C_{GS}	$C_{OX}\,WL_D$	$C_{OX}\,WL_D + \frac{1}{2}\,WLC_{OX}$	$C_{OX}\,WL_D + \frac{2}{3}\,WLC_{OX}$
C_{BG}	$C_{OX}\,WL$	0	0
C_{BD}	C_{BD1}	$C_{BD1} + \frac{C_{BC1}}{2}$	C_{BD1}
C_{BS}	C_{BS1}	$C_{BS1} + \frac{C_{BC1}}{2}$	$C_{BD1} + \frac{2}{3}\,C_{BC1}$

The R_D, R_S are ohmic resistances in source and drain regions from the physical drain and source connections to the actual drain (D´) and source (S´) of the device which will be approximately equal to 50 ohms.

2.8.7. Measurement of MOSFET Model Parameters

Knowledge of process parameters, as well as of statistical and or worst case information, is necessary during the design stage so that trade offs between performance, complexity and yield can be considered by the designer. Experimentally measured process parameters also serve as an interface between the fabrication and design groups. After fabrication it is checked to verify if the process parameters fall within bounds.

Measurement of λ :

In saturation region

$$I_D = \frac{K'W}{2L}(V_{GS} - V_T)^2 (1 + \lambda V_{DS})$$

For $\qquad V_{DS} > V_{GS} - V_T$ and $V_{GS} > V_T$ ... (A)

For different Values of $(V_{GS} - V_T)$ with constant V_{BS} the above equation is plotted and shown in Fig. 2.8.7 (*a*) below.

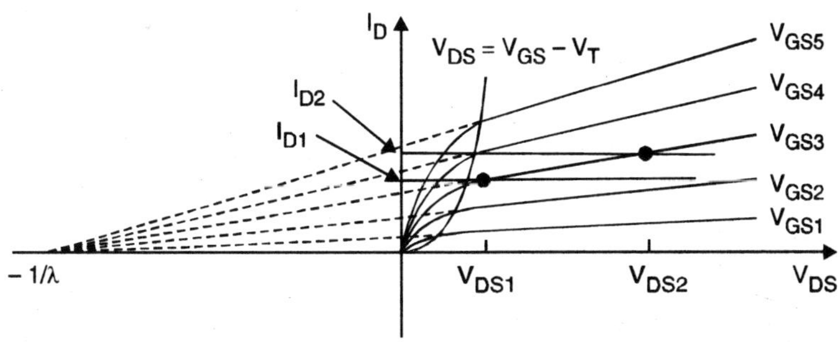

Fig. 2.8.7 (*a*) Measurement of λ.

Projection of these curves into quadrant 2 shows that all curves intersect the V_{DS} axis @ $V_{DS} = -1/\lambda$. therefore the λ can be determined by fixing V_{GS} and V_{BS}, (thus fixing $V_{GS} - V_T$) and determining the $-V_{DS}$ axis intercept of I_D V_S V_{DS}. Let I_{D1} and I_{D2} be the currents corresponding to two distinct values V_{DS1} and V_{DS2}. Then

$$\lambda = \frac{I_{D2} - I_{D1}}{I_{D1} V_{DS2} - I_{D2} V_{DS1}}$$

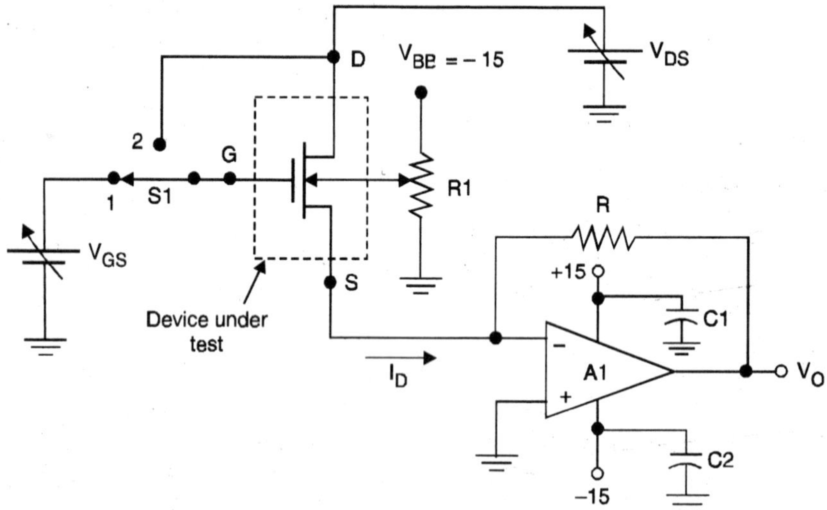

Fig. 2.8.7. (*b*) Circuit for measuring MOSFET parameters.

Put S_1 in position 1 for measuring λ. Fix V_{GS} and V_{BS} so that $V_{GS} > V_T$. Op-amp is used to establish $V_0 = -I_D R$ while maintaining source Vol = OV.

$$\lambda = V_{02} - V_{01} / V_{01} V_{0s2} - V_{02} V_{0s1}$$

where $V_{01}, V_{02}, \rightarrow$ O/P vol. of OP amp

Measurement of V_T and V_{T0}

With V_{BS} and V_{DS} fixed, the values of I_D ($I_{01} - I_{02}$) for V_{GS1} and V_{GS2} can be measured. If for both values of V_{GS} the device is operating in saturation region then

$$V_T = (V_{GS2} (I_{01}/I_{01})^{1/2} - V_{GS1}) / (I_{01}/I_{02})^{1/2} - 1$$

If V_{BS} is set to O_v. V_{TO} can be found. Here also S_I is put in position 1 to measure V_T
Measurement of $\mathbf{K^1}$: Switch S_1 is kept @ position 2. From equation (A)

$$K' = \frac{2LI_D}{W} (V_{GS} - V_T)^2 (1 + \lambda V_{DS})$$

Since λ V_{DS} is much less than 1.

$$I_D \approx \frac{K'W}{2L} (V_{GS} - V_T)^2 \quad \text{for } V_{GS} > V_{T,} V_{DS} > V_{GS} - V_T$$

For depletion devices to establish a large dc bias on V_{DS} keep S_1 in position 1.

$$I_D = \frac{K'W}{2L}(V_{GS} - V_T)^2 \qquad \text{for } V_{GS} > V_T$$

Measurement of δ

We know $V_T = V_{T0} + \gamma\left(\sqrt{\phi - V_{BS}} - \sqrt{\phi}\right)$

Both γ and ϕ can be determined by obtaining values of V_T, from two distinct non zero values V_{BS}. For large values of V_{BS}, the threshold voltage is only weakly dependent upon ϕ. In that case, using a typical value of ϕ, ϕ_{typ}, a good approximation to δ can be obtained with a single large value of V_{BS}.

$$\gamma \approx \frac{V_T - V_{T0}}{\sqrt{\phi_{typ} - V_{BS}} - \sqrt{\phi_{typ}}}$$

2.8.8. Short Channel Devices (Channel Lengths < 3μ)

In this the channel region is more irregular in structure. The effects of the transition region from the drain and source diffusions to the channel become significant, causing need for increased complexity in the device model.

Advantage of Short Channel Devices

1. Reduced area.
2. Improved speed (because of reduced i/p capacitance).

Limitations

Deterioration in the O/P impedence.

Matching is difficult to achieve.

O/P characteristic of a short channel transistor compared with that of a long channel device is shown below Fig. 2.8.8 (*a*).

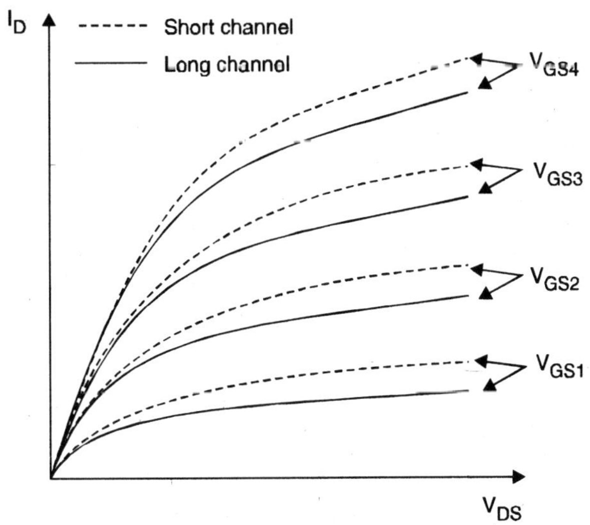

Fig. 2.8.8. Short channel effects. —— 5μ device, - - - - 1μ device.

$I_G = 0$

$I_D = 0,\ V_{GS} < V_T$ (cut off)

$$I_D = \frac{K'W_{eff}}{L_{eff}}(V_{GS} - V_T - V_{DS}/2)\,V_{DS}\left[1+\lambda\left(\frac{1+\theta L_{eff}}{\theta L_{eff}}\right)V_{DS}\right] \quad |\ V_{GS} > V_T,\ V_{GS} - V_T > V_{DS} \text{ [ohmic]}$$

$$I_D = \frac{K'W_{eff}}{L_{eff}}(V_{GS} - V_T)^2\left[1+\lambda\left(\frac{1+\theta L_{eff}}{\theta L_{eff}}\right)V_{DS}\right] \quad |\ V_{GS} > V_T,\ V_{DS} > V_{GS} - V_T \text{ [sat]}$$

θ has units $(\text{length})^{-1} \approx 0.2\ \mu^{-1}$

W_{eff}, L_{eff} denote the effective channel width and length.

$$W_{eff} = W - W_R \qquad L_{eff} = L - 2\,L_D - L_R$$

Where W, L are drawn width and length. W_R and L_R are constants representing width and length reduction due to processing. I_D is the lateral diffusion of the source or drain under the gate. Short channel devices are widely used in digital logic, where a net improvement in speed reduction in circuit area are readily achievable.

2.8.9. Sub-threshold Operation

It has been assumed that when $V_{GS} < V_T$, $I_D = 0$ with positive V_{DS} and when $V_{GS} > V_T$, $I_D \neq 0$ with V_{DS} = +ve. But it is not true experimentally. For $V_{GS} < V_T$ and with V_{DS} = +ve there will be a small I_D. This extremely small current finds application in the field of Bio-medical. If $V_{GS} < V_T$, the devices are said to be operating in weak inversion, or sub threshold region. At room temp, the transition between strong inversion and weak inversion actually occurs around $V_{GS} \cong V_T$ + 100 mV.

A plot of I_D V_S V_{GS} for the MOS with $V_{DS} = V_{GS}$ is shown in Fig. 2.8.9 (*a*).

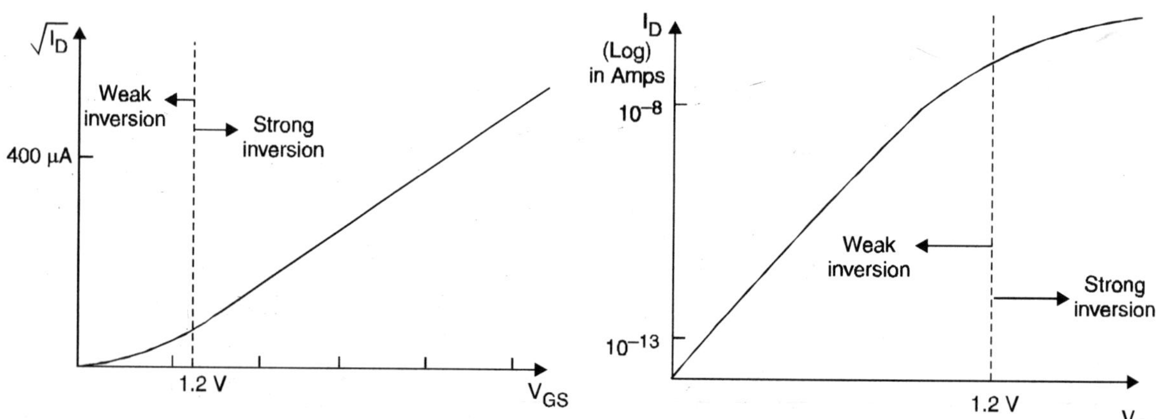

(*i*) Vertical axis – quadratic. (*ii*) Vertical axis – logarithmic.

Fig. 2.8.9. (*a*) $I_D - V_{DS}$ characteristics for MOS in weak inversion.

A dc model for the MOS operating in weak inversion

$V_{GS} = V_T + 2n\,V_t$ *n* is constant between 1 and 2

V_t = KT/q = 26 mV @ room temp.

K – Boltzmann's constant

where T – device temp. in Kelvin

q – charge of an electron.

$$I_G = 0 \quad \text{and} \quad I_D = \frac{W}{L} I_{DO}\, e^{-V_{BS}[(1/nV_t) - (1/V_t)]} (1 - e^{-V_{DS}/V_t})$$

$$= \frac{W}{L} I_{DD}\, e^{-V_{BS}[(1/nV_t) - (1/V_t)]} \left(1 - e^{-V_{DS}/V_t}\right) e^{(V_{GS} - V_t)/nV_t}$$

where V_t – Threshold Voltage

I_{DO} and n are process parameters and $I_{DO} \cong 20$ n A and $n = 2$.

A plot of I_D V_S V_{DS} is shown in Fig. 2.8.9 (*b*).

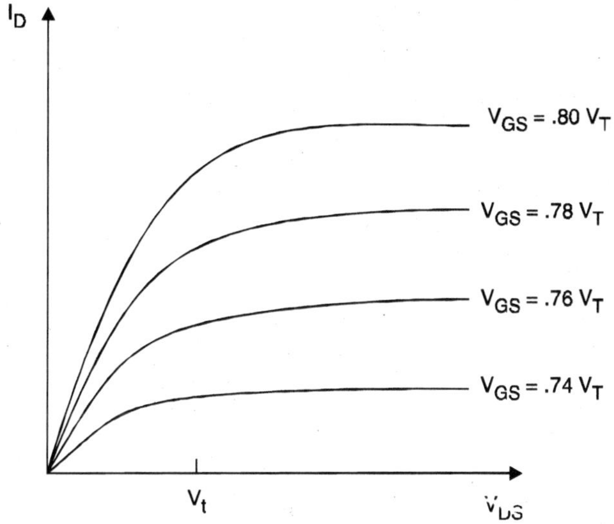

Fig. 2.8.9. (*b*) $I_D - V_{DS}$ characteristics for MOSFET operating in subthreshold.

If $V_{DS} > 3V_t$ the term e^{-V_{ds}/V_t} becomes negligible and it attains strong inversion. If V_{BS} is also equal to 0.

Then $\quad I_G = 0 \quad$ and $\quad I_D \cong W/L\, I_{DO}\, e^{(V_{gs} - V_T)/nV_t}$

$$I_{DO} \cong \frac{K' 2(nV_t)^2}{e^2}$$

Practical limitation of devices operating in weak inversions

1. Frequency Response is poor.

2. The drain and source substrate currents associated with the reverse biased moat-substrate junction are not necessarily negligible compared to subthreshold drain currents.

3. Linearity is poor for $V_{DS} > 3V_t$.

4. Deterioration of matching characteristics of MOS transistors with decreasing drain currents.

2.8.10. Modeling Noise Sources in MOSFET

Two primary noises in MOSFET

1. Thermal noise associated with carriers.
2. Flicker noise associated with trapping and releasing of electrons in $Si\text{-}SiO_2$ interface.
3. These noise sources contribute to the total drain current and can thus be modeled as a current source between the drain and source. The thermal noise current is white noise. It has zero mean. It is most easily characterized by its spectral density.

$$S_{IW} = 4\ KT/R_{FET} \text{ in ohmic region}$$
$$= 8\ KTgm/3 \text{ in saturation region}$$

where T – temp. is * Kelvin

K – Boltzmann's constant

$$R_{FET} - \text{FET resistance} \cong \frac{K'\,2(nV_t)^2}{e^2}$$

$$g_m - \text{small signal trans conductance} = \sqrt{2K'\frac{W}{L}}\ \sqrt{|I_{DQ}|}$$

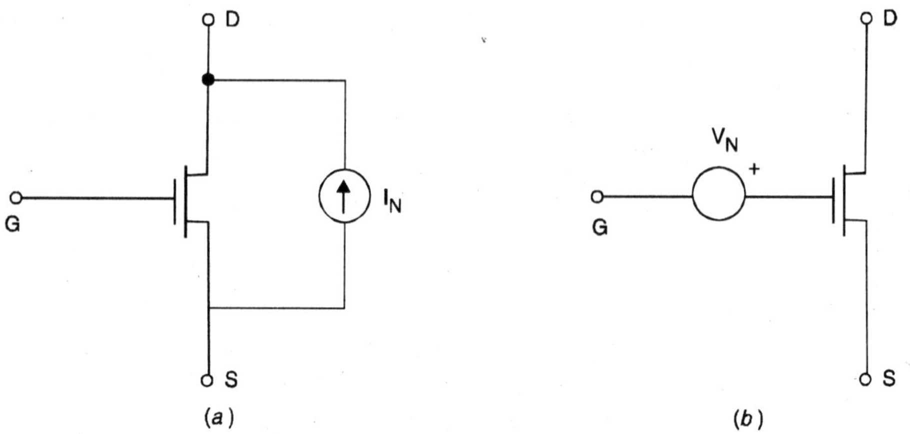

Fig. 2.8.10. Model of noise sources in the MOSFET.

The flicker noise current in both the saturation and ohmic regions is characterized by the spectral density

$$S_{IF} = 2K_f\,K'\,I_{DQ}/C_{OX}\,L^2f$$

where K_f – Flicker noise co-efficient

I_{DQ} – quiescent current

f – frequency. K', C_{OX} and L – model parameters. This is termed as 1/f noise because of the 1/f dependence. The white noise and 1/f noise are statistically uncorrelated and hence the spectral current density of the noise current source, I_N can be obtained by adding S_{IW} and S_{IF} to get $S_N = S_{IW} + S_{IF.}$ The RMS noise current source in the freq. Band $[f_2, f_2]$ can be obtained from the spectral density and is given by —

$$I_{NB} = \sqrt{\int_{f_2}^{f_1} S_N \, df}$$

Also I_{WB} – RMS white noise current = $\sqrt{\int_{f_2}^{f_1} S_{1W} \, df}$

and I_{FB} – RMS flicker noise current = $\sqrt{\int_{f_2}^{f_1} S_{1F} \, df}$

$$I_{NB} = \sqrt{I^2{}_{WB} + I^2{}_{FB}}$$

An equivalent small signal model that characterizes the noise performance can be obtained by replacing the noise current source from drain to source with a noise voltage source in series with the gate. (V_N). This is termed an input referred noise source. The spectral density of the input referred noise voltage source relates to the noise spectral density of the noise current source by the expression

$$S_{VN} = \frac{S_{IN}}{g_m^2}$$

Correspondingly the i/p referred noise voltage relates to the noise current source by the expression

$V_N = I_N / g_m$ Where g_m – small signal transconductance gain

2.8.11. MOSFET Model for Digital Application

Many digital application involve large numbers of gates and transistors. But analysis using the device models studied so far is totally impractical. Therefore a simple model is needed. But this will not provide highly accurate results. Models that are widely used for this purpose are shown below :

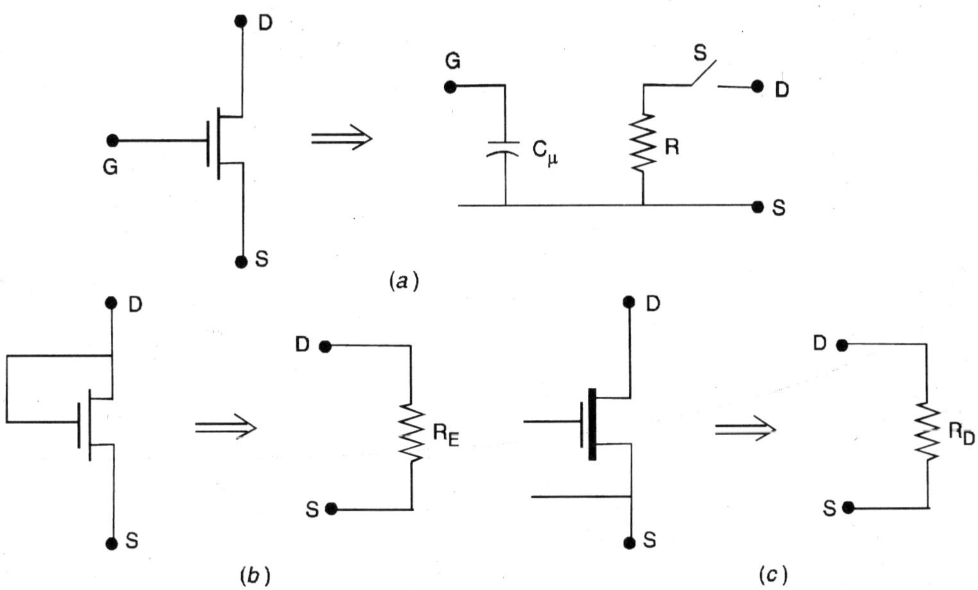

(*a*) Three terminal MOSFET. (*b*) Two terminal enhancement. (*c*) Two terminal depletion.

Fig. 2.8.11. Models for the MOSFET used for simple analysis of digit circuits.

In Fig. (*a*) S_1 is open if $V_{GS} < V_H/2$ and closed if $V_{GS} > V_H/2$. (V_H = logical high in digital logic.)

$$C\mu = C_{ox} WL$$

$$R = \frac{L}{K' W (V_H - V_T)}$$

The two terminal enhancement and depletion configuration of Fig. (*b*) and (*c*) are characterized by

$$R_E = \frac{L}{K' W (V_H - V_{TE}) R_D} = \frac{L}{K' W |V_{TD}|}$$

where V_{TE}, V_{TD} denote threshold voltage of the enhancement and depletion transistors.

3

Basic Circuits for Digital System

INTRODUCTION

The Logic circuits are implemented using transistors and a number of different technologies exist viz. NMOS, CMOS, PMOS, BiCMOS. This chapter explains how various circuits used in digital systems can be implemented using MOS transistors. Let us first consider how logic variables can be represented as signals in electronic circuits. Here we are concentrating on binary values '0' and '1'. These values can be represented either as levels of voltage or current. But most popular representation is using voltage levels. The minimum voltage is represented as V_{SS} and maximum voltage is represented as V_{DD}. We will assume that V_{SS} is '0'V, corresponding to electrical ground, denoted as Gnd. The voltage V_{DD} represents the power supply voltage and the most common level for V_{DD} is 5 V, but 3.3V is also popular. In this chapter, we will assume that V_{DD} = 5V.

3.1. TRANSISTOR SWITCHES

Logic circuits are built with transistors. Now we will see how a transistor operates as a simple switch. The figure below shows the symbol of NMOS transistor.

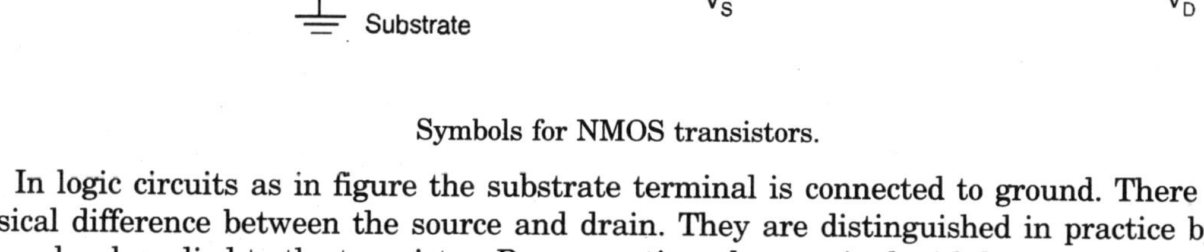

Symbols for NMOS transistors.

In logic circuits as in figure the substrate terminal is connected to ground. There is no physical difference between the source and drain. They are distinguished in practice by the voltage level applied to the transistor. By convention, the terminal with lower voltage level is referred as the source. The voltage levels applied to the gate controls the transistor. If V_G is low, then there is no connection between source and drain and we say that the transistor is turned off. If V_G is high then the transistor is turned on and it acts as a closed switch that connects the source and drain. In other way when S = 1 a '0' from '*b*' is passed to '*a*' since nMOS is a good conductor of logic '0'.

The figure below illustrate this.

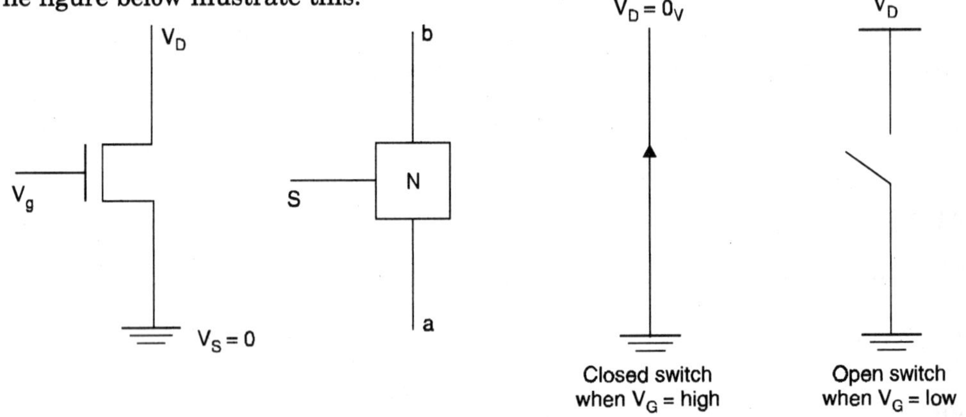

Similarly in logic circuits the substrate of PMOS is always connected to V_{DD}. The figure below shows symbols for PMOS.

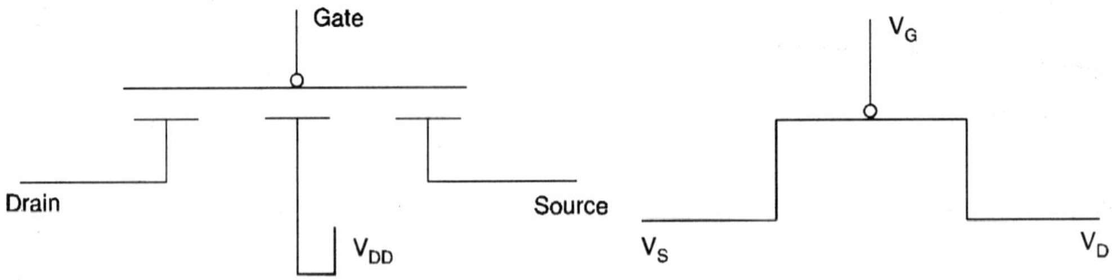

Symbols for PMOS.

If V_G is high, PMOS is turned off and acts as a open switch. If V_G is low transistor is turned on and acts as a closed switch that connects the source and drain. In the PMOS transistor the source is the node with the higher voltage. In other way when S = 0 a '1' is passed from '*b*' to '*a*' since PMOS is a good conductor of logic 1.

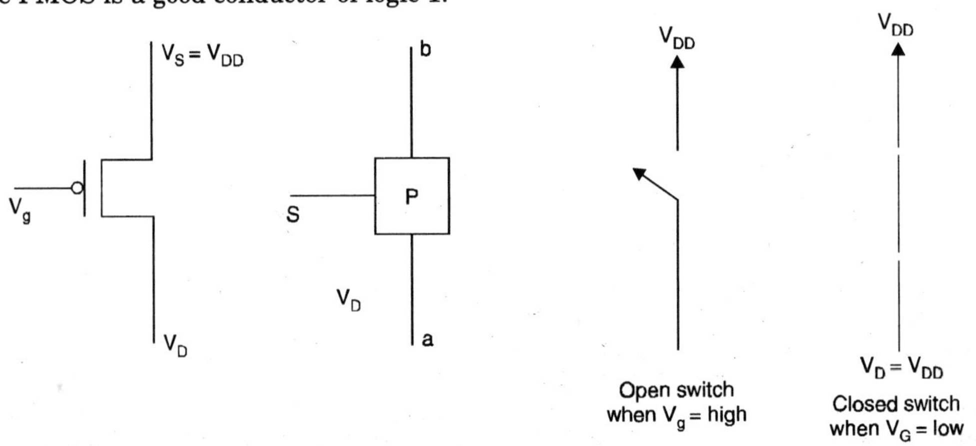

Summarizing

A NMOS transistor is turned on when its gate terminal is high. And when it is turned on the drain is pulled down to ground. While a PMOS transistor is turned on when the gate is low and when it is turned on the drain is pulled to V_{DD}. By the way the transistors operate, we can understand the NMOS transistor cannot be used to pull its drain complete up to V_{DD}. Similarly a PMOS transistor cannot be used to pull its drain terminal completely to ground.

A N switch is almost a perfect switch when a '0' is to be passed from i/p to o/p. However N switch is an imperfect switch when passing a '1'. In doing this the '1' voltage level is reduced by a small amount. The *p*-switch functions in a opposite way. It is almost a perfect switch when a '1' is to be passed from i/p to o/p. However it acts as an imperfect switch when passing a '0'.

3.2. REALISATION OF GATES USING NMOS TECHNOLOGY

Logic circuits or gates can be built with NMOS or PMOS or with both. First we will see the circuit implementation using NMOS only.

(*i*) **Inverter.** The figure below shows Logic symbol, circuit implementation using NMOS and Truth table of an Inverter.

Logic symbol

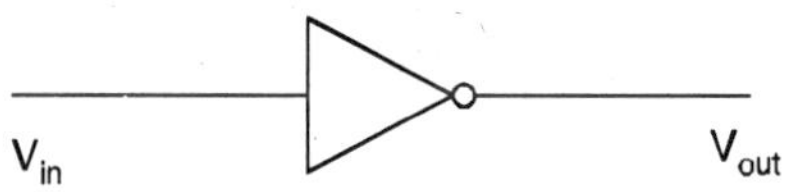

Truth Table

V_{in}	V_{out}
0	1
1	0

Circuit Implementation

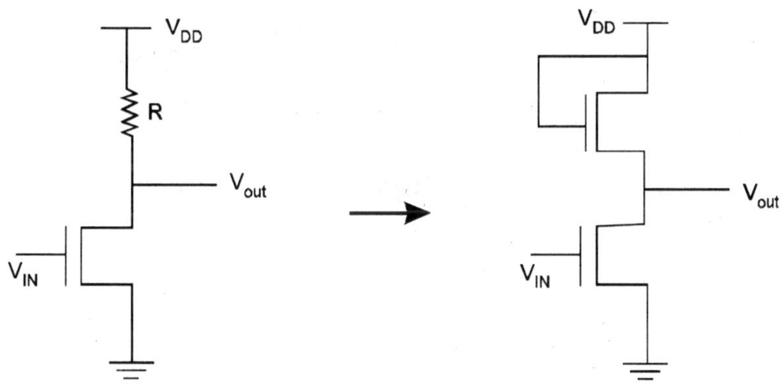

The purpose of the resistor in NOT gate is to limit the current flow when V_{IN} = 5 V. When V_{IN} = 0V, NMOS is turned off and no current flows through resistor and V_{out} = 5 V = V_{DD}. *i.e.,* o/p is connected to V_{DD} and no path or connection between o/p and gnd. When V_{in} = 5 V, NMOS is turned on, and pulls V_{OUT} to a low voltage level. Now o/p is connected to Gnd terminal. This acts as an inverter. Rather than using a resistor to limit the current flow, an NMOS transistor with its gate shorted to V_{DD} is normally used.

(*ii*) **NAND gate :** The figure below shows logic symbol, circuit implementation and truth table of NAND gate.

Logic Symbol

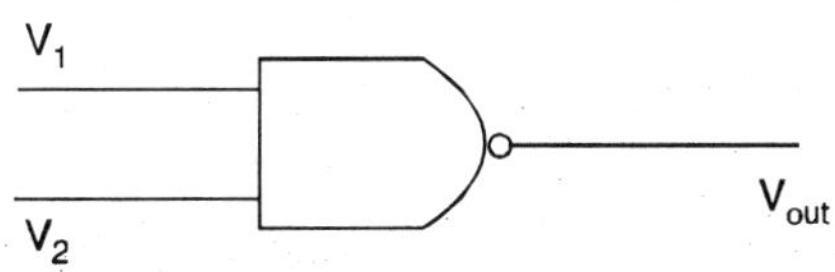

Circuit Implementation

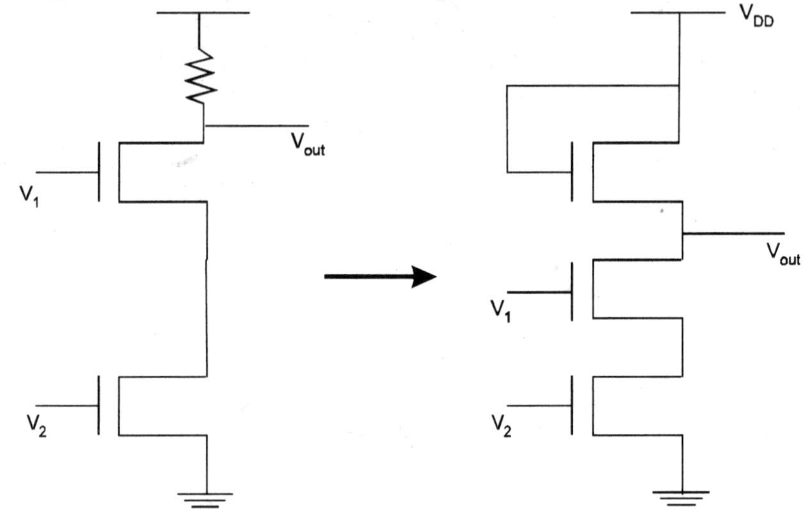

Truth Table

V_1	V_2	V_{out}
0	0	1
0	1	1
1	0	1
1	1	0

When $V_1 = V_2 = 5$ V both transistors will be ON and put V_{out} to a low voltage. If either V_1 or V_2 is '0' then there is current flow through the series connected transistors and $V_{out} = 5$ V. The realized function is a NAND function

(*iii*) **NOR gate.** The figure below shows logic symbol, circuit implementation and truth table of NOR gate.

Logical Symbol

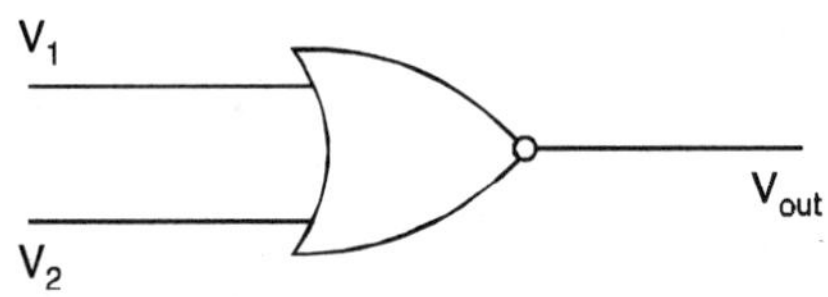

Circuit Implementation

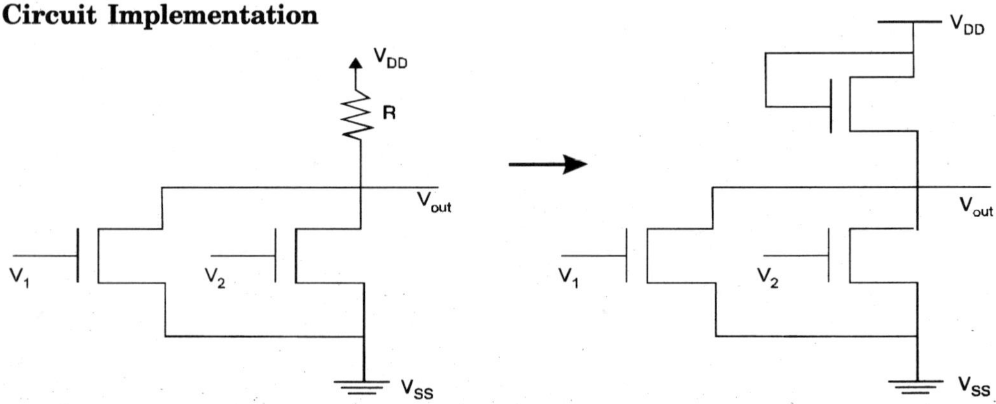

Truth Table

V_1	V_2	V_{out}
0	0	1
0	1	0
1	0	0
1	1	0

When $V_1 = V_2 = 5$ V then both the transistors will conduct and pull down the V_{out} to '0'. Even if one of the gate i/p is at 5 V there is a pull down path for V_{DD} to ground and the output will be only at '0'. Only when both the transistor gate inputs are at the '0' level, both the transistor will be off and output is high. Thus a NOR function has been realized.

***(iv)* AND gate :** NAND gate followed by an inverter acts as an AND gate. The Logic symbol, circuit implementation and TruthTable are given below logic symbol.

Logic Symbol

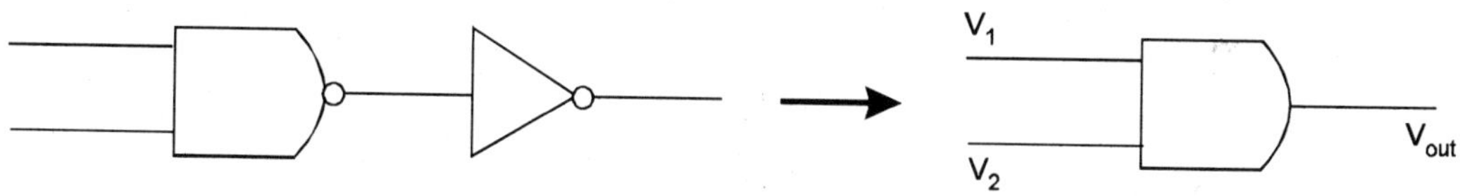

Circuit Implementation

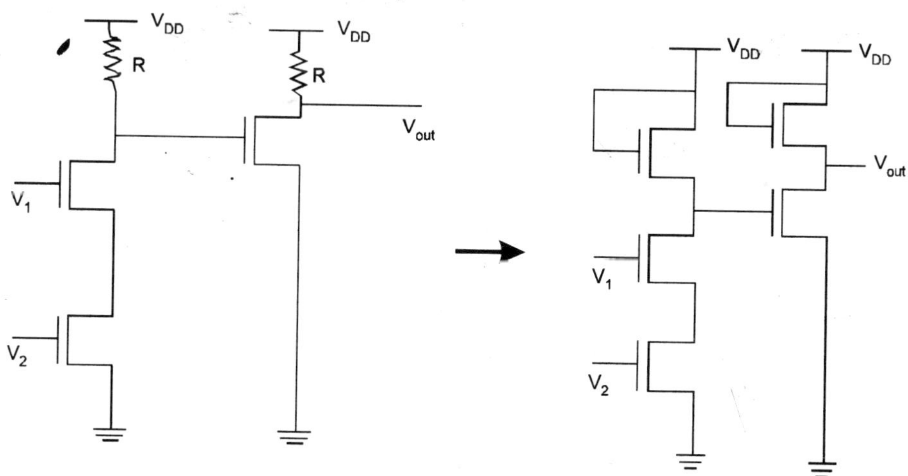

Truth Table

V_1	V_2	V_{out}
0	0	0
0	1	0
1	0	0
1	1	1

(v) OR gate : A NOR gate followed by an inverter acts as an OR gate. The Logic symbol, circuit implementation and TruthTable are given below.

Logic Symbol

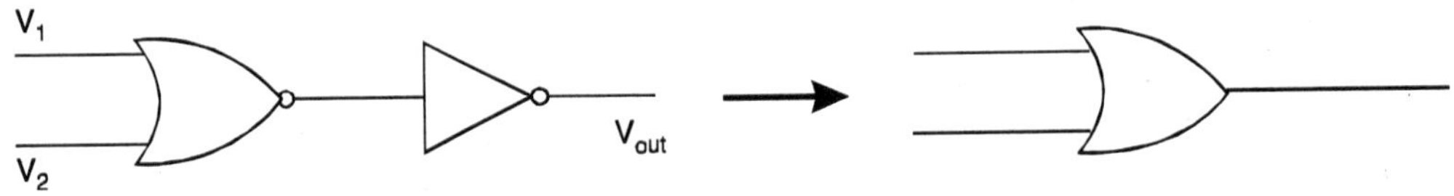

Circuit Implementation

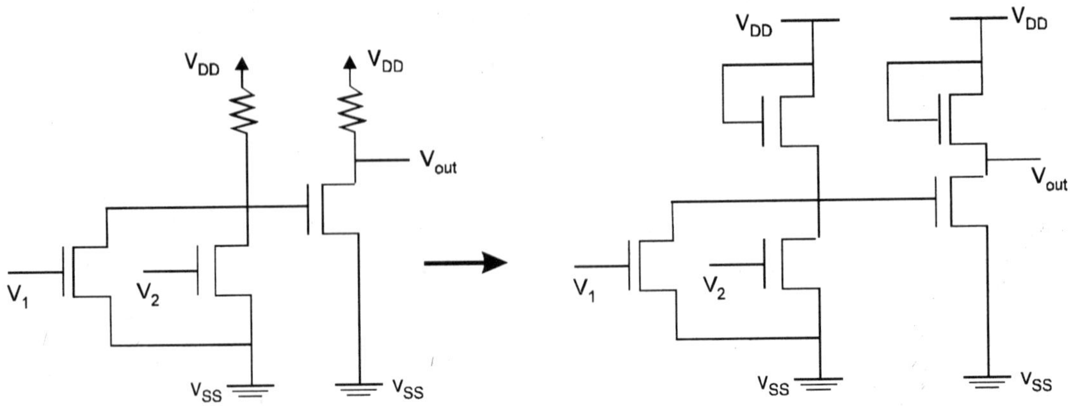

Truth Table

V_1	V_2	V_{out}
0	0	0
0	1	1
1	0	1
1	1	1

3.3. REALISATION OF GATES USING PMOS TECHNOLOGY

Under this topic we will see how the Basic gates can be implemented using PMOS only. As the readers are now very well aware of functionality of PMOS and NMOS transistors and also as it is hoped that the readers have sufficient knowledge about the functionality of all basic gates only the logical symbol and circuit implementation of all gates using PMOS are given in figures below. The operation is self explanatory. The only difference from NMOS implementation is that the resistor to limit current is replaced here with a PMOS transistor whose gate is always connected to G_{ND}.

(*i*) PMOS NOT GATE

Logical Symbol

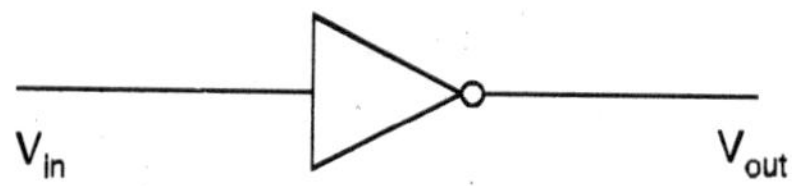

Circuit Implementation

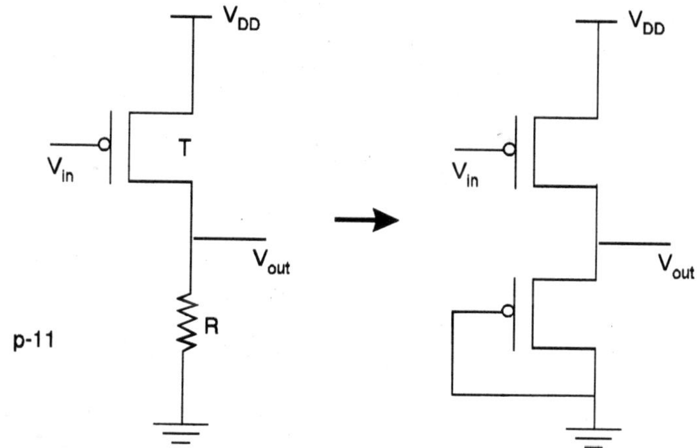

When $V_{in} = 0$, T is switched on and acts as closed switch and $V_{out} = 1$ *i.e.*, Output is connected to V_{DD}. When $V_{in} = 1$, T is switched off and act as open circuit and $V_{out} = 0$, as now o/p is connected to G_{nd} through resistor.

(*ii*) AND GATE

Logical Symbol

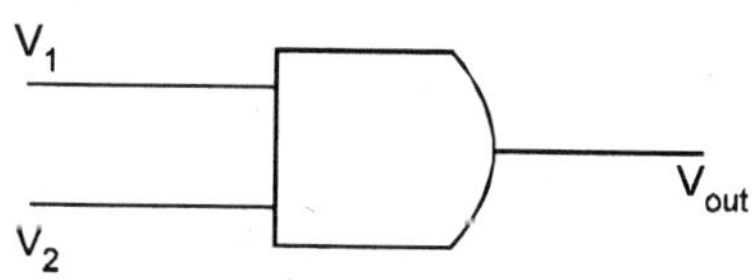

Circuit Implementation

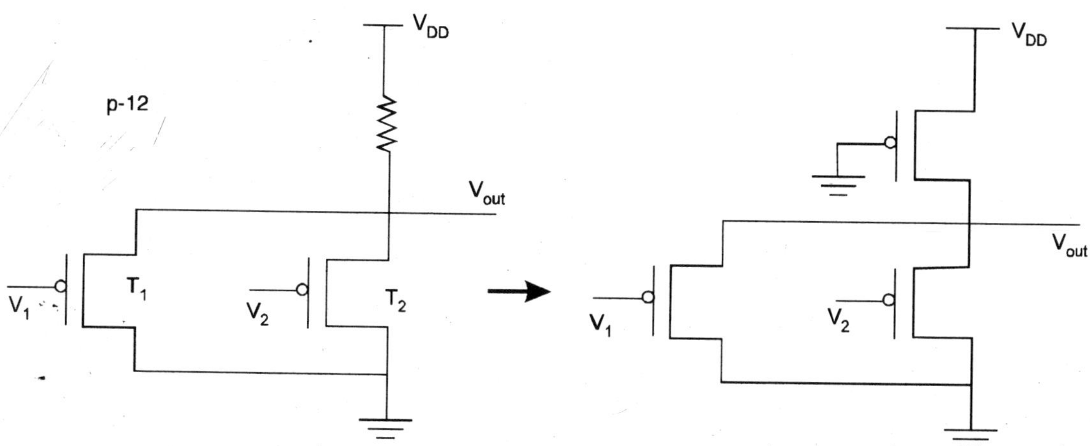

(*iii*) OR GATE USING PMOS

Logical Symbol

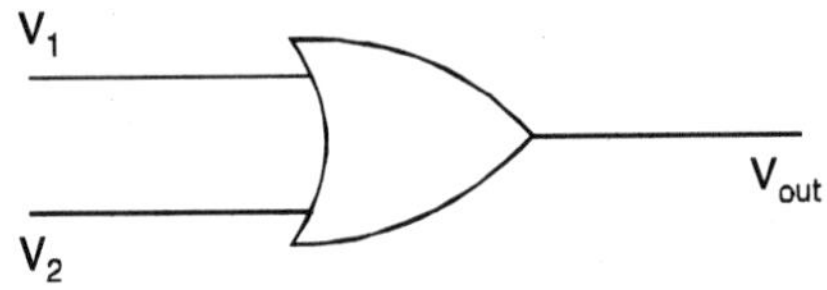

Circuit Implementation

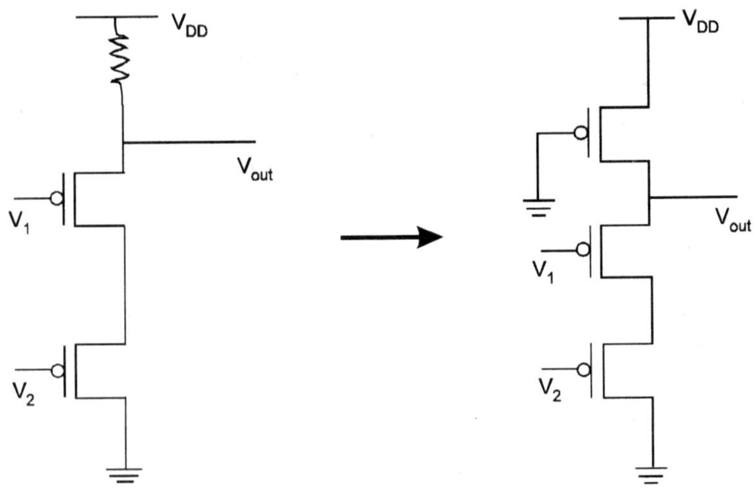

Truth Table

V_1	V_2	V_{out}
0	0	0
0	1	1
1	0	1
1	1	1

(*iv*) NAND GATE

Logical Symbol

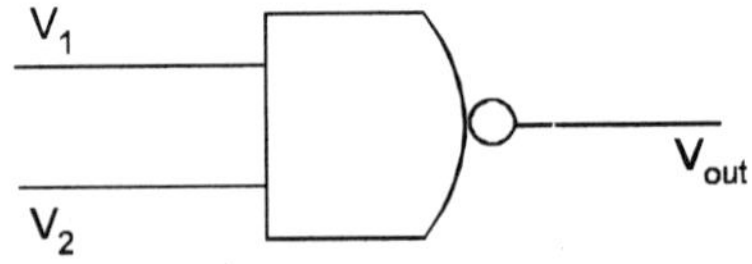

Truth Table

V_1	V_2	V_{out}
0	0	1
0	1	1
1	0	1
1	1	0

Circuit Implementation

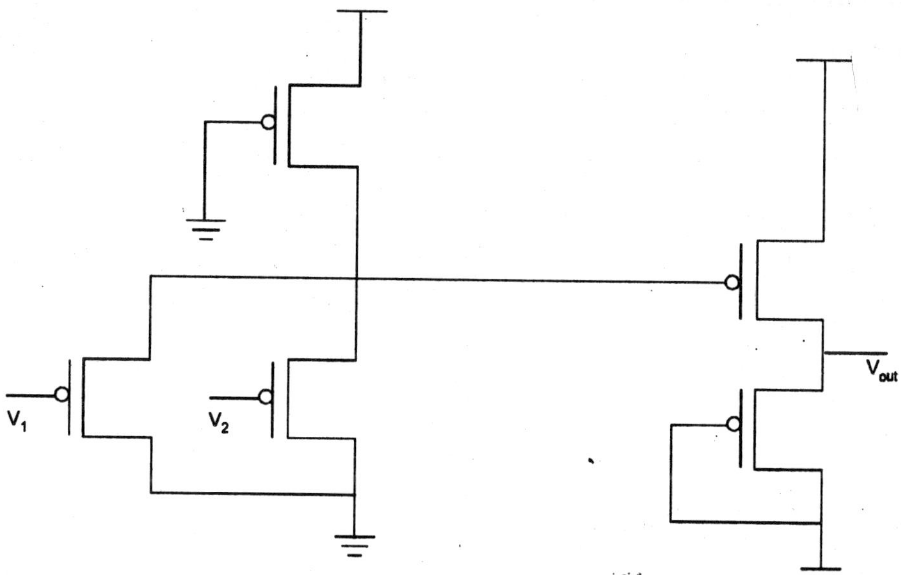

(v) NOR GATE

Logical Symbol

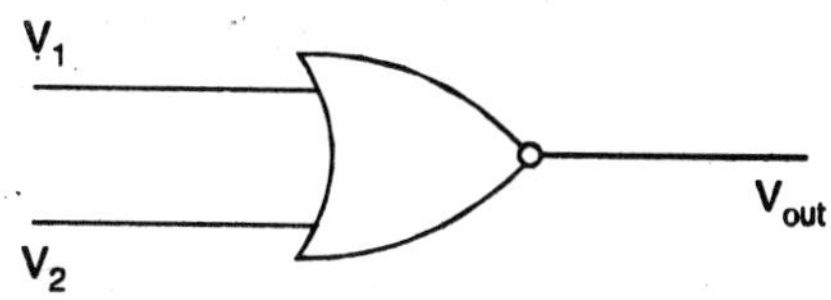

Circuit Implementation

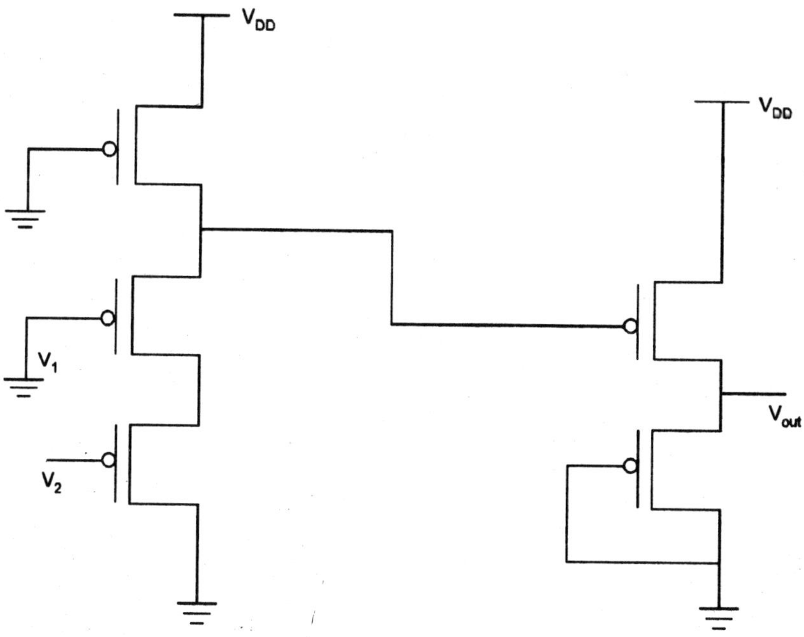

Truth Table

V_1	V_2	V_{out}
0	0	1
0	1	0
1	0	0
1	1	0

3.4.1 Combinational Logic

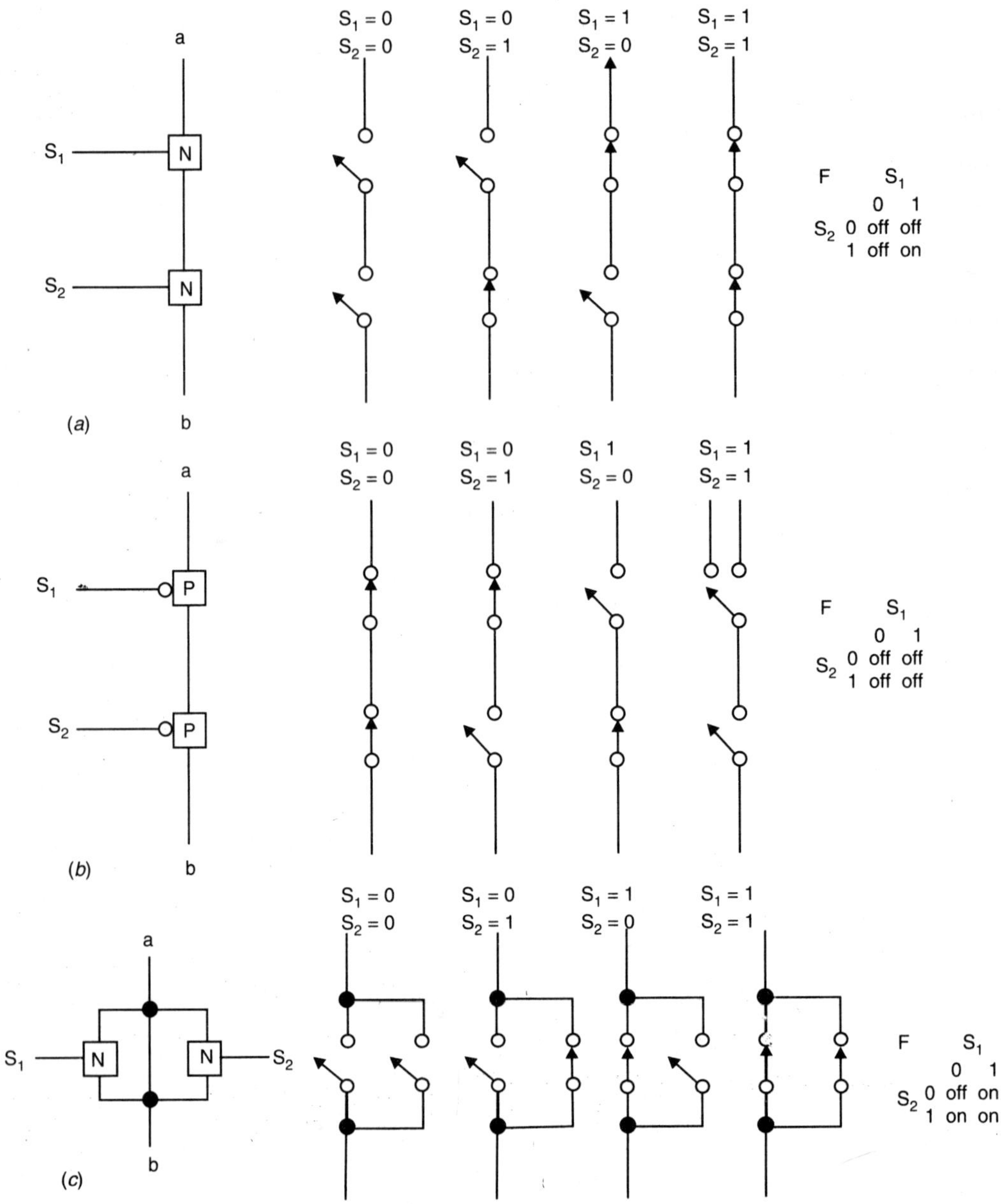

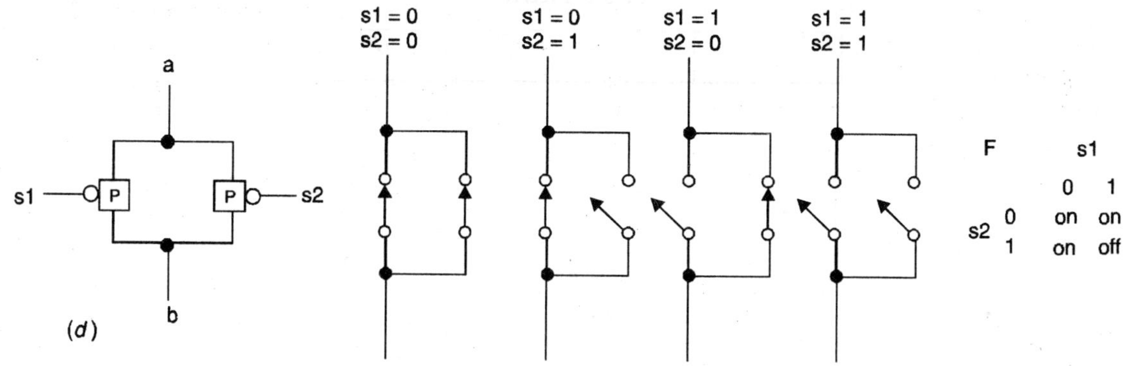

Fig. 3.4. *n*-Switches and *p*-Switches in series and parallel.

If two *n*-switches are placed in series the composite switch obtained is closed only when both switches are closed *i.e.*, When $S_1 = S_2 = 1$. This functionality resembles to that of an AND gate function. If two *p*-switches are placed in series the composite switch obtained is closed only when $S_1 = S_2 = 0$. This is similar to a NOR gate function. So, a function similar to an AND gate is obtained when two *n*-switches are connected in series. Fig. (*a*). A function similar to a NOR gate is obtained when two p-switches are connected in series. Fig. (*b*)

When two *n*-switches are connected in parallel the composite switch is closed if either switch is closed *i.e.* either S_1 or S_2 is '1'. Thus an OR function is created. In a similar way if two p-switches are connected in parallel it will bring out a NAND function. So a function similar to an OR gate is obtained when two *n*-switches are connected in parallel. Fig. (*c*). A function similar to a NAND gate is obtained when two p-switches are connected parallel. Fig. (*d*).

3.4.2. Realisation of Logic Gates Using CMOS

Structure of the CMOS CKT

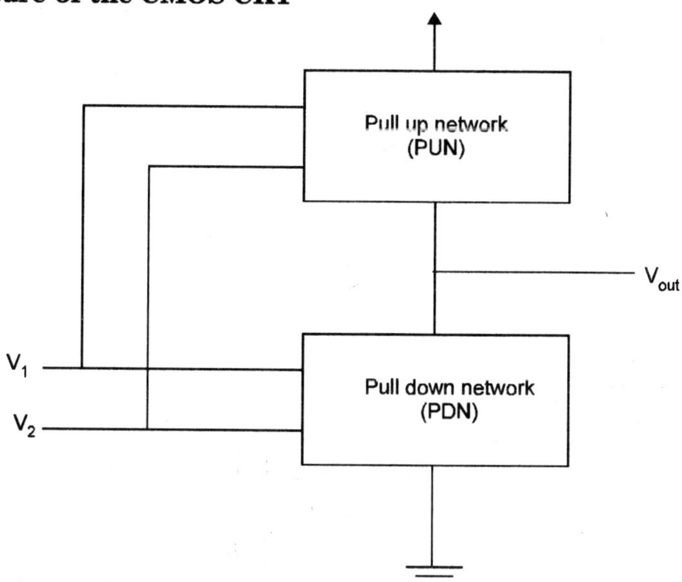

Now let us consider both NMOS and PMOS transistors in realizing logic gates. This approach is known as CMOS Technology. In NMOS circuits the logic functions are realized by the arrangements of NMOS transistors, combined with pull up device that acts as a resistor. The part of the circuit that involves NMOS transistors are referred as pull down network(PDN).

The concept of CMOS circuit is based on replacing the pull-up device with pull up network [PUN] that is built up using PMOS transistors such that the functions are realized by the PDN and PUN. These networks are complement to each other.

PDN and PUN have equal number of transistors, which are arranged so that two networks are duals of one another. Whenever the PDN has NMOS transistors in series, the PUN has PMOS transistors in parallel and vice versa.

***(i)* CMOS NOT Gate.** The Fig. below shows Logical symbol, circuit implementation and Truth table of an inverter. When V_{in} = 0V, T_2 is off and T_1 is ON. As T_1 is ON o/p is connected to V_{DD} and this makes V_{out} = V_{DD} = 5V. When V_{in} = 5V, T_2 is ON and T_1 is OFF, the o/p is now connected to V_{ss} or GND. This makes V_{out} = 0V. From the above description it is understood that how an inverter can be realized using CMOS. A key point is that no current flows in a CMOS inverter (between V_{DD} and V_{SS}), when the input is either low or high. This is true for all CMOS circuits and hence no power is dissipated under steady state condition. This property brings CMOS the most popular technology for building logic circuits.

Logic Symbol

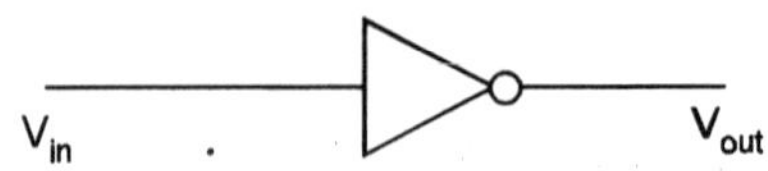

Circuit implementation

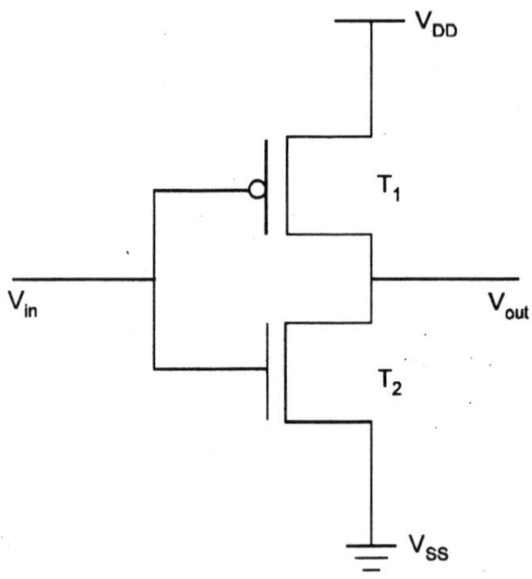

Truth Table

V_{in}	T_1	T_2	V_{out}
0	ON	OFF	1
1	OFF	ON	0

(ii) CMOS NAND Gate.

Logic symbol

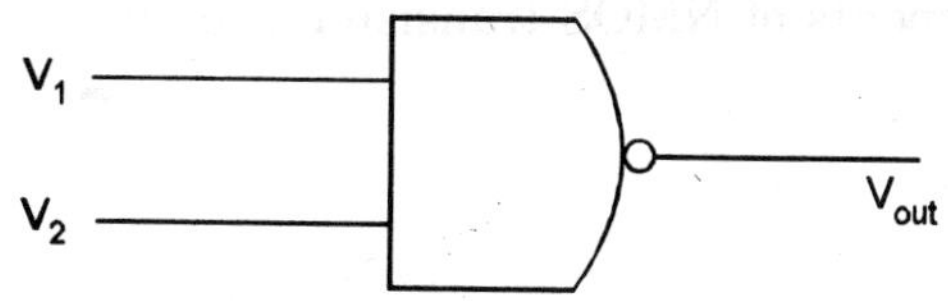

Circuit implementation

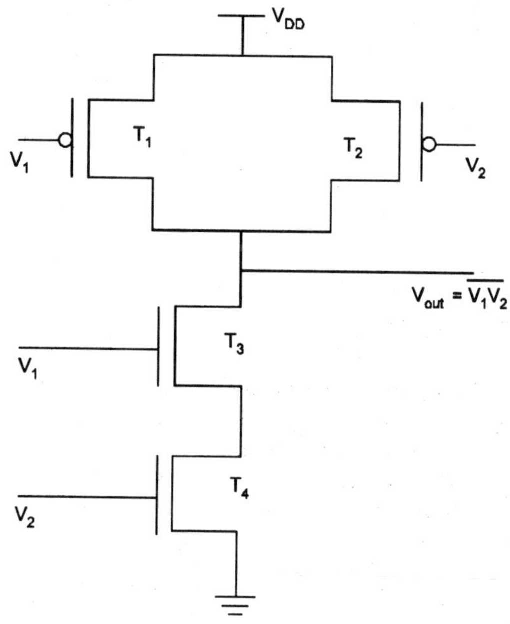

Truth Table

V_1	V_2	T_1	T_2	T_3	T_4	V_{OUT}
0	0	ON	ON	OFF	OFF	HIGH
0	5	ON	OFF	OFF	ON	HIGH
5	0	OFF	ON	ON	OFF	HIGH
5	5	OFF	OFF	ON	ON	LOW

***(iii)* CMOS Realization of AND Gate :** AND gate can be realized by a NAND followed by NOT gate

Logic symbol

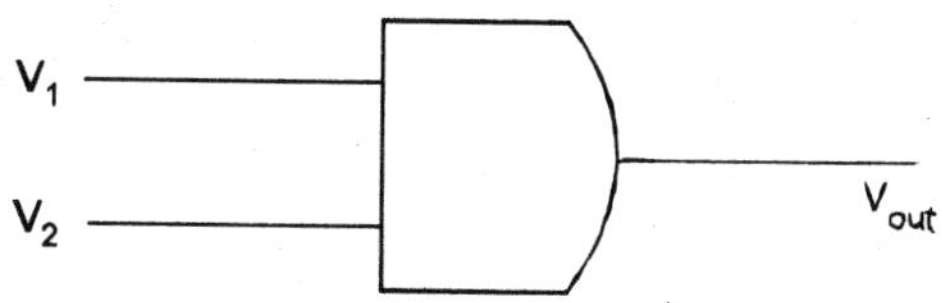

Circuit implementation

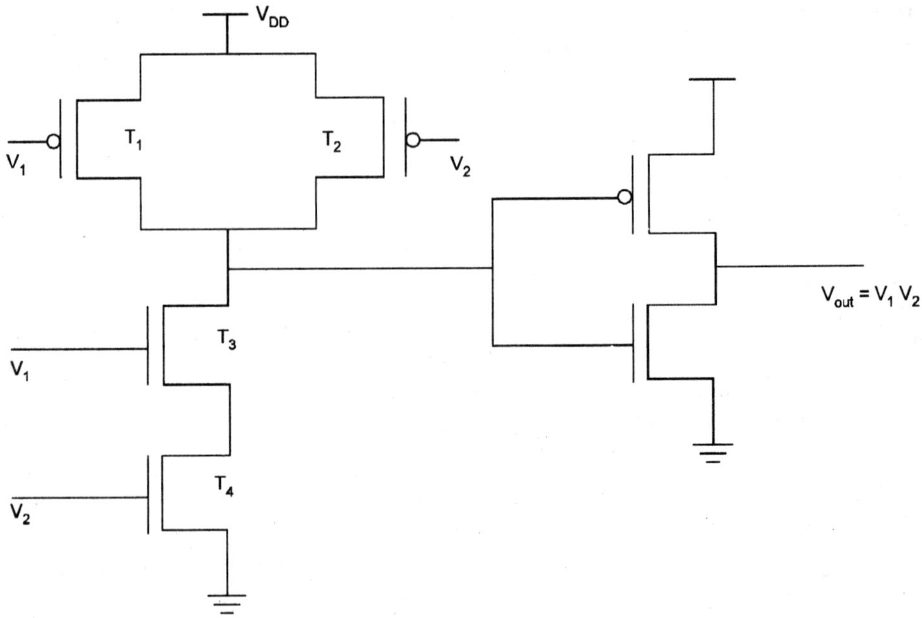

(*iv*) CMOS Realization of NOR Gate

Logic symbol

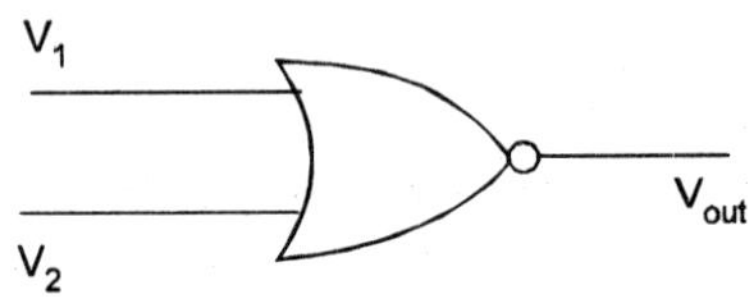

Circuit implementation

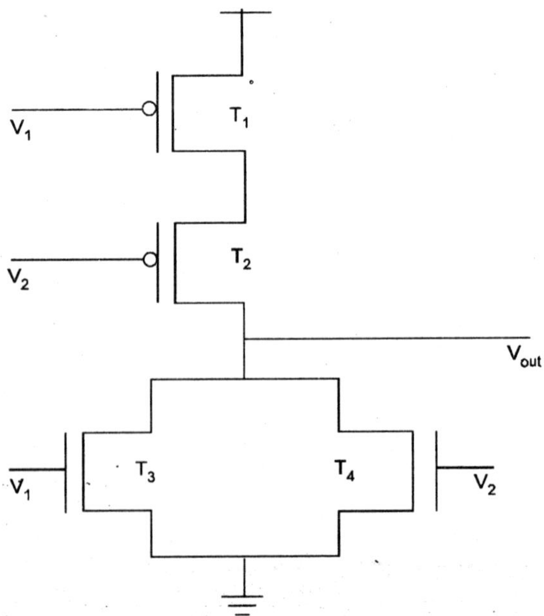

Truth Table

V_1	V_2	T_1	T_2	T_3	T_4	V_{OUT}
0	0	ON	ON	OFF	OFF	5
0	5	ON	OFF	OFF	ON	0
5	0	OFF	ON	ON	OFF	0
5	5	OFF	OFF	ON	ON	0

(v) CMOS Realisation of OR Gate : An OR gate can be realized by a NOR gate followed by an inverter.

Logic symbol

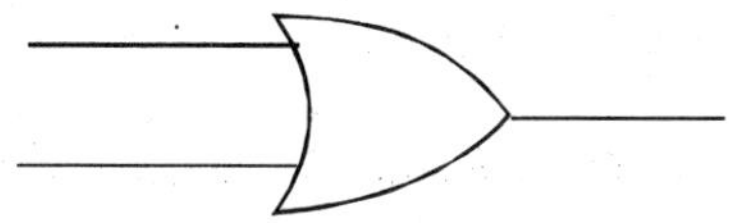

Circuit implementation

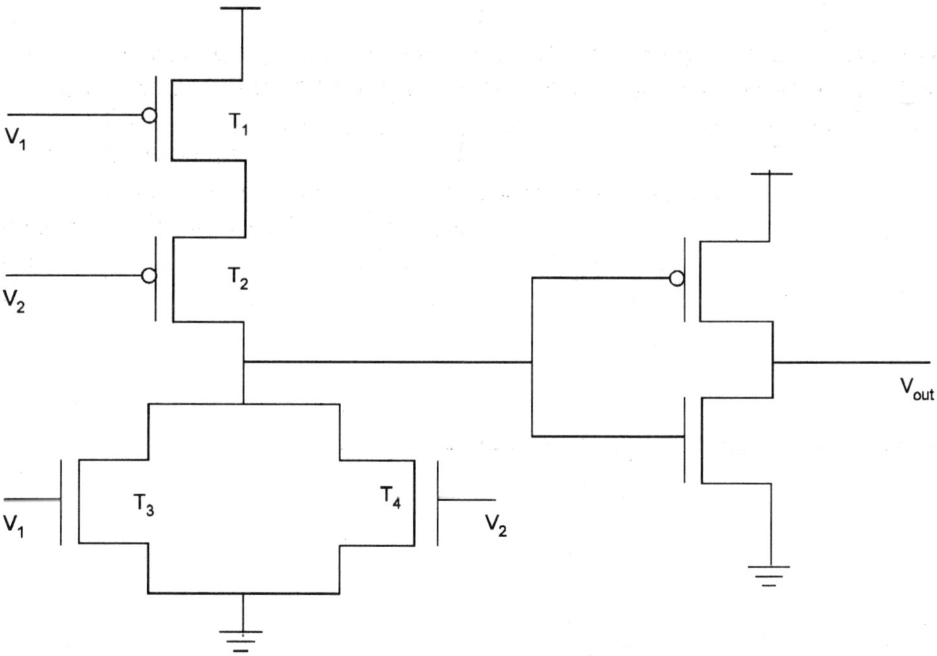

3.5. COMPOUND GATES

Using a combination of series and parallel switch structures a compound gate can be formed.

Example 1. Consider the function

$$Y = \overline{(A + B)(C + D)}$$

The decomposition of this function and generation of the diagram can be approached as follows. For the n-side, take the uninverted expression ((A + B). (C + D)) .The OR function requires the parallel connection of switches as shown below :

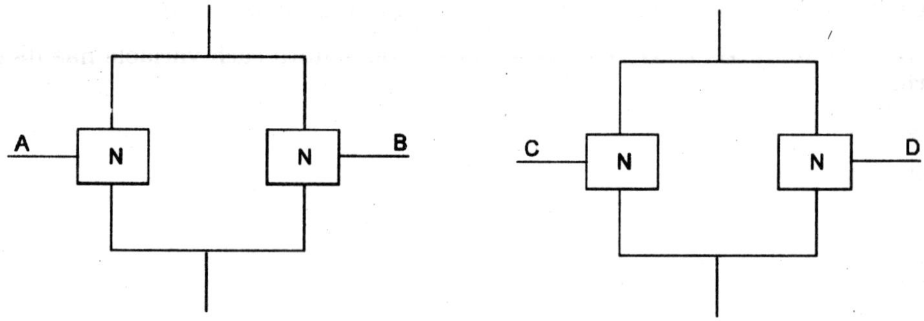

AND ing of (A + B) and (C + D) can be implemented by series connection. As shown below :

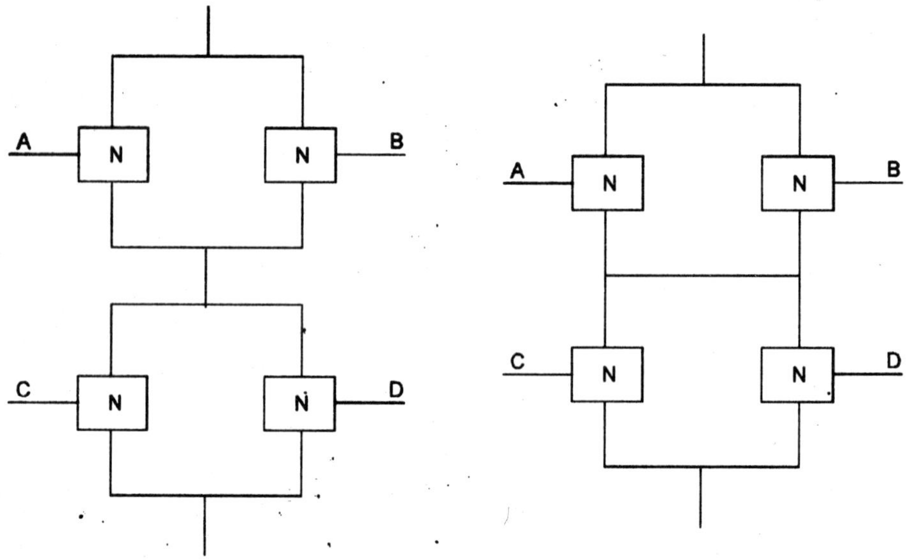

The equivalent transistor implementation is shown below :

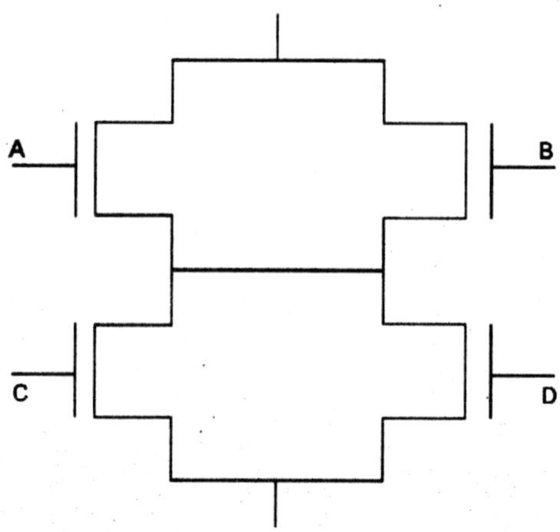

For *p*-side, invert the expression used for *n*-type. Which gives

Y= (A + B) (C + D). Using Morgan's theorem it can be split as $\overline{(A+B)}+\overline{(C+D)}$ which can be further reduced to $\overline{A}\ \overline{B}+\overline{C}\ \overline{D}$. The Reduction is done until each variable has its complemented form.

Now, we have to realize for $\overline{A}\ \overline{B}+\overline{C}\ \overline{D}$. For *p*-type, NAND function is realized by parallel combination of two transistors and NOR function is realized by series connection. $\overline{A}\ \overline{B}$ is realized by series connection of two p-transistors. Similarly $\overline{C}\ \overline{D}$ is realized by series connection of two *p*-transistors. Finally $\overline{A}\ \overline{B} + \overline{C}\ \overline{D}$ is obtained by the parallel connection of above two series structures.

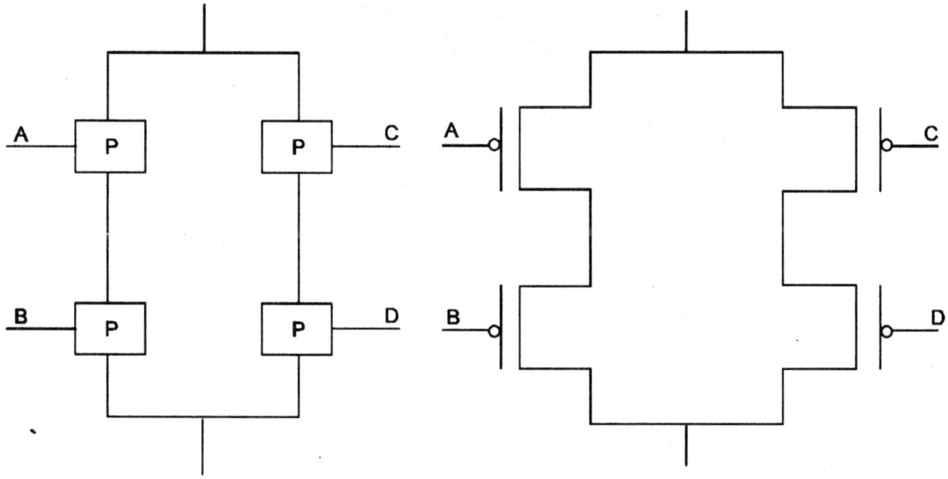

Finally, realization for y = $\overline{(A + B).(C + D)}$ is shown below :

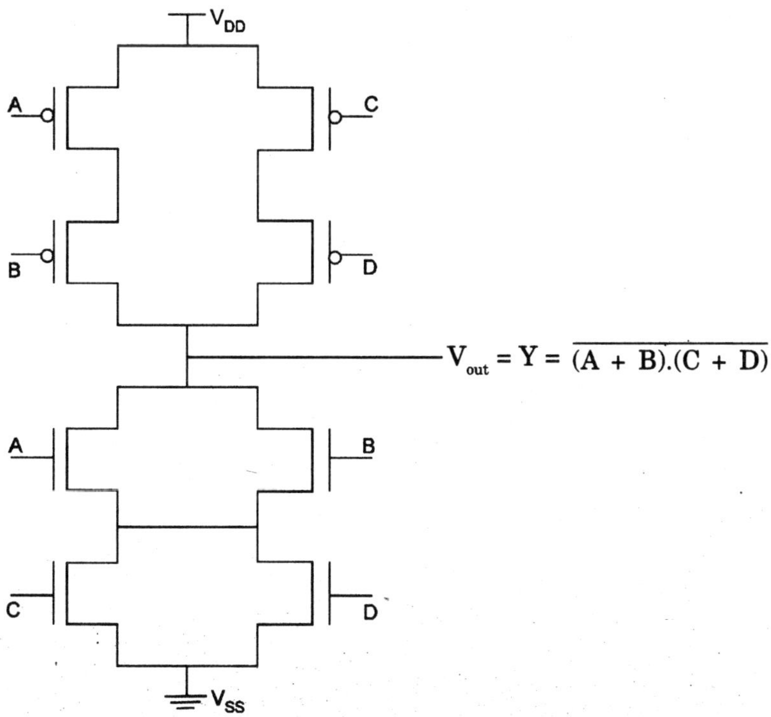

Example 2. Realize the function $y = \overline{AB + CD}$ using compound gates

For *n*-type : Y = A.B + C.D

AND function is realized by series combination of NMOS. OR function is realized by parallel combination of NMOS. This is shown in Fig below :

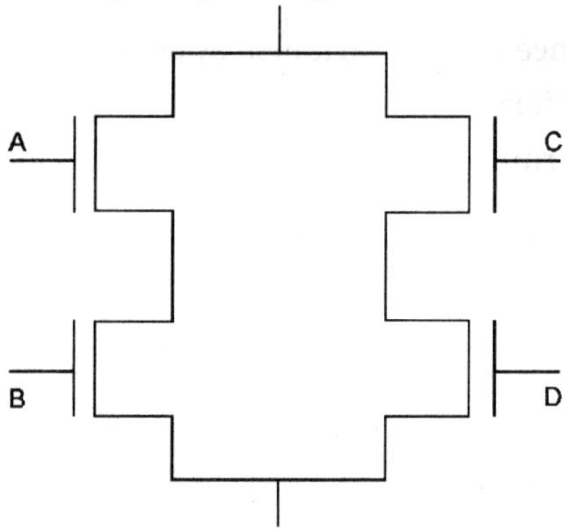

For *p*-type :

$$Y = \overline{AB + CD}$$

$$Y = \overline{AB} \cdot \overline{CD}$$

$$Y = (\overline{A} + \overline{B}) \cdot (\overline{C} + \overline{D}).$$

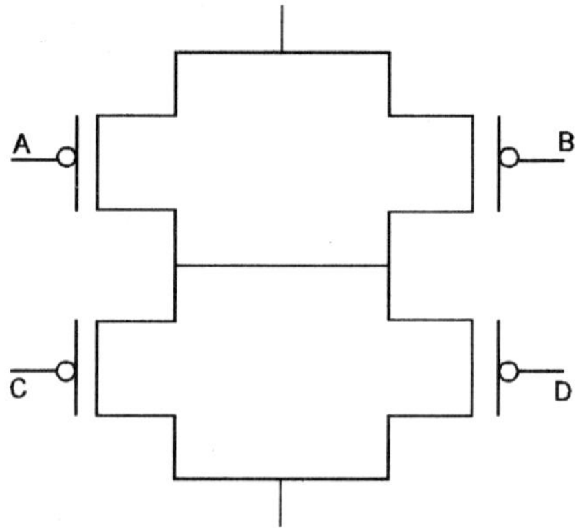

The final structure for

$Y = \overline{AB + CD}$

is shown below :

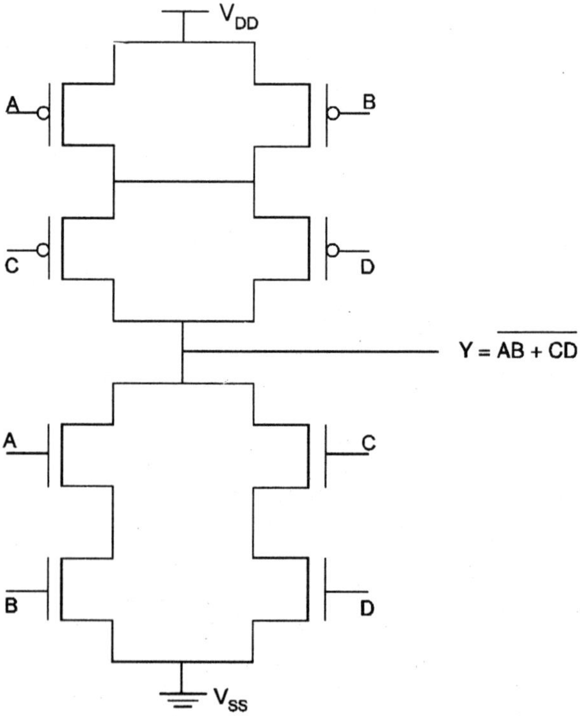

Example 3. Realize the function; $y = \bar{x}_1 + \bar{x}_2 \; \bar{x}_3$

Since all the variables appear in their complemented form, we can directly realize the equation using *p*-transistors

$\overline{x_2 + x_3} = \bar{x}_2 \; \bar{x}_3$ (*i.e.*, NOR function of x_2 and x_3)

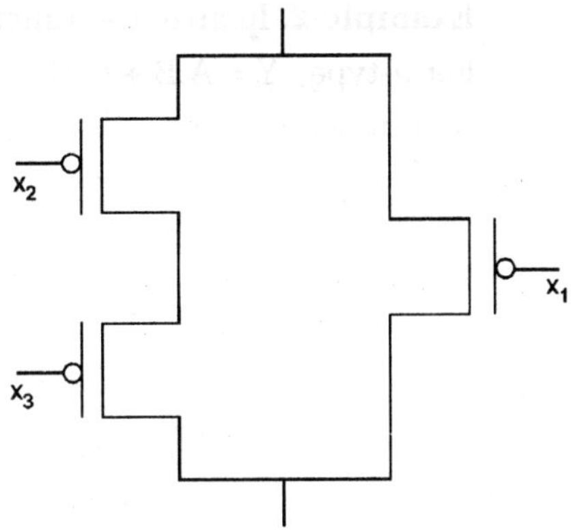

For n-type, take the inversion.

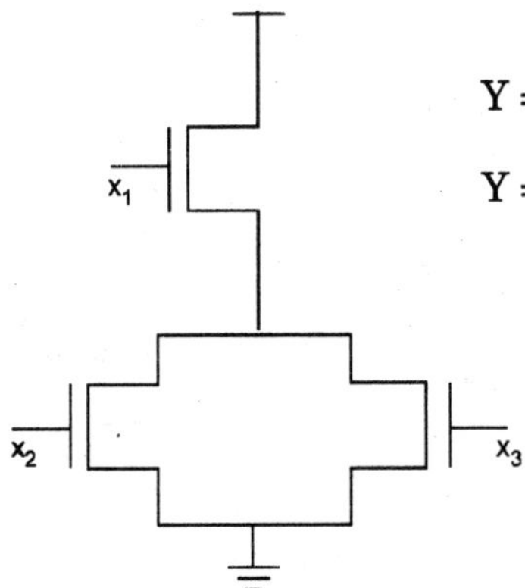

$Y = \overline{\bar{x}_1 + \bar{x}_2 \, \bar{x}_3}$

$Y = x_1 . (\bar{\bar{x}}_2 + \bar{\bar{x}}_3)$

$Y = \bar{\bar{x}}_1 \cdot \overline{\bar{x}_2 \, \bar{x}_3}$

$Y = x_1 . (x_2 + x_3)$

Final structure;

$Y = \bar{x}_1 + \bar{x}_2 \; \bar{x}_3$

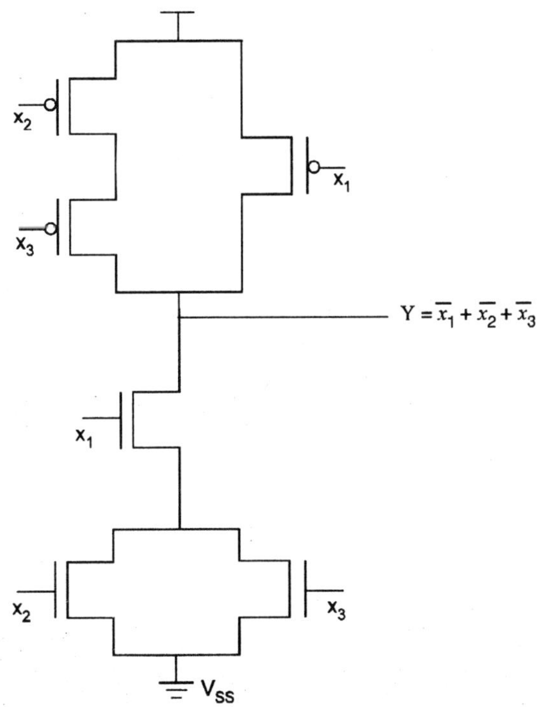

Example 4. Realize the function;

$Y = \overline{x_1} + (\overline{x_2} + \overline{x_3})\,\overline{x_4}$

As every variable has its complement realize it for *p*-type directly.

$(\overline{x_2} + \overline{x_3})$ is nothing but NAND function of x_2 and x_3. *i.e.*, x_2 and x_3 should be connected in parallel.

$(\overline{x_2} + \overline{x_3}).\ \overline{x_4}$ is nothing but NOR function of $(\overline{x_2} + \overline{x_3})$ and x_4 *i.e.*, x_4 should be connected in series with parallel combination of x_2 and x_3.

Finally x_1 is in parallel with $\overline{x_4}\ (\overline{x_2} + \overline{x_3})$

For n-type; Invert / complement the above expression.

$$Y = \overline{\overline{x_1} + (\overline{x_2} + \overline{x_3})\,\overline{x_4}}$$

$$Y = \overline{\overline{x_1}} \cdot \overline{(\overline{x_2} + \overline{x_3})\,\overline{x_4}}$$

$$Y = x_1 . \overline{(\overline{x_2} + \overline{x_3})} + x_4$$

$$Y = x_1 . (\overline{\overline{x_2}} \cdot \overline{\overline{x_3}} + x_4)$$

$$Y = x_1 . (x_2 x_3 + x_4)$$

Final realization structure;

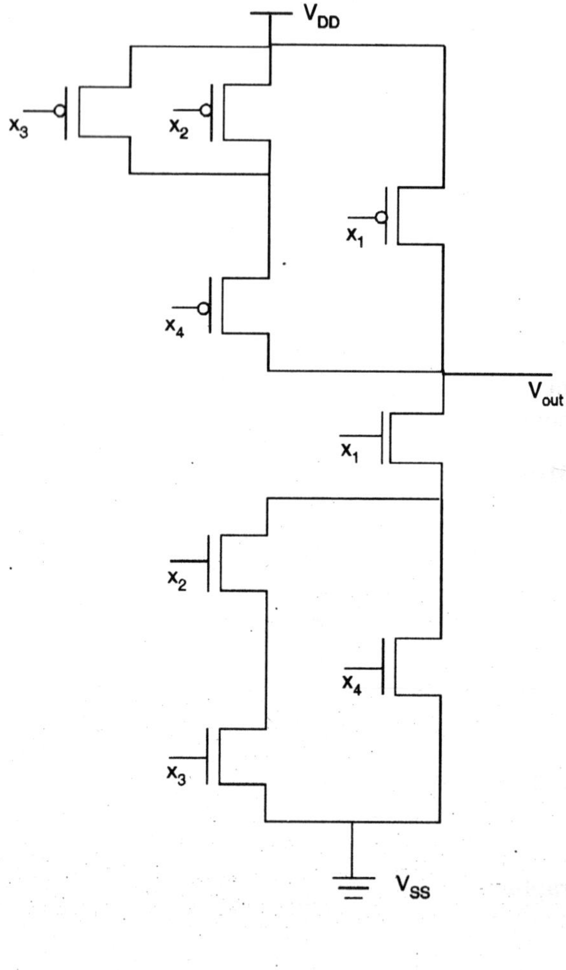

3.6. TRANSMISSION GATE IMPLEMENTATION

By combining an *n*-switch and a *p*-switch in parallel (sources and drains of n-transistor and *p*-transistor are connected) a switch in which both '0's and '1's are passed in an acceptable fashion can be obtained. This is termed as a complementary switch because the control signal is applied to the gate of *n*-transistor and the complement of control signal – S or $\overline{\text{S}}$ is applied to gate of *p*-transistor. This complementary switch is also called as a transmission gate. The following Fig. shows schematic ions commonly used for representing transmission gate.

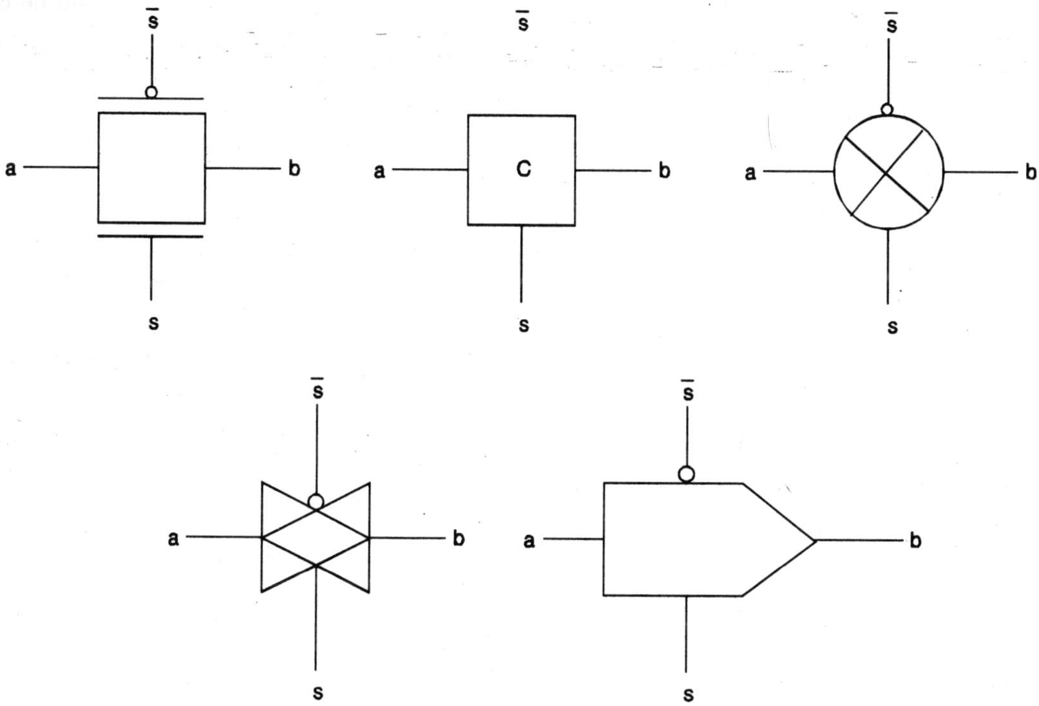

Fig. 3.6. Schematic icons of complementary switch or transmission gate.

The transmission gates can be used in a variety of applications. They lead to efficient implementations of XOR logic gates, multiplexer circuits, etc.

3.6.1. Multiplexer

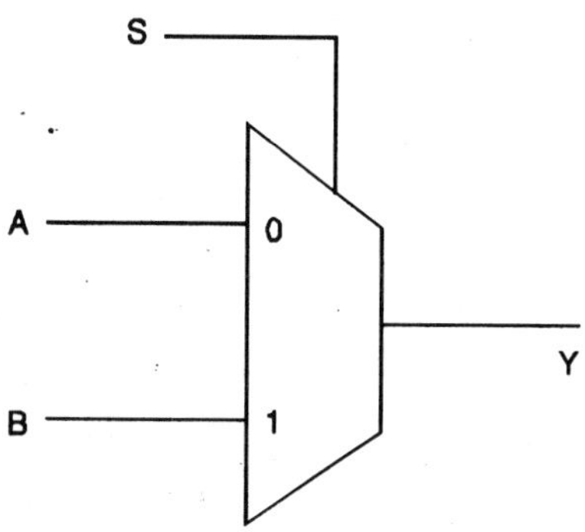

Graphical symbol of 2:1 mux.

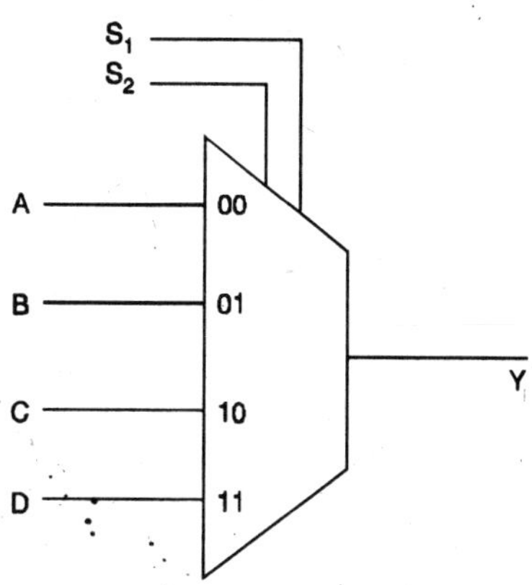

Graphical symbol of 4:1 mux.

A Multiplexer circuit has a number of data inputs, one or more select input and one output. It passes the signal value on one of the data inputs to the output. The data input to be passed to output is selected by the select inputs. The select input '*s*' chooses as the o/p of the multiplexer either inputs A or B. The following figure shows graphical symbol for a 2 : 1 mux and 4 : 1 mux.

2 : 1 mux using transmission gate.

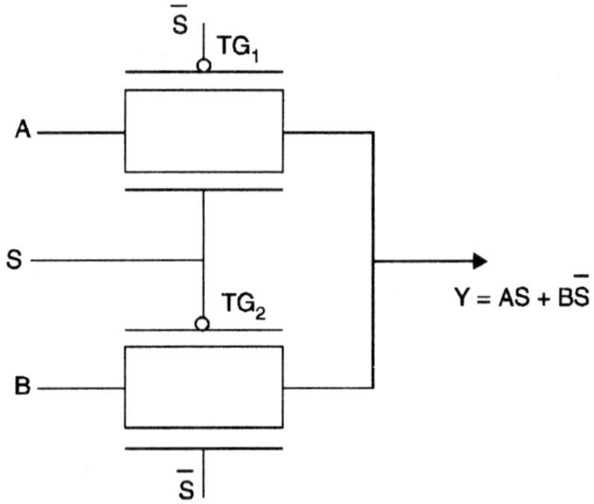

(*i*) When S = 1, $\bar{S}$ = 0. Now TG_1 is ON and TG_2 is OFF. The output Y = A.

(*ii*) When S = 0, $\bar{S}$ = 1. Now TG_2 is ON and TG_1 is OFF. The output Y = B.

4 : 1 mux using TG.

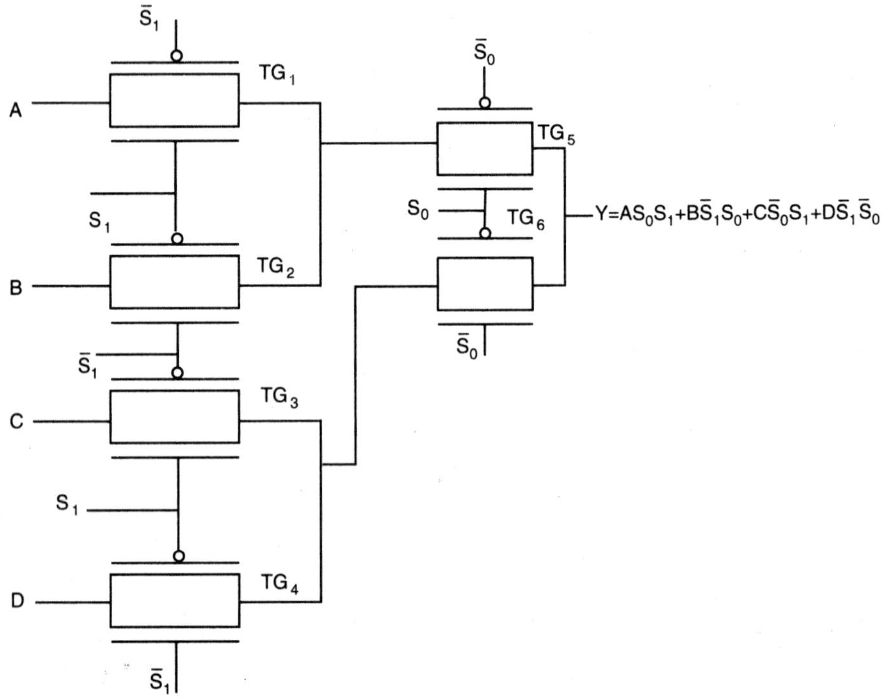

1. A will be passed to o/p when $S_0 = S_1 = 1$. Because Now $\overline{S_0} = 0$ and TG_1 and TG_5 will be ON.
2. B will be passed to o/p when $S_1 = 0$ and $S_0 = 1$. This condition will switch ON TG_2 and TG_5.
3. C will be passed to o/p when $S_0 = 0$ and $S_1 = 1$. This condition will switch ON TG_3 and TG_6.
4. D will be passed to o/p when $S_1 = 0$ and $S_0 = 0$. This condition will switch TG_4 ON and TG_6.

3.6.2. Demultiplexer

Placing the value of a single data input onto multiple data outputs is called a Demux. The below figure shows graphical symbol for a 1 : 2 and 1 : 4 Demux.

Symbol for 1 : 2 Demux.

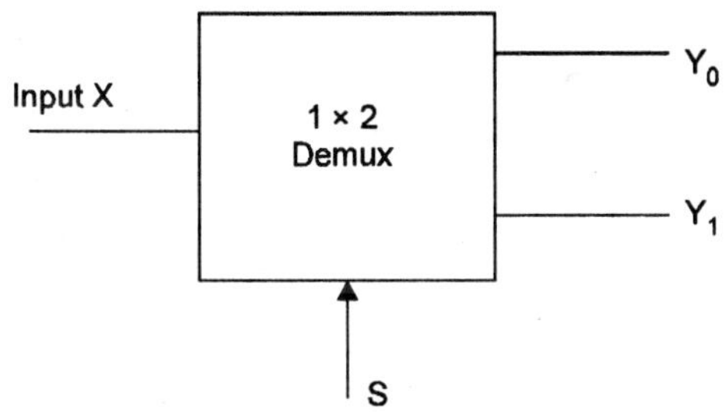

Symbol for 1 : 4 Demux.

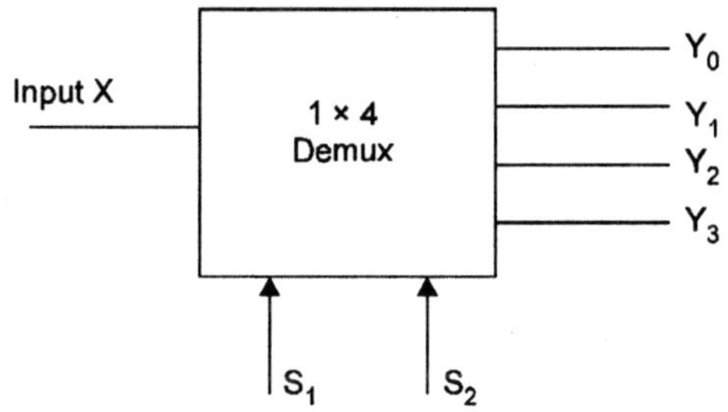

TG implementation of Demux.

(*i*) 1 : 2 Demux.

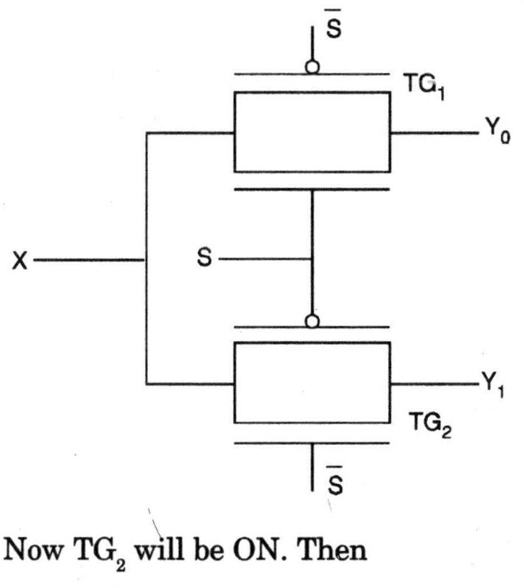

When $S = 0$, $\overline{S} = 1$. Now TG_2 will be ON. Then input 'X' is passed to Y_1. When S = 1, S = 0. The TG_1 will be ON and input 'X' is passed to Y_0.

(*ii*) 1 : 4 De mux

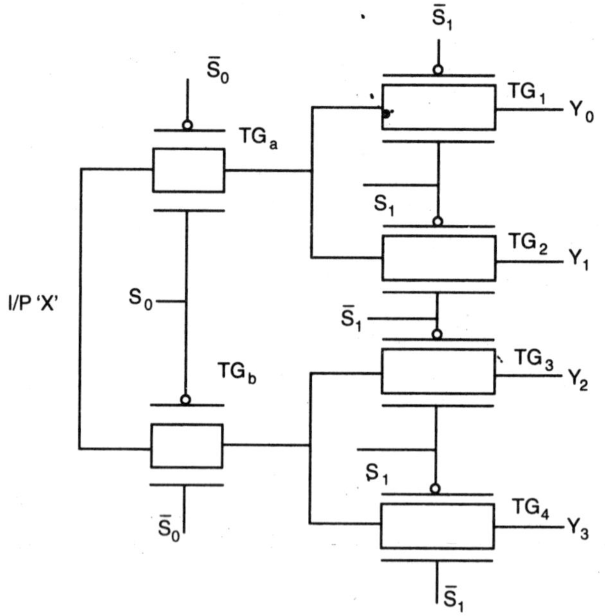

(*i*) When S_0 = 1 and S_1 = 1TG_a and TG_1 is ON. X is passed to Y_0.
(*ii*) When S_1 = 0 and S_0 = 1TG_a and TG_2 is ON. X is passed to Y_1.
(*iii*) When S_0 = 0 and S_1 = 1TG_b and TG_3 is ON. X is passed to Y_2.
(*iv*) When S_1 = 0 and S_0 = 0TG_b and TG_4 is ON. X is passed to Y_3.

3.6.3 (*a*) Exclusive OR

Graphical Symbol

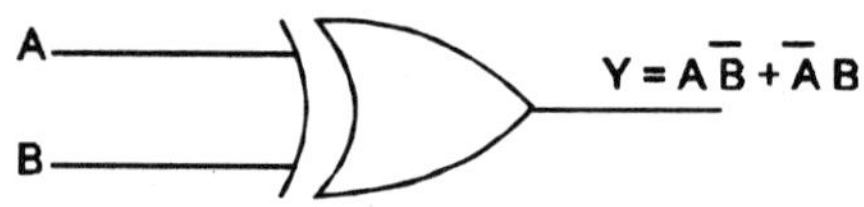

Implementation of EX-OR gate using Transmission Gate

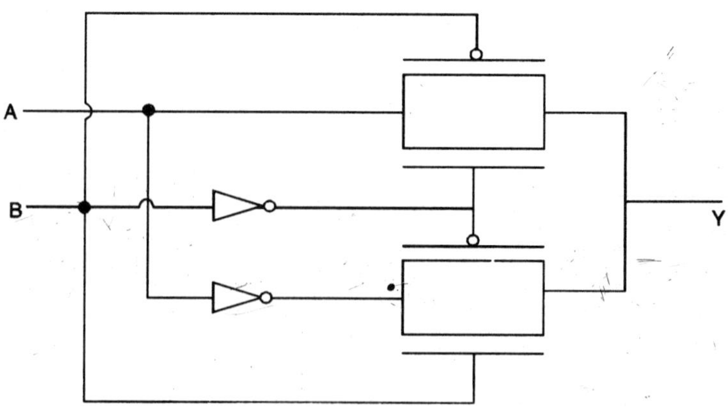

Truth Table

A	*B*	$Y = A + B$
0	0	0
1	1	1
1	0	1
1	1	0

3.6.3 (*b*) Exclusive NOR

Graphical Symbol

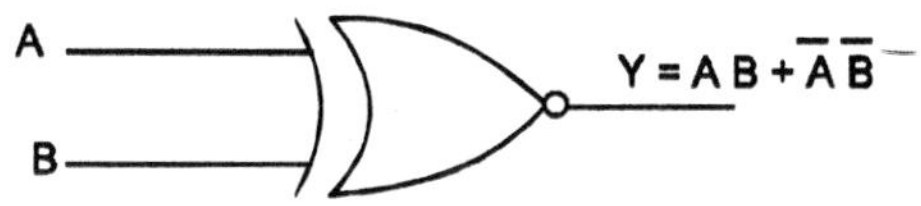

Truth Table

A	*B*	*Y*
0	0	1
0	1	0
1	0	0
1	1	1

Implementation of EX-NOR using Transmission gate

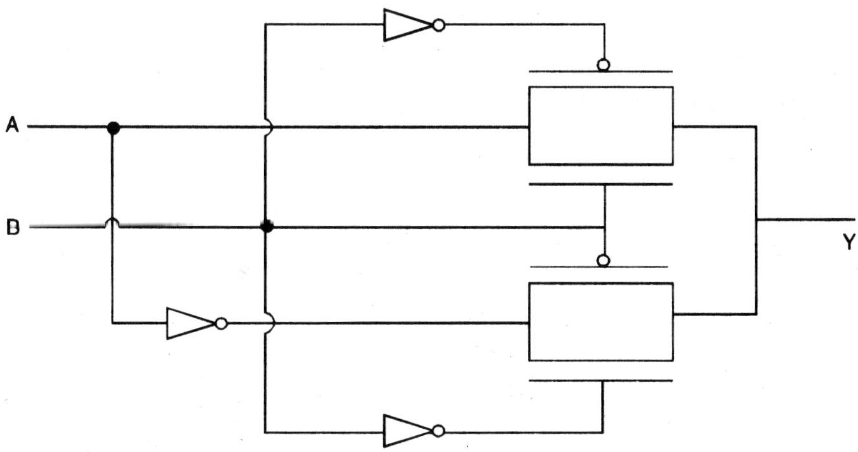

3.6.4. Half Adder Using Mux (2-bit)

A	*B*	*SUM*	*CARRY*
0	0	0	0
0	1	1	0
1	0	1	0
1	1	0	1

Sum = A XOR B Carry = AB

Implementation using Transmission Gate

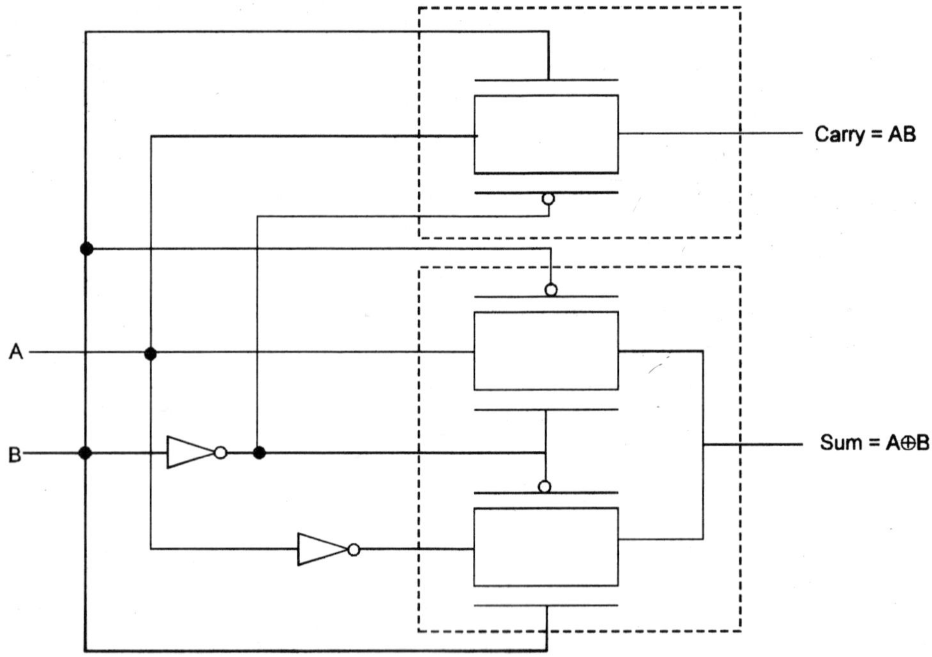

3.6.5. 4 : 2 Encoder Using Transmission Gates

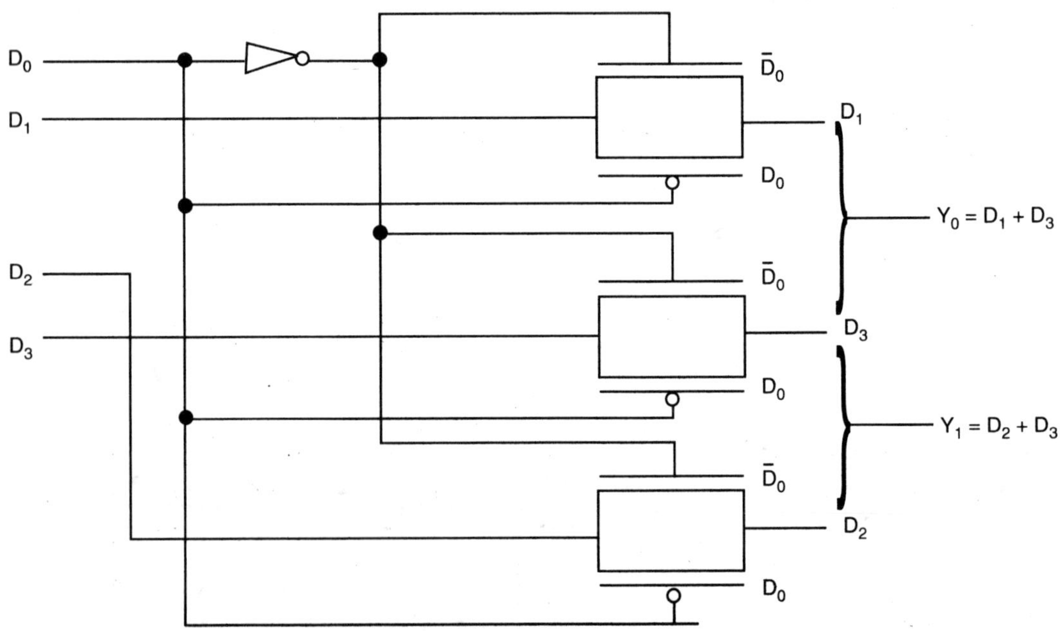

Truth Table

D_3	D_2	D_1	D_0	y_1	y_0
0	0	0	1	0	0
0	0	1	0	0	1
0	1	0	0	1	0
1	0	0	0	1	1

3.6.6. 2 : 4 Decoder Using Transmission Gates

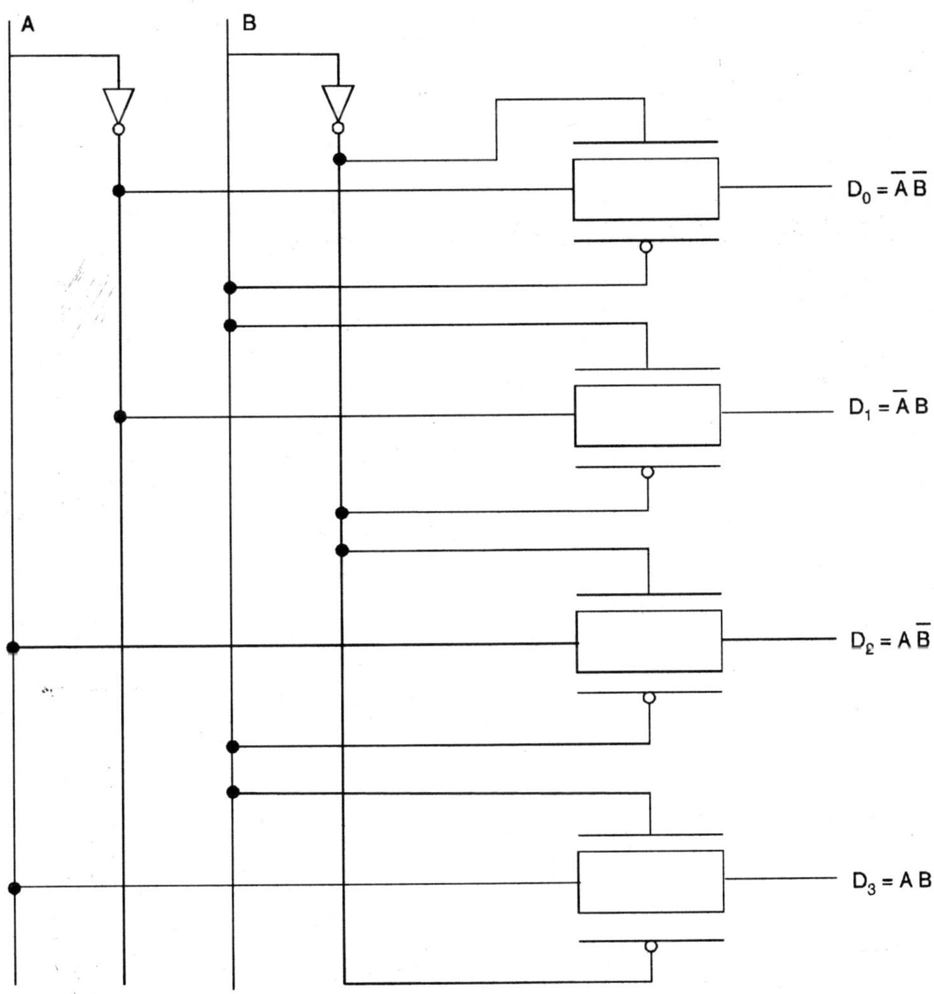

Truth Table

A	B	D_3	D_2	D_1	D_0
0	0	0	0	0	1
0	1	0	0	1	0
1	0	0	1	0	0
1	1	1	0	0	0

3.7. LATCHES AND REGISTERS

Latch. It is a feedback connection of two NOR-gates or two NAND-gates, which can store one bit of information.

Gated latch. It is a basic latch that includes input gating and a control input signal. The latch retains its existing state when the control input is equal to 0. Its state may be changed when the control signal is equal to 1.

Control input = clock.

Flip-flop. It is a storage element based on gated latch principle, which can have its output state changed only on the edge of the controlling clock signal.

Register. A FF stores one bit of information. When a set of n-flip flops is used to store n-bits of information, such as an n-bit number, we refer to these flip-flop as a register. A common clock is used for each flip-flop in a register.

Fan-in. The fan-in of a logic gate is the number of inputs the gate has in the logic path being exercised.

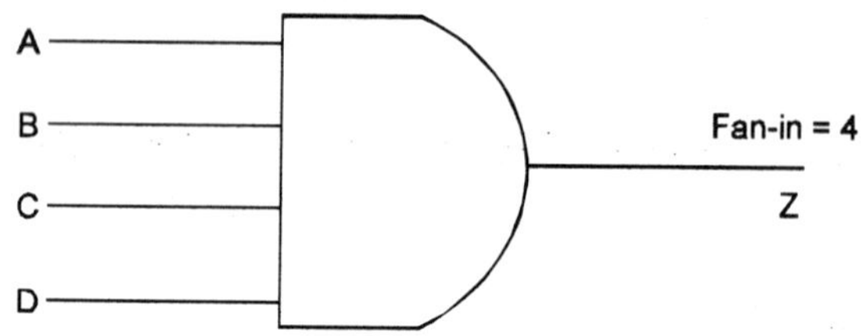

Fan-out. The fan out of a logic gate is the total number of gate inputs that are driven by the gate output.

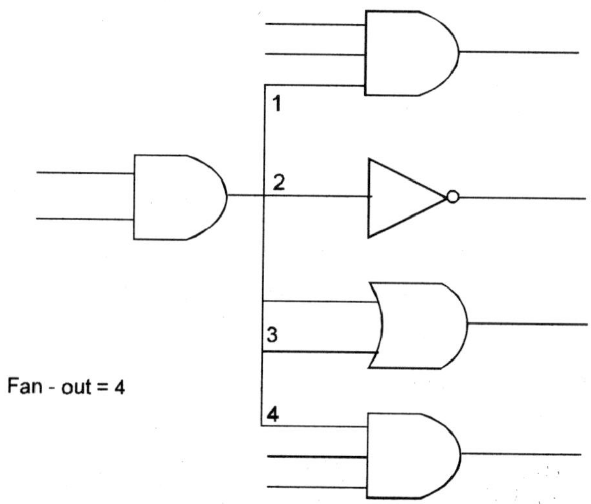

3.7.1. Level Sensitive Latch

A D latch can be constructed using 2 complementary switches (Transmission gates) and 2 inverters. This is shown Fig. 3.7.1. It consists of a data input, D, a clock input, CLK, and outputs Q and $\overline{Q}$. When CLK = '1', switch C_1 is ON and C_2 is off. Therefore Q = D and $\overline{Q} = \overline{D}$. When CLK = '0', Switch C_1 is open and C_2 is closed. So the 'D' input is ignored. Now the inverter pair forms a closed circuit and this causes the present state of Q to be stored. From the above, it is understood that the output is dependent on the level of the clock signal. Hence the name level sensitive latch. Also since this responds to a positive clock (*i.e.*, the D input is passed to o/p when CLK = 1) this is a positive level sensitive latch. By reversing the control connections to switches C_1 and C_2 a negative level sensitive latch may be constructed.

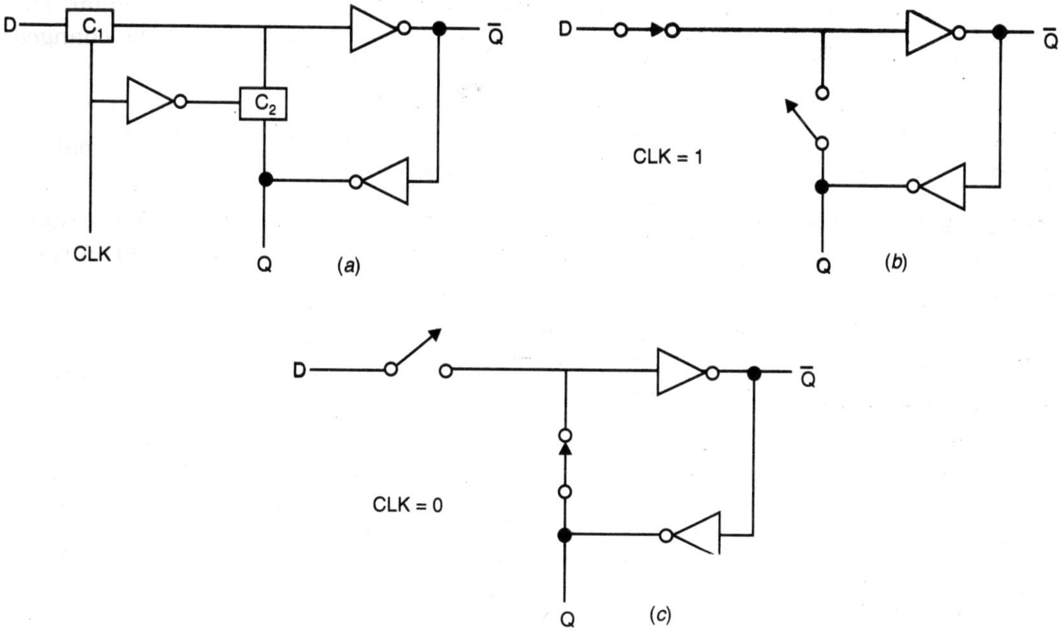

Fig. 3.7.1. (*i*) A positive level sensitive D-latch.

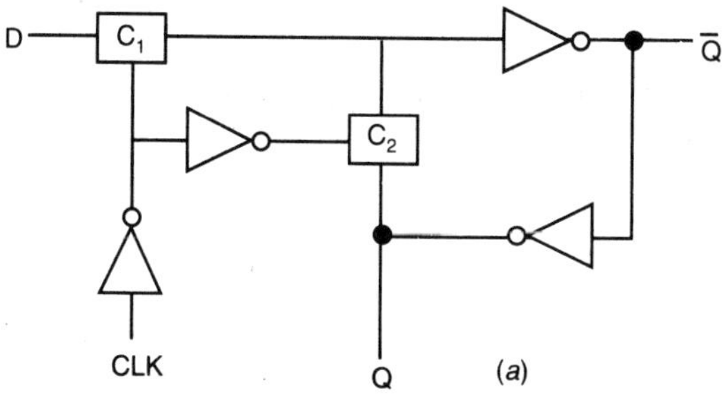

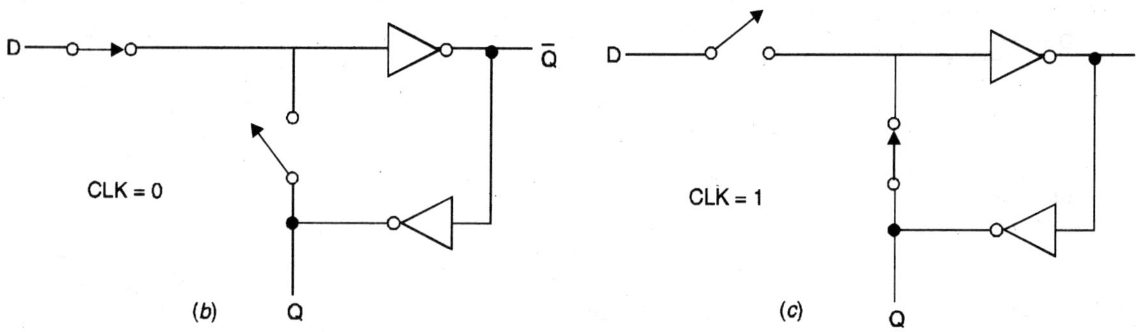

Fig. 3.7.1. (*ii*) A negative level sensitive D-latch.

For a negative level sensitive latch when CLK = 0, C_1 is on and C_2 is off. $\overline{Q} = \overline{D}$ and Q = D. when CLK = 1, C_1 is off and C_2 is on, so a closed path is established about inverter pair and now the Q holds the current state D. Since the latch responds to a negative clock (*i.e.*, D input is passed to the o/p when CLK is low) it is named as a negative level sensitive latch.

3.7.2. Edge Triggered D Register

By combining two level sensitive latches, one positive level sensitive and one negative level sensitive an edge triggered register may be constructed. Let us refer the first latch stage (a –ve level sensitive latch) as master and succeeding one (a +ve level sensitive latch) as slave. When CLK = 0, C_1 is closed, C_2 is open and C_3 is open and C_4 is closed. The master level sensitive latch output $\overline{QM}$ follows the D input. Now $\overline{QM} = \overline{D}$. As C_3 is open the slave holds the previous value. When a clock changes from '0' to '1', the master latch ceases to follow or sample the input and D value is stored at the time of clock transition. During clock transition the slave latch also opens, and the stored master value ($\overline{QM}$) is passed to the output of slave latch Q. Now Q follows $\overline{QM}$. The D input is prevented from affecting the output as the master is disconnected from the D input. When the clock changes from 1 to 0, the cycle begins *i.e.*, master will start following the input D and the slave latch locks with the master latch output. As the input is sampled at the positive edge of the clock this is a positive edge triggered register. A negative edge triggered register can be constructed by reversing the latch polarities. The Fig. 3.7.2 (*i*) and 3.7.2 (*ii*) below illustrates this. Also Fig. 3.7.2 (*a*), (*b*) shows the clocked D flip-flop using gates and implementation of clocked D using CMOS.

Truth Table of D-FF

D	*Q*	$\overline{Q}$
0	0	1
1	1	0

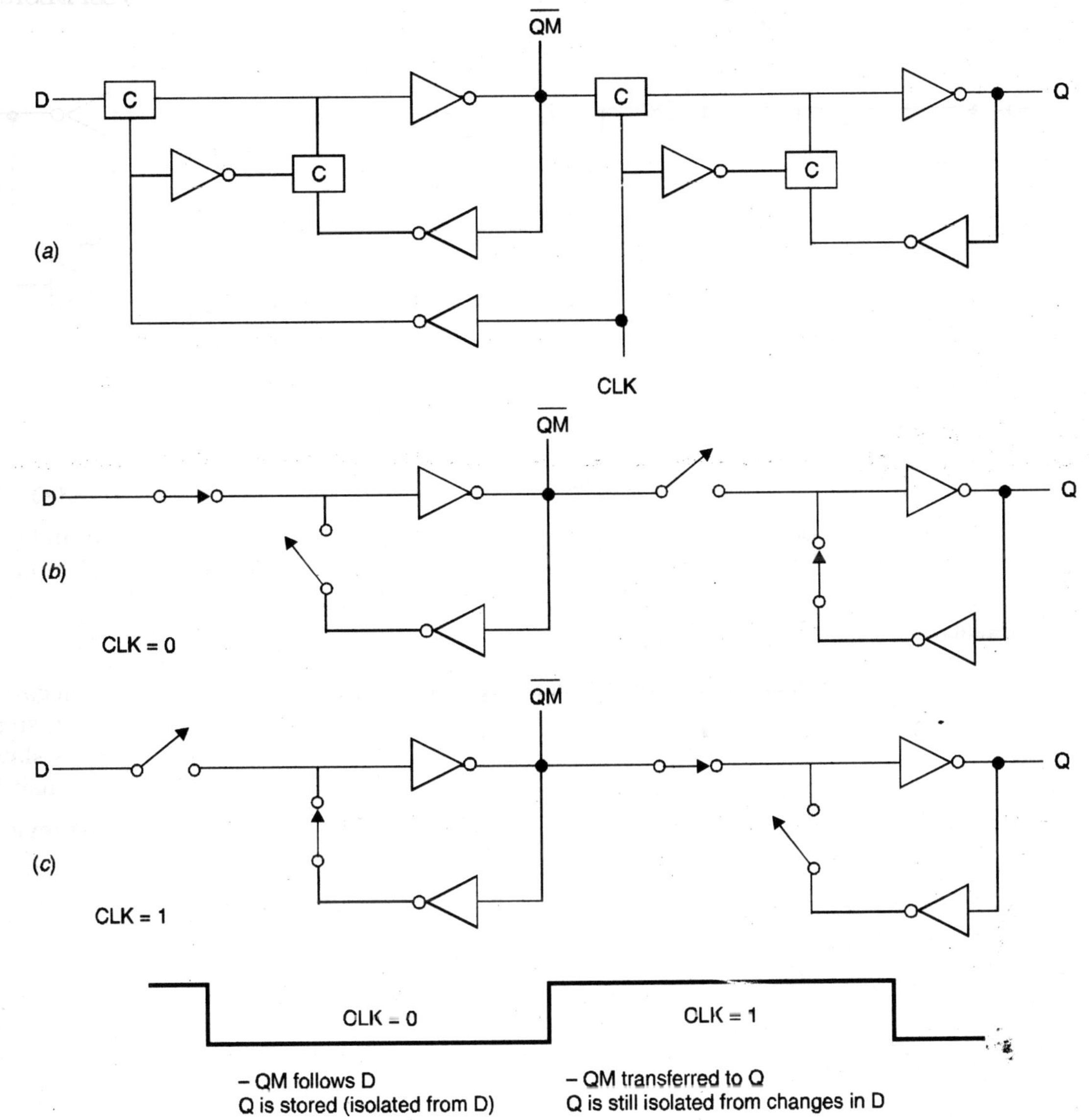

Fig. 3.7.2. (*i*) Positive edge triggered D register.

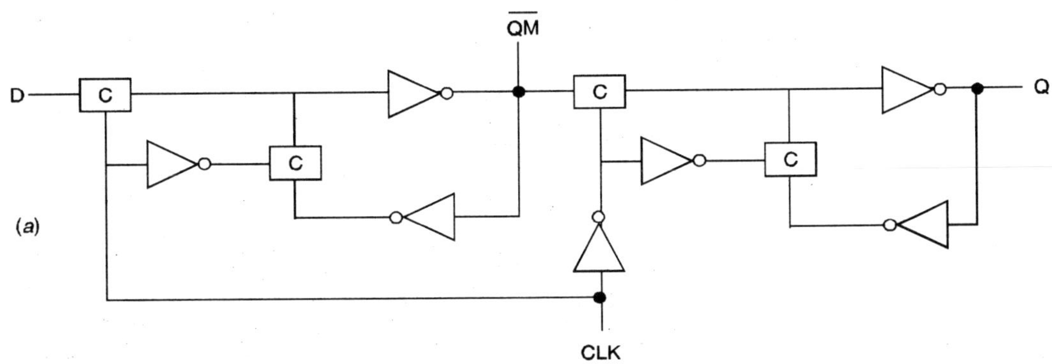

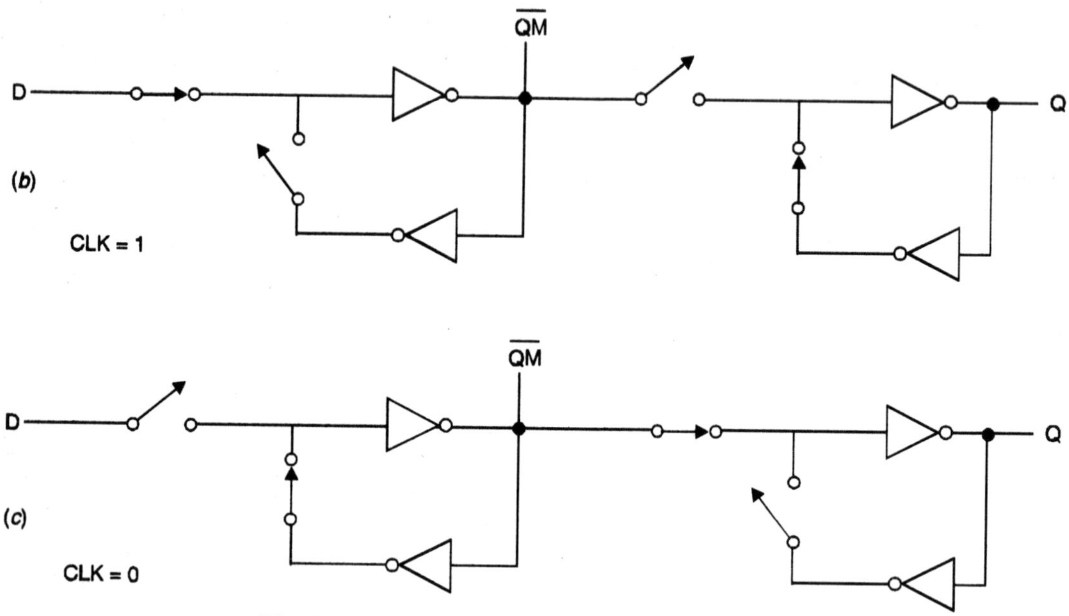

Fig. 3.7.2. (*ii*) Negative edge triggered D register.

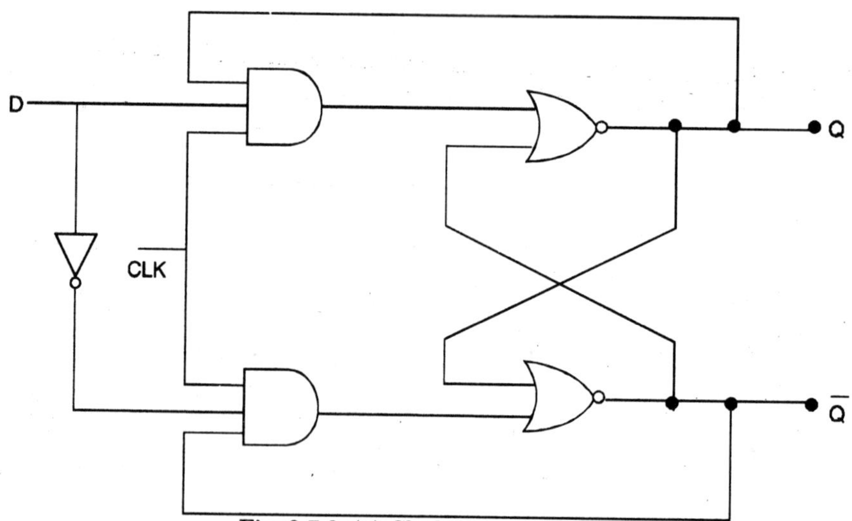

Fig. 3.7.2. (*a*) Clocked D flip-flop.

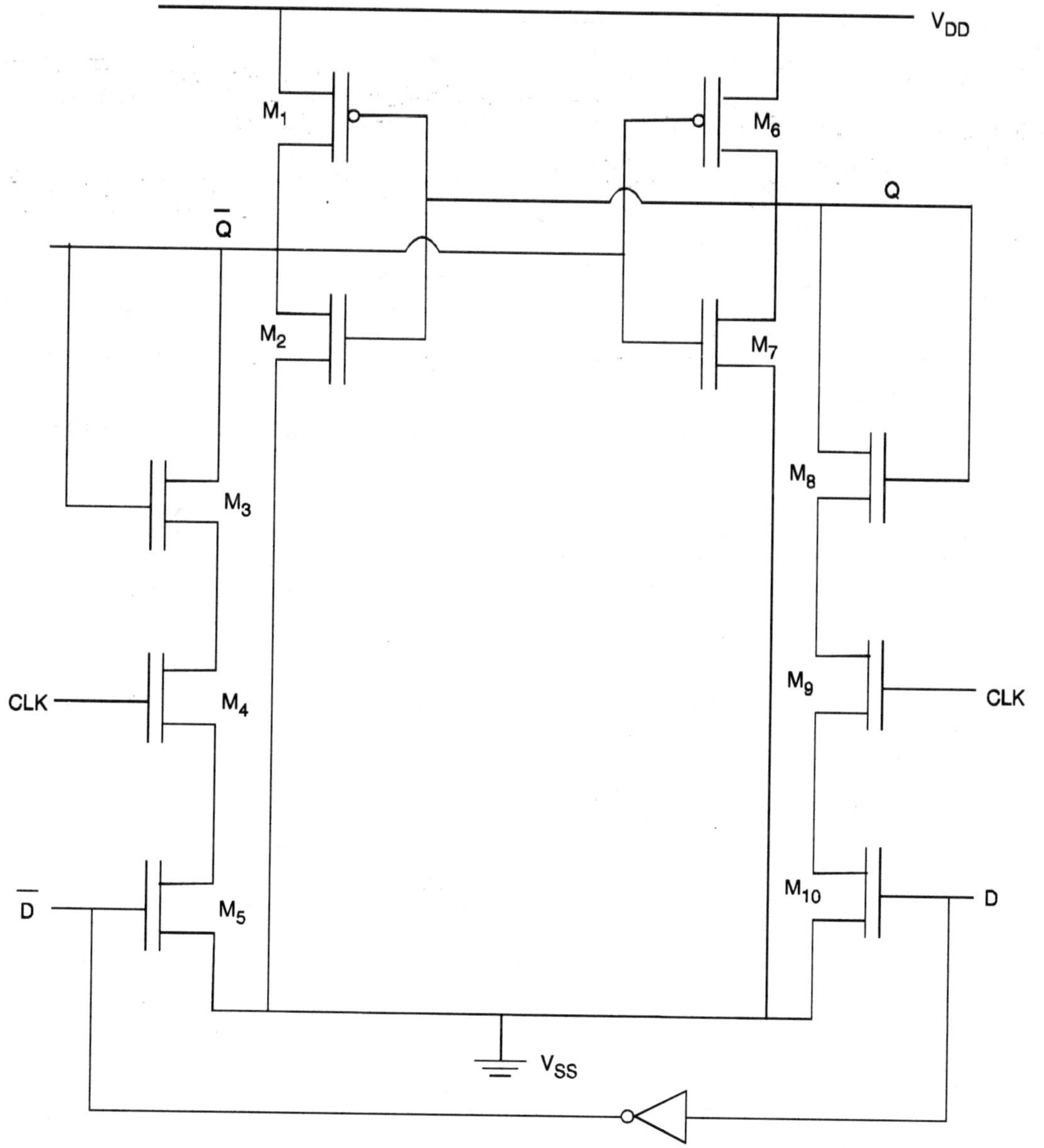

Fig. 3.7.2. (*b*) CMOS implementation for clocked D.

3.7.3. CMOS Clocked SR Flip-Flop

The SR flip-flop is mainly formed by a cross-coupled inverter pair. The remaining transistors are used to drive the FF from one stable state to other, upon application of set and reset pulses and a high clock signal. A simple RS latch with CMOS implementation is shown in Fig. 3.7.3 (*a*) (*i*), (*ii*) below :

The Fig. 3.7.3 (*c*) shows the CMOS implementation of clocked SR flip-flop. The truth table is shown below :

Truth table of SR flip-flop

S	R	Q	$\overline{Q}$	
0	0	Q	$\overline{Q}$	No Change
0	1	0	1	Reset
1	0	1	0	Set
1	1	Indeterminate		

Consider the situation when CLK Q is high, S = 1 and R = 0. This makes M_6, M_8 and M_5 to be switched ON. And M_7 is switched off. Since M_6 and M_5 are switched ON, a closed path is established between o/p of inverter 1 and ground. In other way $\overline{Q}$ is shorted to GND. This $\overline{Q}$ is applied as input to the inverter 2. And this makes Q to become 1. The Fig below shows the gated SR latch.

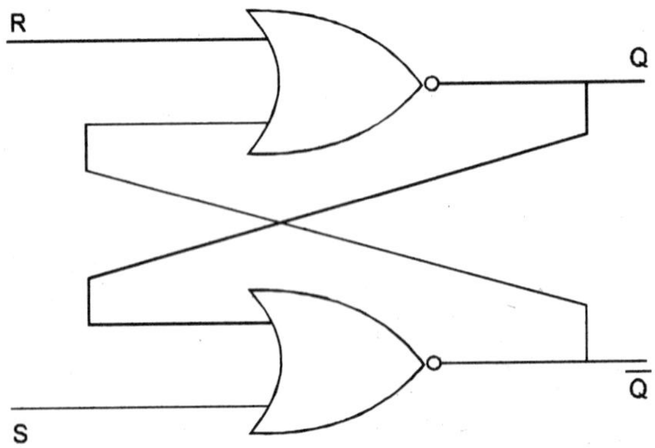

Fig. 3.7.3. (*a*) (*i*) A simple SR latch.

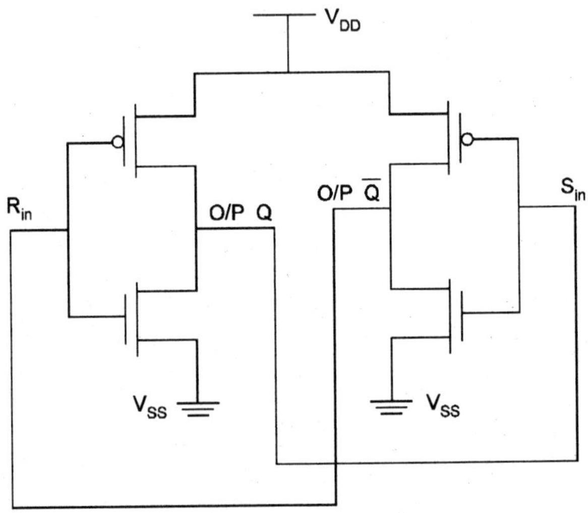

Fig. 3.7.3. (*a*) (*ii*) CMOS implementation of simple SR latch.

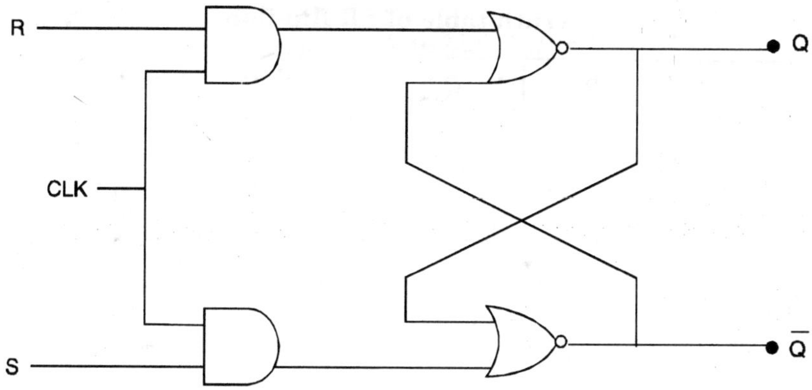

Fig. 3.7.3. (*b*) Gated RS latch.

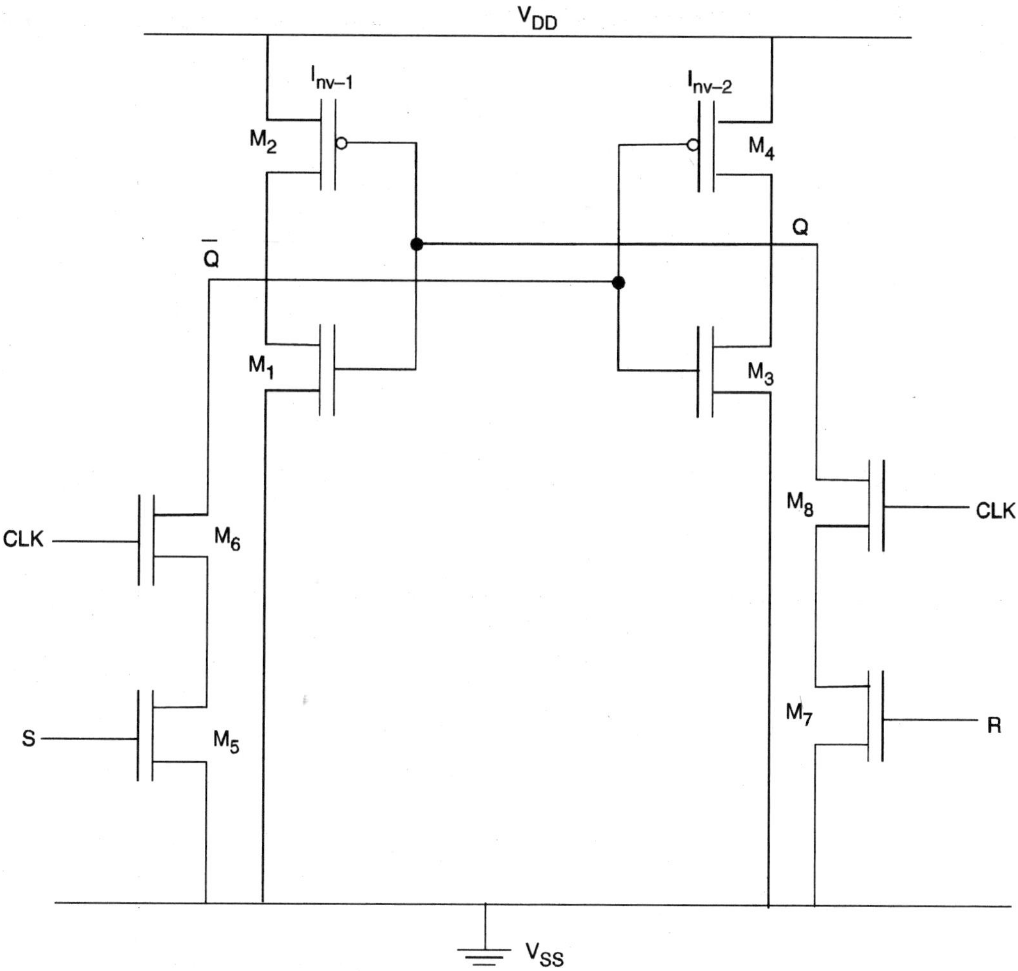

Fig. 3.7.3. (*c*) CMOS implementation of Clocked RS latch.

3.7.4. T-register

A toggle register (or) "T-latch" toggles the Q-output when T = 1. When CLK = 1, the Q output is complemented whereas, When CLK = 0, Q is unchanged

Truth Table of T-register

T	Q $(t+1)$	
0	Q (t)	No change
1	$\overline{Q(t)}$	Toggle

The Fig. 3.7.4 (*b*) shows implementation of T-register using TG and Fig. 3.7.4 (*c*) shows implementation using CMOS.

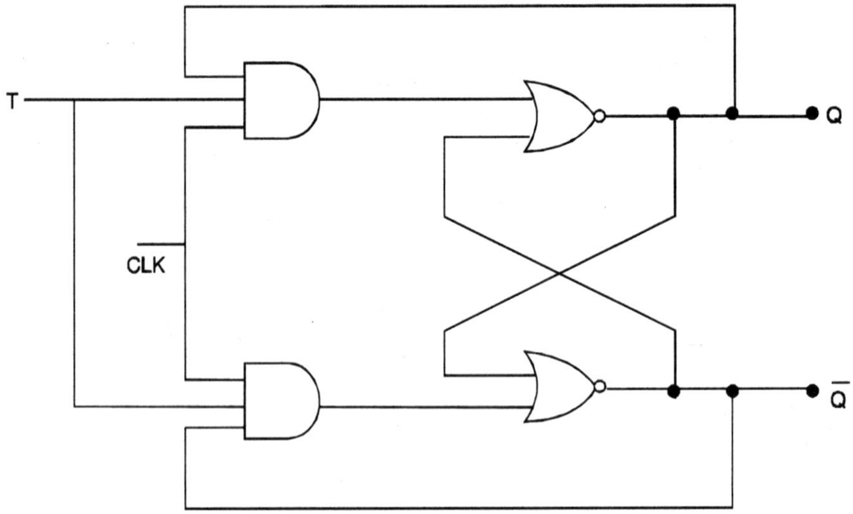

Fig.3.7.4. (*a*) T flip-flop using gates.

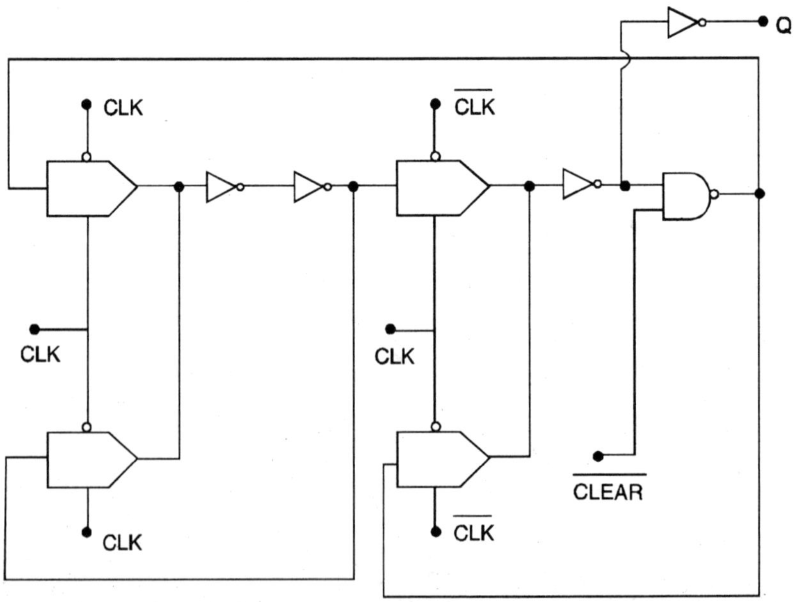

Fig. 3.7.4. (*b*) T register using TG.

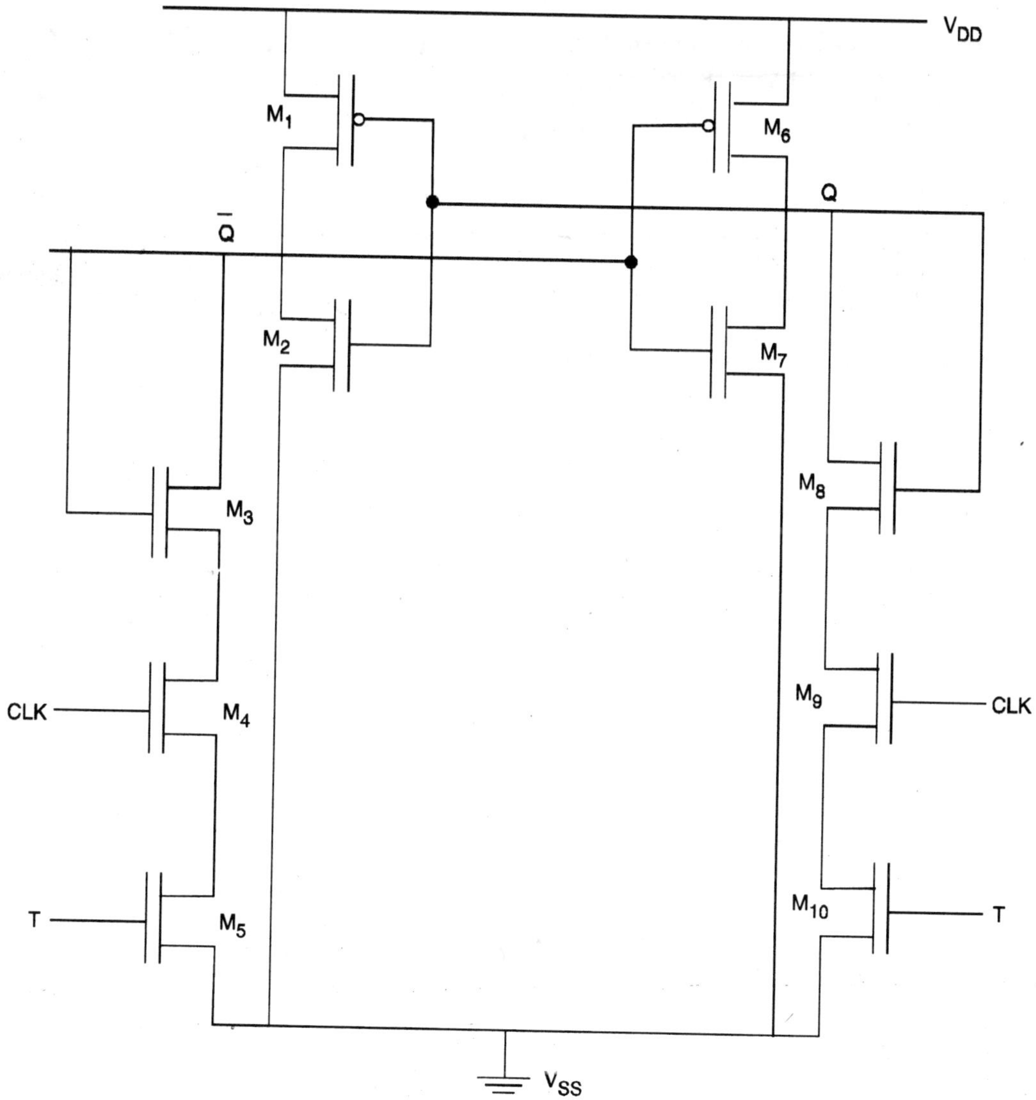

Fig. 3.7.4. (*c*) T register using CMOS.

Assume that the present state of Q = 0 and $\overline{Q}$ = 1. Let T = 0 so that M_5 and M_{10} are switched off (*i.e.*, transistors M_5 and M_{10} are open) As T = 0 there will be no change in the next state of Q and $\overline{Q}$. It maintains as Q = 0 and $\overline{Q}$ = 1. Now make T = 1. Since Q = 0, M_8 is off and as $\overline{Q}$ = 1, M_3 is ON. Transistors M_4, M_5, M_9 and M_{10} all are switched ON by respective clocks and state of T. Since M_3, M_4 and M_5 are switched ON, a short circuit path is established between the o/p of inverter1 (M_1 and M_2 forms inverter1) and ground. So maximum current flows through the short circuit and the voltage drop is zero. Therefore $\overline{Q}$ = 0 as it is connected to ground. As $\overline{Q}$ is applied as input to inverter 2 (M_6 and M_7 forms Inverter 2) output Q = 1. Thus the truth table is verified.

3.7.5. Clocked JK Flip-Flop

The Fig. 3.7.5 (*a*) shows JK-FF using Gates.

The Fig. 3.7.5 (*b*) shows the CMOS implementation of JK flip-flop.

The Truth Table of JK Flip-Flop

J	K	Q	$\overline{Q}$	
0	0	Q	$\overline{Q}$	No change
0	1	0	1	
1	0	1	0	
1	1	$\overline{Q}$	Q	Complement of previous state.

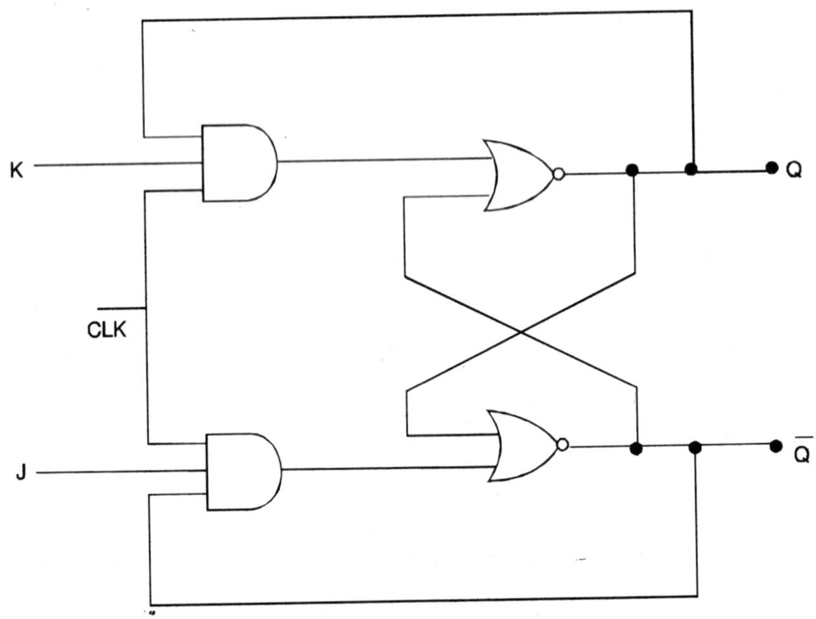

Fig. 3.7.5. (*a*) JK Flip Flop.

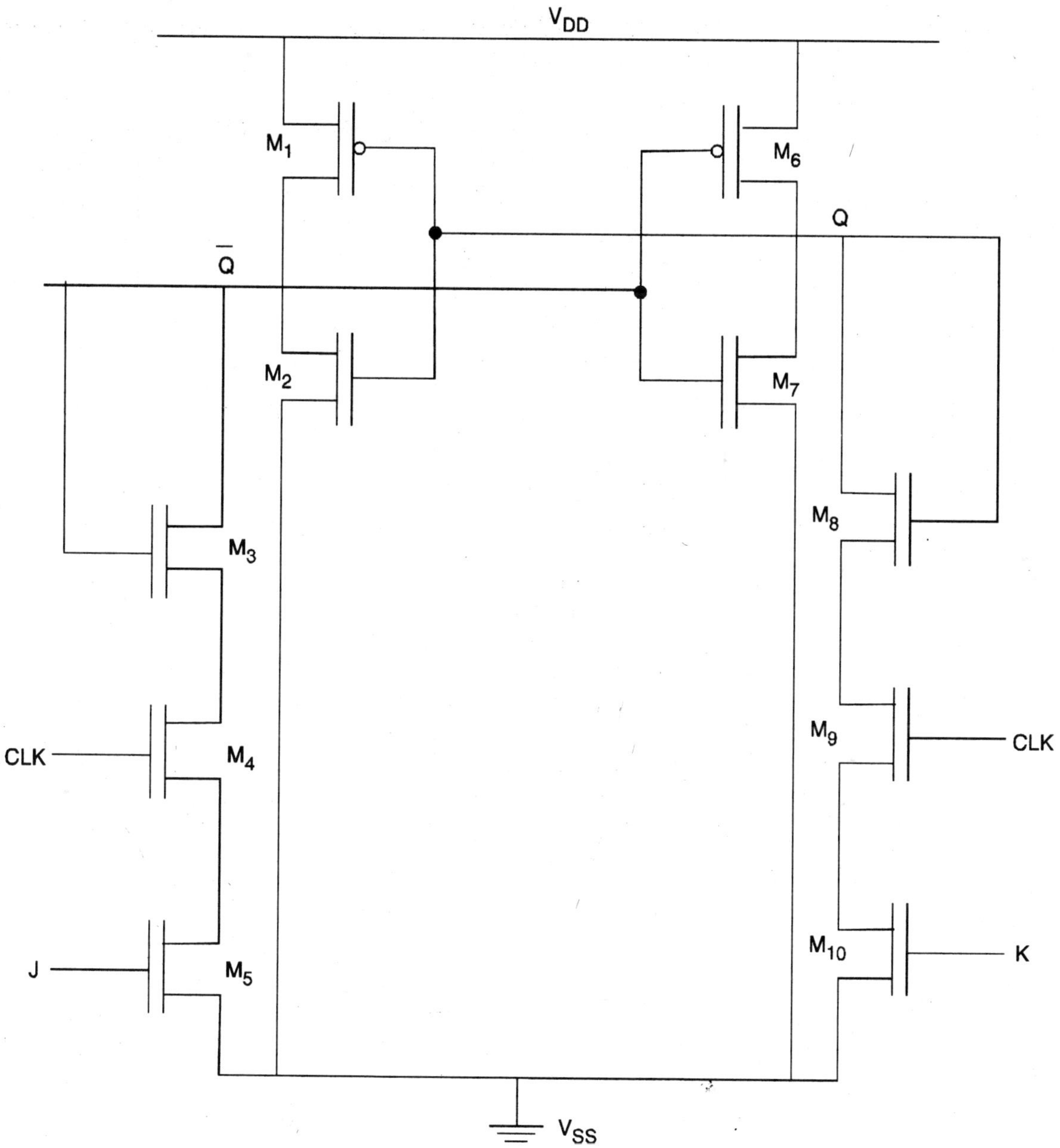

Fig. 3.7.5 (*b*). CMOS implementation of JK FF.

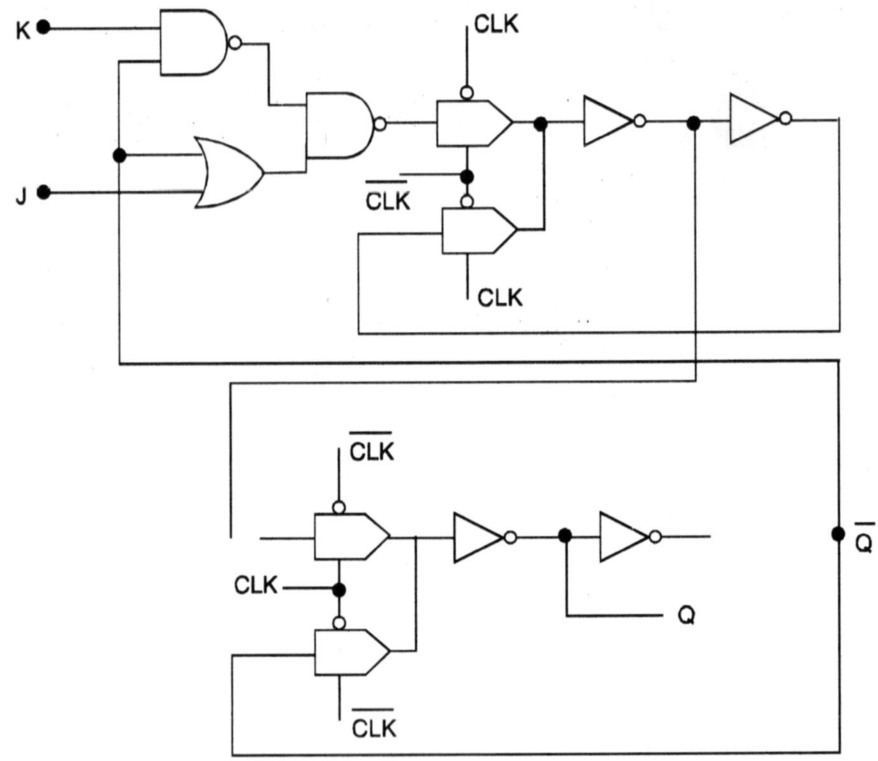

Fig. 3.7.5 (*c*) TG implementation of JK-Register.

Initially let us assume Q = 1 and $\overline{Q}$ = 0 and let CLK = high. When input J = 0 and K = 1, transistor M_5 and M_3 are open and M_4 is closed. In a similar way M_8, M_9 and M_{10} all are closed. Now a short circuit path is established between Q (output of inverter 2) and GND. So the voltage drop across this path is '0'. *i.e.,* Q = 0. Now since Q is applied as input to inverter-1. $\overline{Q}$ will be '1'. Thus the truth table can be verified. The Fig. shows the gated JK-Flip flop

3.8. FAN-IN AND FAN-OUT IN LOGIC GATES

The fan-in of a logic gate is defined as the number of inputs to the gate. It may be impractical to increase the number of inputs beyond a small number. Because we have to consider the effect of n(number of i/p's of NAND gate or equivalently '*n*' number of NMOS transistors) on the propagation delay t_p through the gate.

$$Tp \simeq \frac{1.7C}{K'_p \frac{W}{L} V_{DD}} .n \qquad \text{... (1)}$$

where C is the equivalent capacitance at the output of the gate including the parasitic capacitance contributed by each of the *n*-transistors. The Fig. 3.8.1 shows high fan-in NMOS NAND gate.

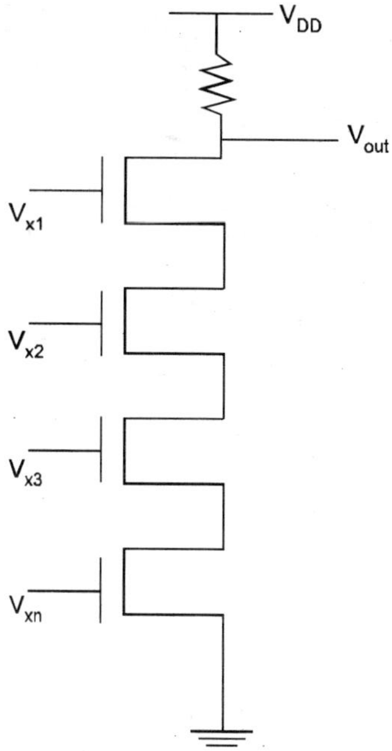

Fig. 3.8.1. High fan-in NMOS NAND gate.

K'_n → Process transconductance parameter-constant.

N → Number of transistors in series

Since the transistors are connected in series, we can consider then being equivalent to one long transistor with length $n \times L$ and width W. The performance of the gate can be improved somewhat by increasing 'W' for each NMOS-transistor. But this change further increases 'C' and comes at the expense of chip-area. (So, when we try to reduce the propagation delay, the chip area increases). Another drawback of the circuit is that each NMOS transistor has the effect of increasing V_{OL} volt (V_{OL} low output voltage level), hence reducing the noise margin. So NAND gates can be build in this manner only if the fan-in is small. Fig. 3.8.2 shows high fan-in NMOS NOR gate.

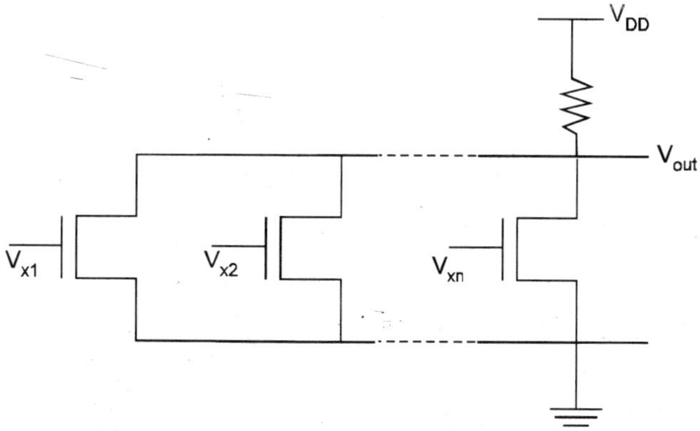

Fig. 3.8.2. High fan-in NMOS NOR gate.

In this case, NMOS transistors are connected in parallel. They can be viewed as one large transistor with width $n \times w$ and length L. According to Eq. (1) the propagation delay should be decreased by the factor-n. However, the parallel connected transistors increase the load capacitance – C at the gate's output. For simplicity assume that C – is increased by the factor 'n' hence t_p – is unaffected by the fan-in. It is thus possible to build high fan-in NOR gates in NMOS technology. In CMOS, we can construct a high fan-in gate by using 2 (or) more lower fan-in gates. As an example, a six – input AND gate can be realized as 2 three – input AND gates that connect to a two – input AND gate.

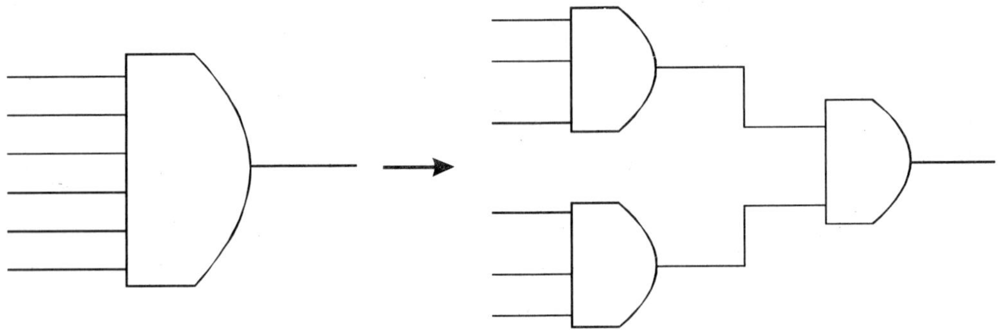

Fig. 3.8.3. Realisation of six I/P AND gate using 2 three I/P AND gate.

Fan-Out. The fan-out of logic – gate is the total number of gate inputs that are driven by the gate output.

That is the number of other gates that a specific gate drives is called its fan-out. In Example-1 Fig. 3.8.4, G_1 drives four other gates.

Example 1

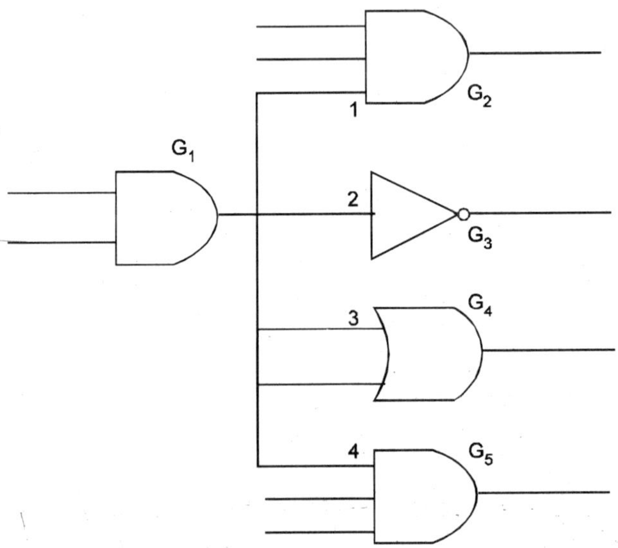

Fig. 3.8.4.

Example 2

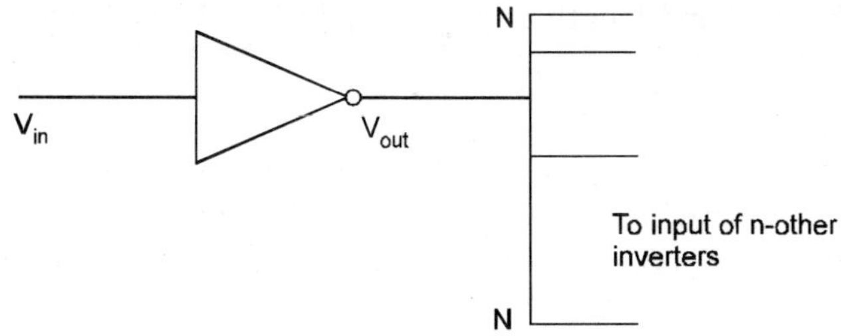

Fig. 3.8.5 (*a*)

Example 2, Fig. 3.8.5 (*a*) shows an inverter N that drives the inputs of n- other inverters. Equivalent circuit of Example 2 is shown in Fig. 3.8.5 (*b*) for timing purpose. Each of the other inverters contributes to the total capacitive loading on node V_{out}.

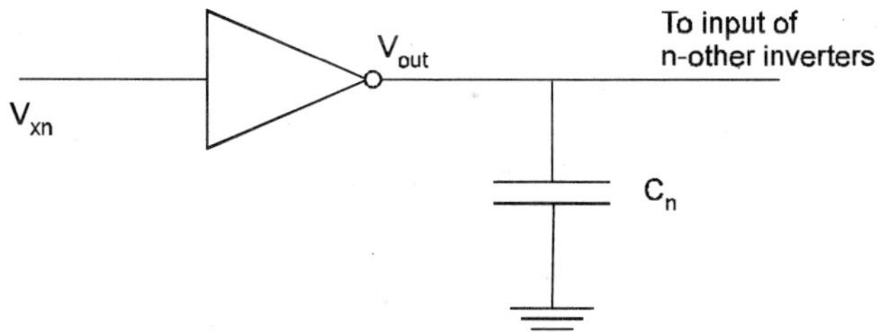

Fig. 3.8.5. (*b*) Equivalent circuit of Fig. 3.8.5 (*a*).

From the above figure the n-inverters are represented by one large capacitor – C_n. For simplicity, assume that each inverter contributes a capacitance C and that $C_n = n \times C$. We know that the propagation delay increases in direct proportion to '*n*' as given by the equation.

$$tp \simeq \frac{1.7C}{K'_n \frac{W}{L} V_{DD}}$$

The Fig. 3.8.6 shows how '*n*' (fan-out) affects the propagation delay.

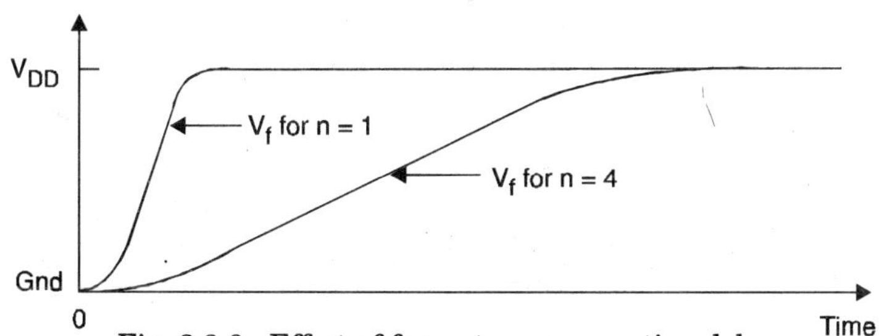

Fig. 3.8.6. Effect of fan-out on propagation delay.

The propagation delay can be reduced by decreasing *n* and also by increasing W/L ratios of the transistors.

$$\text{Propagation delay} = tp \cong \frac{1.7C}{K'_n \frac{W}{L} V_{DD}}$$

K'_n = process transconductance parameter, It is a constant that depends on the technology being used and has the units A/v². (For PMOS – K'_p). This expression specifies that the speed of the circuit depends both on the value of 'C' and on the dimensions of the transistor. The delay can be reduced by making 'C' smaller (or) by making the ratio W/L – larger. The above expression shows the propagation time when the output changes from a high level to low level. The low to high propagation time is given by the same expression but using K′p and W/L of the PMOS transistor. 'L' is usually set to the minimum value that is permitted according to the specification of fabrication technology used. The value of W-is chosen depending on the amount of current flow, hence propagation delays that are desired.

3.9. BUFFERS

A buffer circuit is a one which is used to avoid loading effect. In circuits in which a logic gate has to drive a large capacitive load; buffers are often used to improve performance. A buffer is a logic gate with one input $-V_{in}$ and one output $-V_{out}$ which produces $V_{in} = V_{out}$. The simplest implementation of buffer uses two inverters.

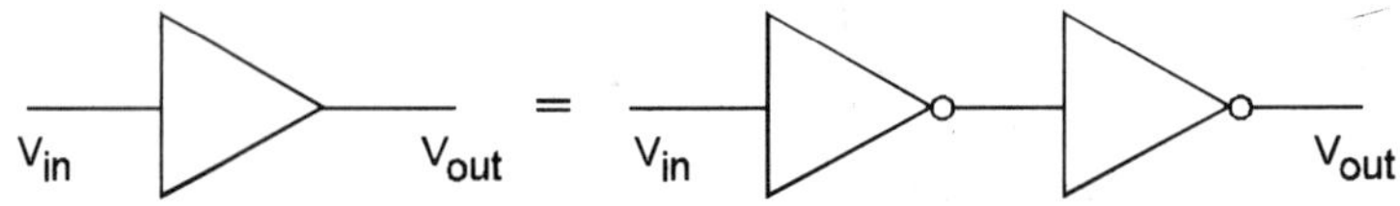

Fig. 3.9. (*a*) Graphical symbol.

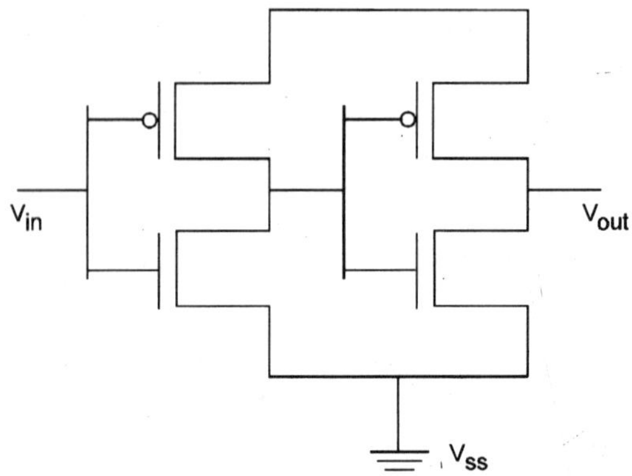

Fig. 3.9. (*b*) Implementation of buffer.

Buffers can be created with different amounts of drive capability, depending on the sizes of the transistors. In general, because they are used for driving higher capacitive loads than normal, buffers have transistors that are larger than those in typical logic gates.

3.9.1. Tri-state Buffers

There is a type of buffer called tri-state buffer. It has one input V_{in}, one output V_{out} and a control input called enable-e.

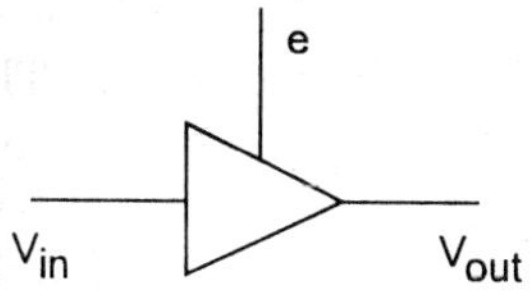

Fig. 3.9.1 (*a*) Graphical symbol.

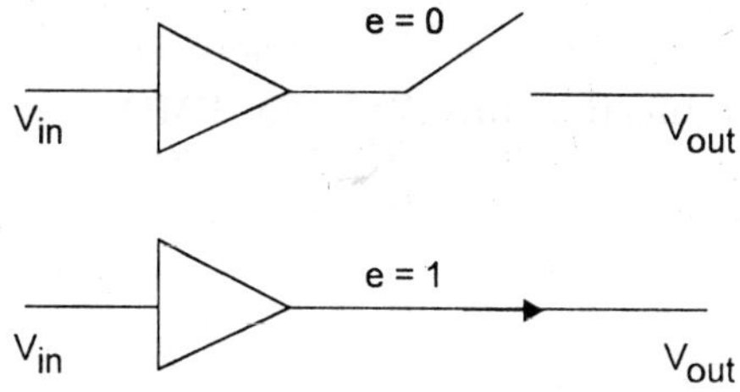

Fig. 3.9.1 (*b*) Equivalent circuit.

Truth Table

e	V_{in}	V_{out}
0	0	Z
0	1	Z
1	0	0
1	1	1

When e = 0, the output is denoted by the logic – value – Z, which is called High – impedance state (i/p is disconnected from output). The name tri-state derives from the fact that there are two normal states for a logical signal – 0 and 1 and Z – represents a third state that produces no-output signal. When e = 1, the buffer drives the value V_{in} onto V_{out}, causing $V_{out} = V_{in}$.

3.10. NOISE MARGIN

This parameter allows us to determine the allowable noise voltage on the input of a gate so that the output will not be affected

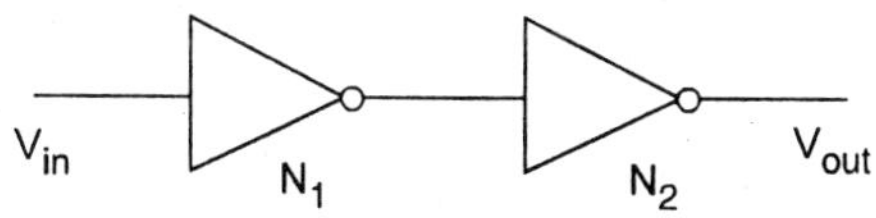

The above Fig shows a NOT- gate driving another NOT- gate. Electronic circuits are constantly subjected to random perturbations, called Noise, which can alter the output voltage levels produced by the gate N_1. It is essential that this noise not cause the gate N_2 to misinterpret a low logic value as a high one (or) vice-versa. Consider the case where N_1 produces its low voltage level v_{OL}. The presence of noise may alter the voltage level but as long as it remains less than V_{IL}, it will be interpreted correctly by N_2. The ability to tolerate noise without affecting the correct operation of the circuit is known as noise margin.

For the low output voltage, we define the low noise margin as the difference in magnitude between the maximum LOW output voltage of the driving gate and the maximum input low voltage accepted by the driven gate.

$$N_{ML} = V_{ILmax} - V_{OLmax}$$

A similar situation exists when N_1 produces its high output voltage V_{on}, Any existing Noise in the circuit may alter the voltage level, but it will be interpreted correctly as N_2 as long as the voltage is greater than V_{IH}. The high noise margin is defined as the difference in magnitude between the minimum HIGH output voltage of the driving gate and the minimum input HIGH voltage accepted by the receiving gate.

$$N_{MH} = V_{OH}\,\text{min} - V_{IH}\,\text{min}$$

V_{IH} min = minimum HIGH input voltage
V_{IL} max = maximum LOW input voltage
V_{OH} min = minimum HIGH output voltage
V_{OL} max = maximum LOW output voltage

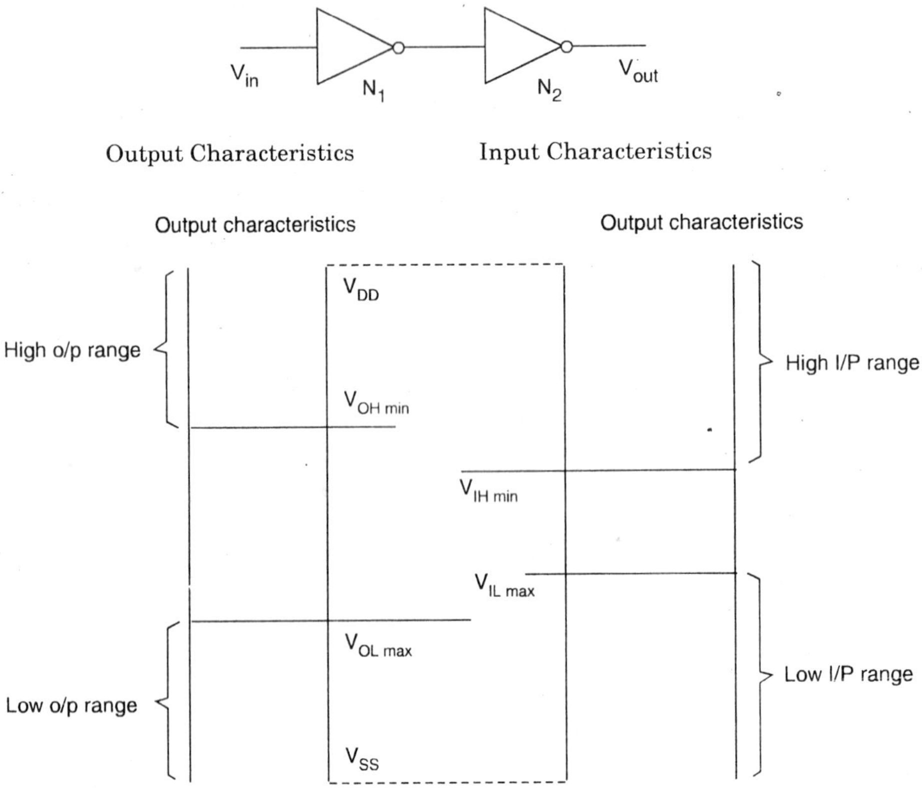

Fig. 3.10. Noise margin.

3.11. SETUP AND HOLD TIME

The setup and hold time of a register are deviation from an ideal register caused by finite circuit delays. The hold time is defined as the delay between the clock input to the register and the storage element. That is, the data has to be held for this period while the clock travels to the point of storage. The setup time is the delay between the data input of the register and the storage element. As the data takes a finite time to travel to the storage point, the clock can not be changed until the correct data value appears. In CMOS registers these delays normally are very small. In a synchronous system, if the data input to a register does not obey the setup and hold time contraints the clock race problems may occur which will result in erroneous data being stored in registers. With hold time violation the data changes to a new value before the clock can change. With a setup time constraint the clock changes before the data assumes the correct value.

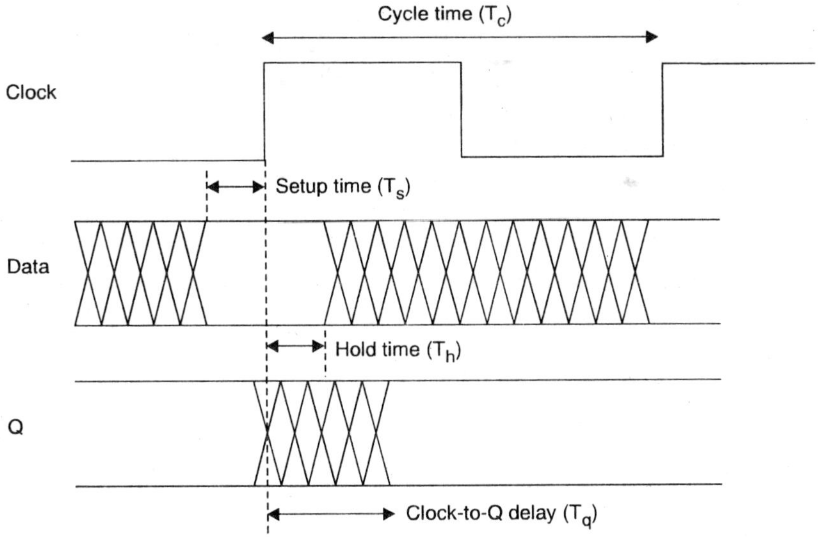

Fig. 3.11. A single phase clock showing setup time, hold time, cycle time and clock to Q delay.

The Fig. 3.11 shows the parameters of interest in single phase clock. The corresponding definitions are as follows.

Set up time (T_s). The time before the clock edge that the D-input has to be stable is called the set up time (T_s).

Hold time (T_h). The time after the clock edge that the D-input has to remain stable is called the hold time (T_h).

Clock-to-Q delay (T_q). The delay from the positive clock input to the new value of the Q-output is called the clock-to-Q delay (T_q).

Cycle Time (T_c). The time between successive positive clock transitions is called the cycle Time (T_c).

3.11.1. System Timing

Latches and registers may be used in variety of ways to implement clocked system.

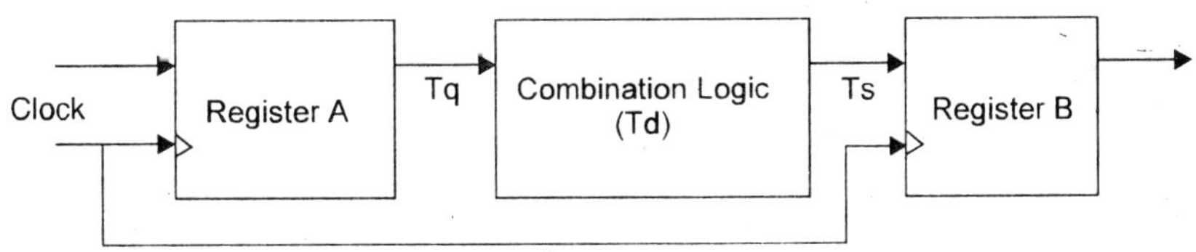

Fig. 3.11.1 (*a*)

The above Fig. 3.11.1 (*a*) shows a pipelined structure with input and output registers separated by combinational logic. Here, the cycle time is given by,

$$\boxed{T_c = T_q + T_d + T_s} \quad \text{... (1)}$$

where $T_d \rightarrow$ worst – case delay through the combinational logic block.

$T_c \rightarrow$ cycle time.

$T_s \rightarrow$ set up time

$T_q \rightarrow$ clock-to-Q-delay

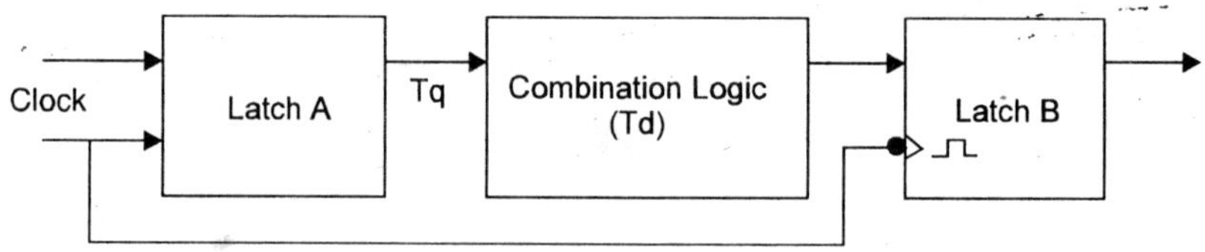

Fig. 3.11.1 (*b*)

An alternate structure is shown in Fig. 3.11.1 (*b*) where the combinational logic is bounded by latches. This structure can be extended as shown in Fig. 3.11.1 (*c*) by alternating positive and negative level-sensitive latches with combinational logic. A saving has been made in the number of latches per combinational block compared with the registered configuration as in Fig. 3.11.1 (*a*).

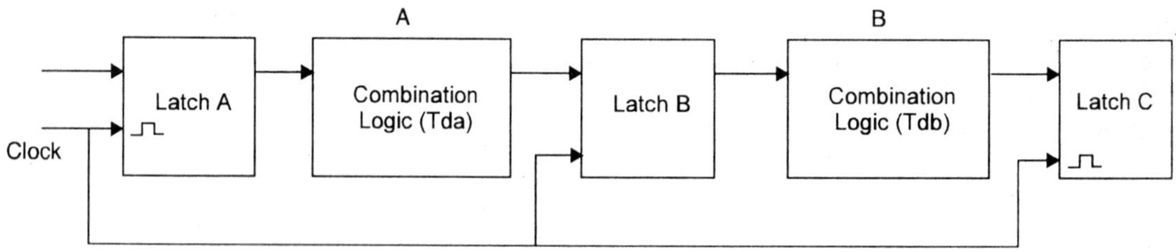

Fig. 3.11.1 (*c*)

However, in the latch case, the logic at the output of latch – A receives data T_q later than 0 – 1 transition must deliver it to latch B, T_s before 1 – 0 transition of the clock. Thus the logic delay of block – A (T_{da}) must satisfy the following inequality.

$$\boxed{T_{da} < T_{c1} - T_{qa} - T_{sb}} \text{ (assuming 50\% duty cycle)} \quad \text{... (2)}$$

where T_{qa} = clock – to – Q time of latch – A

T_{sb} = the setup time of latch – B.

Similarly,

$$\boxed{T_{db} < T_{co} - T_{qb} - T_{sa}} \quad \text{... (3)}$$

(Assuming $T_{sa} = T_{sb} = T_{sc}$ and also assuming A, B, C to be identical)

From 2, In the limit, $T_{da} = (T_c/2) - T_{qa} - T_{sb}$

From 3, $T_{db} = (T_c/2) - T_{qb} - T_{sa}$

Also From 2, $T_{c1} = T_{da} + T_{qa} + T_{sb}$ ... (4)

From 3, $T_{c0} = T_{db} + T_{qb} + T_{sa}$ ... (5)

Adding equation 4 and 5 we get

$$\boxed{T_C = T_{C1} + T_{C0} = T_{da} + T_{db} + 2(T_q + T_s)} \text{ (Assuming a and b as identical)}$$

All state changes occur at the rising clock edge.

4

MOS Circuits Design Process

4.1. INTRODUCTION TO MOS LAYERS

Generally in MOS design, first a specification is turned into masks. Then the substrate (silicon) is processed in accordance with the mask to meet the specification. The MOS circuits are formed on basic-layers *n*-diffusion, *p*-diffusion, polysilicon and metal. Each layer is isolated from one another by thick or thin (thinox) SiO_2 insulating layers. The thinox mask region includes *n*-diffusion, *p*-diffusion and transistor channels. A transistor is formed where polysilicon and thinox regions cross one another and in some processes a second metal layer and a second polysilicon layer may also be used. Layers may be deliberately joined together where contacts are formed.

In order to capture the topology and layer information of the actual circuit in silicon, simple diagrams such as

1. Stick diagrams
2. Layout diagrams can be used.

4.2. STICK DIAGRAMS

Stick diagrams may be used to convey layer information through the use of a colour code or through monochrome encoding scheme.

For different processes different colour codes and encoding scheme will be used. We will discuss only NMOS and CMOS processes. The colour encoding is made compatible with colour terminals, printers and plotters having quite simple colour palettes.

In this book after having a brief look on colour encoding scheme, a detailed look on monochrome encoding is done. Finally stick diagrams are drawn with monochrome encoding scheme only. The encoding schemes are discussed for the following two cases.

1. Single metal NMOS design
2. Double metal CMOS P-well process.

Table 4.1. Encoding for a Simple Single Metal NMOS Process

Sl. No.	*Layers*	*Stick encoding*	
		Colour	*Monochrome*
1.	*n*-diffusion Thinox (*n*-diff and transistor channels)	Green	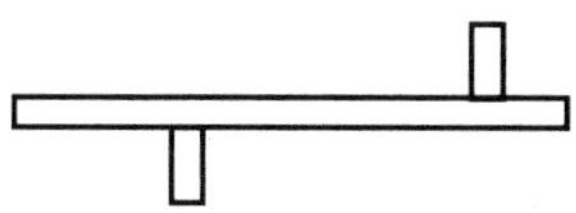
2.	Polysilicon	Red	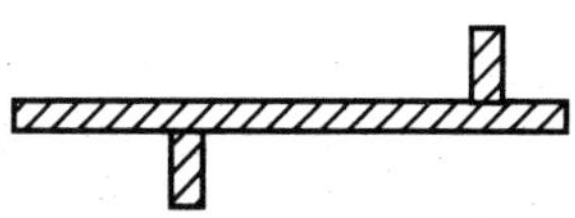

3.	Metal 1	Blue	
4.	Contact Cut	Black	
5.	Overglass	Gray	Not applicable
6.	Implant	Yellow (for NMOS only)	
7.	Buried Contact	Brown (for NMOS only)	

Table 4.2. Encodings for a Double Metal CMOS P-well Process

Sl. No.	*Layers*	*Stick encoding*	
		Colour	*Monochrome*
1.	*n*-diffusion Thinox		
2.	Polysilicon		
3.	Metal 1	→ Encoding as in Table 4.1	
4.	Contact Cut		
5.	Overglass		
6.	P-diffusion (P^+active)	Yellow	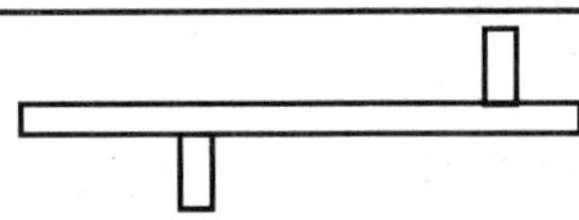
7.	Metal 2	Dark Blue or Purple	
8.	Via	Black	
9.	P-Well	Brown	P-Well edge is shown as a demarcation line in stick diagram
10.	V_{DD} or V_{SS} Contact	Black	

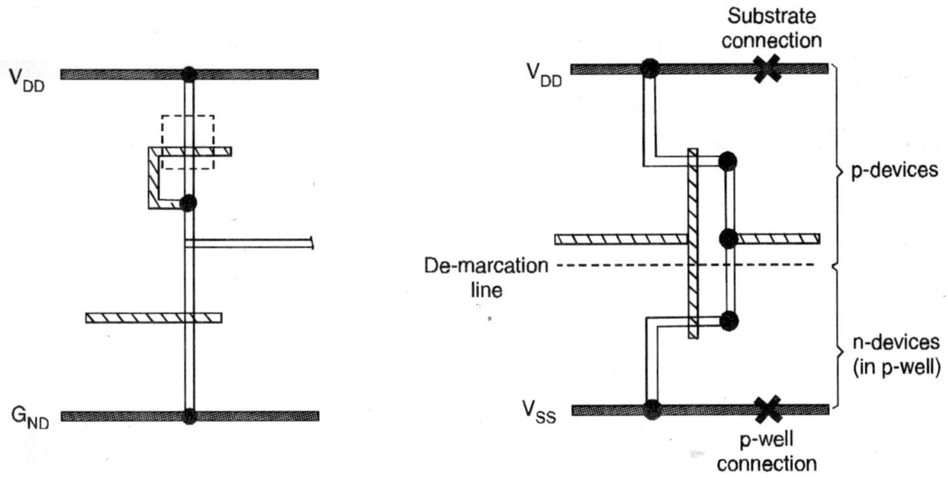

Fig. 4.2.1. Illustration of stick diagram (*i*) NMOS inverter (*ii*) *p*-well CMOS Inverter.

4.2.1. NMOS Design Style

Single metal Single Polysilicon NMOS technology. The stick layout of NMOS involves.

- *n*-diffusion and other thinox regions
- Polysilicon 1
- Metal 1
- Implant
- Contacts

A transistor is formed wherever poly crosses *n*-diffusion. All diffusion wires (interconnections) are *n*-type.

Steps in drawing stick diagram

1. Draw the metal V_{DD} and G_{ND} rails in parallel allowing enough space between them. Next thinox paths may be drawn between the rails for inverters and inverter based logic. Also make appropriate contacts. (See Fig. 4.2.2 (*a*)).

2. Inverters and inverter based logic comprise a pull up structure, usually a depletion mode transistor, connected from the OUTPUT point to V_{DD} and a pull down structure of enhancement mode transistors suitably interconnected between the OUTPUT point and G_{ND}. [See Fig. 4.2.2 (*b*)]. Remember that poly crosses thinox wherever transistors are required. For depletion mode transistors implants are also to be drawn as in Fig. For each transistor write length to width (L : W) ratio.

3. Signal paths may also be switched by pass transistors and long signal paths may often require metal buses. The stick diagram may well represent only a small section of circuit, which will be replicated many times. (Replication-repetition).

Therefore power rails and buses are run in parallel in metal. The control signals (input) are propagated at right angles on polysilicon. At this stage of design **'leaf cell'** boundaries are conveniently shown on the stick diagram and these are placed so that replicated cells may be directly interconnected by direct abutment on a side by side and/or top to bottom basis.

Leaf cell : The layout of IC, *i.e.,* floor plan of IC can be represented heirarchically; cells are built from other cells, except for those cells that are at the lowest level of hiararchy. These lowest-level cells are called leaf cells.

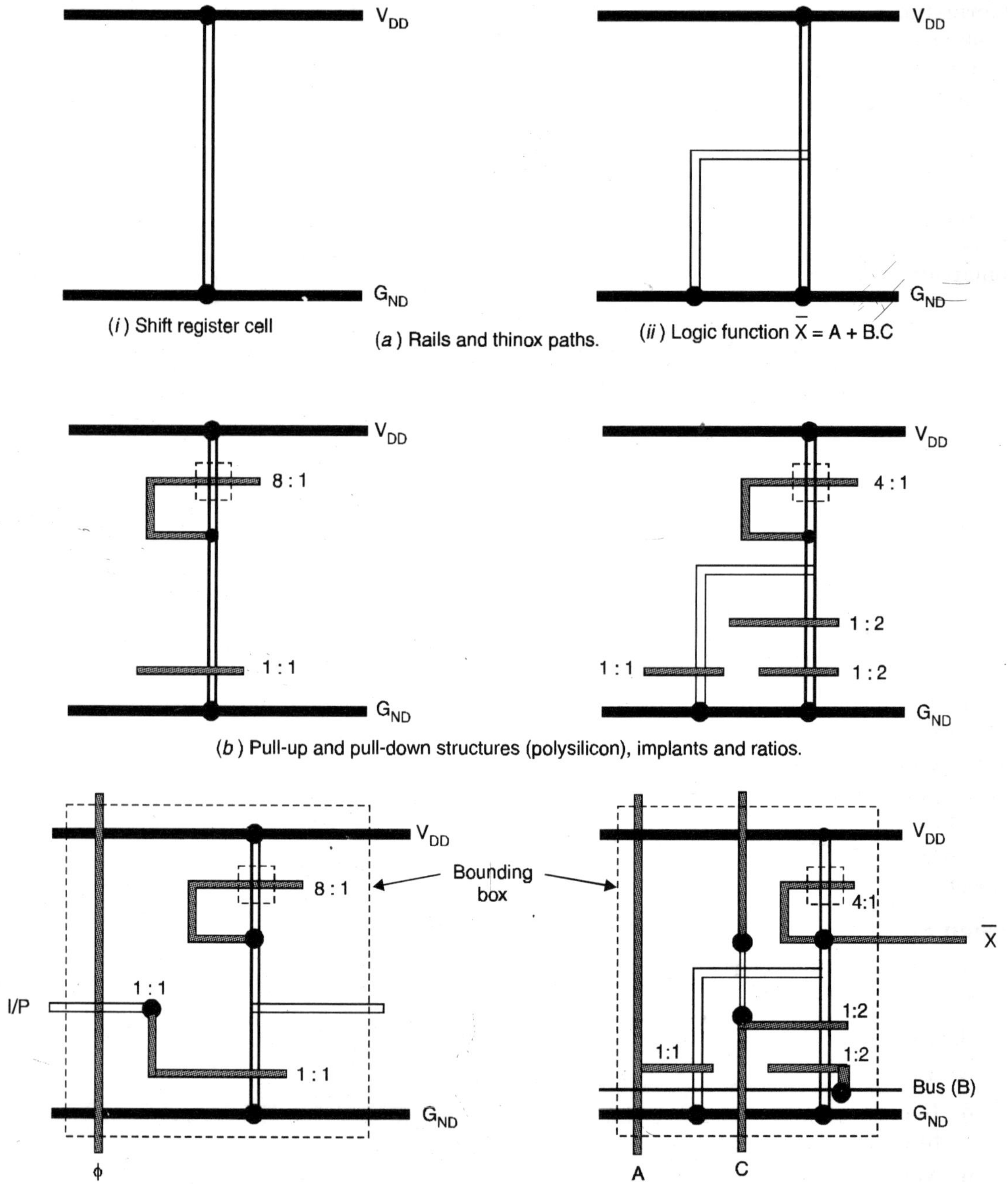

(*i*) Shift register cell (*a*) Rails and thinox paths. (*ii*) Logic function $\overline{X} = A + B.C$

(*b*) Pull-up and pull-down structures (polysilicon), implants and ratios.

(iii) Buses, control signals, interconnections, and 'leaf-cell' boundaries.

Fig. 4.2.2. Examples of NMOS stick layout design style.

From the beginning a design style should encourage the concepts of 'regularity' (through the use of replication) and generality so that design effort can be minimized and the interconnection of leaf cells, subsystems and systems is facilitated. Using stick diagrams layer information and topology at the actual circuit is conveyed. The next step is to turn these into mask layouts. In order that the mask layouts produced during design will be compatible with the fabrication processes, a set of design rules are set out for layouts so that, if obeyed, the rules will produce layouts which will work in practice.

4.2.2. CMOS Design Style

In this depletion mode devices are not utilised. The two types of transistor used, '*n*' and '*p*' are separated in the stick layout by the demarcation line (representing the *p*-well boundary) above which all *p*-type devices are placed. The n-devices are consequently placed below the demarcation line and are thus located in the *p*-well.

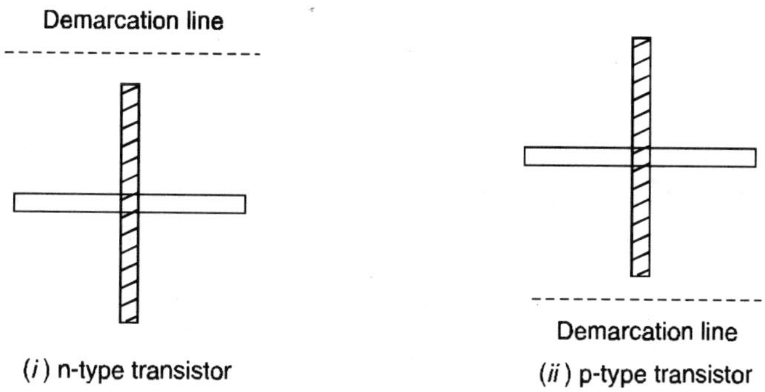

Fig. 4.2.3. *n*-type and *p*-type transistors in CMOS design.

Diffusion paths must not cross the demarcation line and *n*-diffusion and *p*-diffusion wires must not join. Wherever a connection is needed the '*n*' and '*p*' features are normally joined by metal. Apart from demarcation line, there is no indication of the actual *p*-well topology at this (stick diagram) level of abstraction. Their geometry will appear when the stick dia is translated into a mask layout.

To represent the substrate and *p*-well connection place crosses on V_{DD} and V_{SS} rails respectively.

Design Steps

1. Draw V_{DD} and V_{SS} rails in parallel in metal.
2. Draw imaginary demarcation line in between rails. (Fig. 4.2.4 *a*).
3. *n*-transistors are placed below this line and close to V_{SS}
4. *p*-transistors are placed above the line and below V_{DD}. (Fig. 4.2.4 *b*).
5. In both cases, the transistors are conveniently placed with their diffusion paths parallel to the rails.
6. Now interconnect the n with the *p*-transistors as required, using metal and connect to the rails as shown in Fig. (4.2.4 *c*).
7. Remember that only metal and polysilicon can cross the demarcation line, with the restriction, that wires can run in diffusion also.

8. Finally the remaining interconnections are made as appropriate and the control signals and data inputs are added. Fig. (4.2.4 *d*).
9. Make the V_{SS} and V_{DD} contact crosses – one on the V_{DD} line for every four *p*-transistors and one on the V_{SS} line for every four *n*-transistors.
10. The bounding box for the entire leaf-cell may shown.

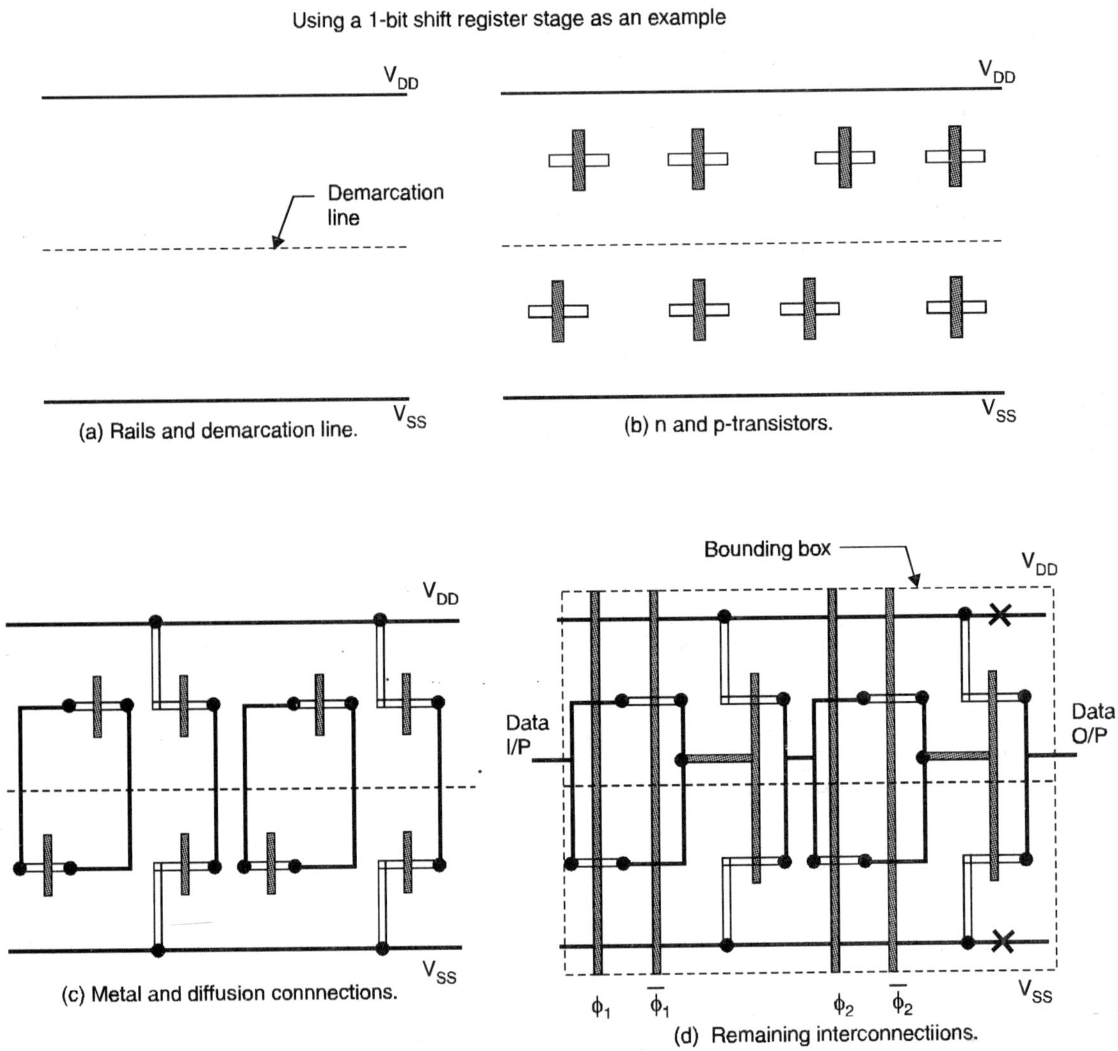

Fig. 4.2.4. Examples of CMOS stick layout design style.

The following Figs. show circuit symbol and the corresponding stick diagrams of basic gates. (*i*) inverter (*ii*) NAND (*iii*) NOR in nMOS and CMOS logic

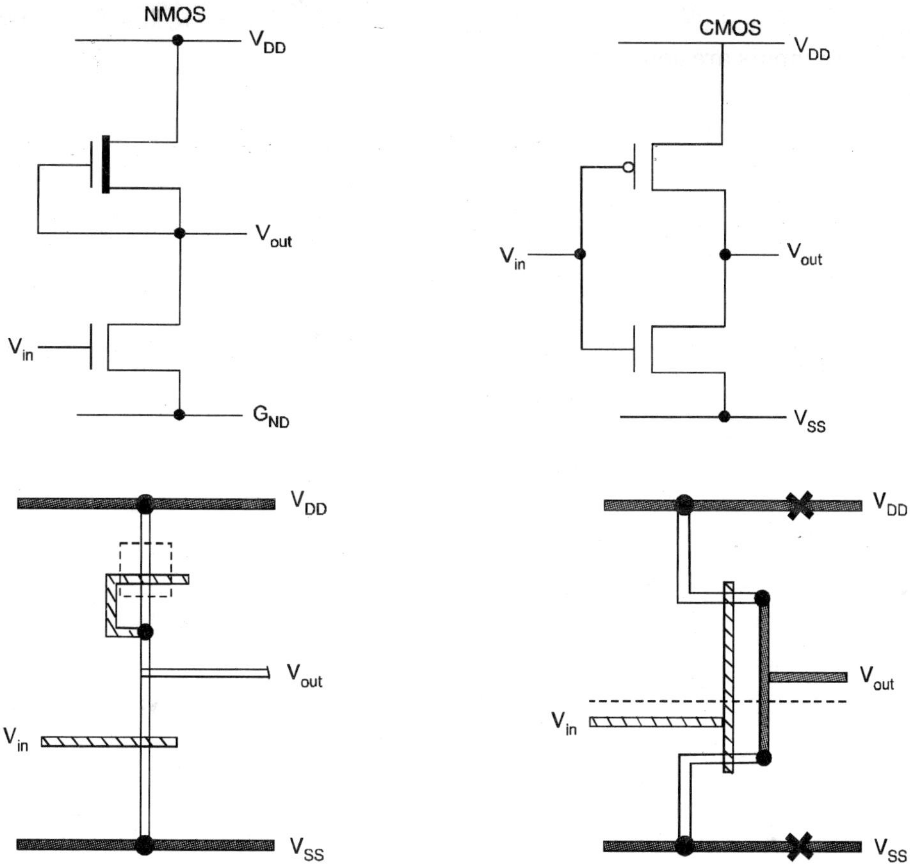

Fig. 4.2.5. Stick diagrams for inverter.

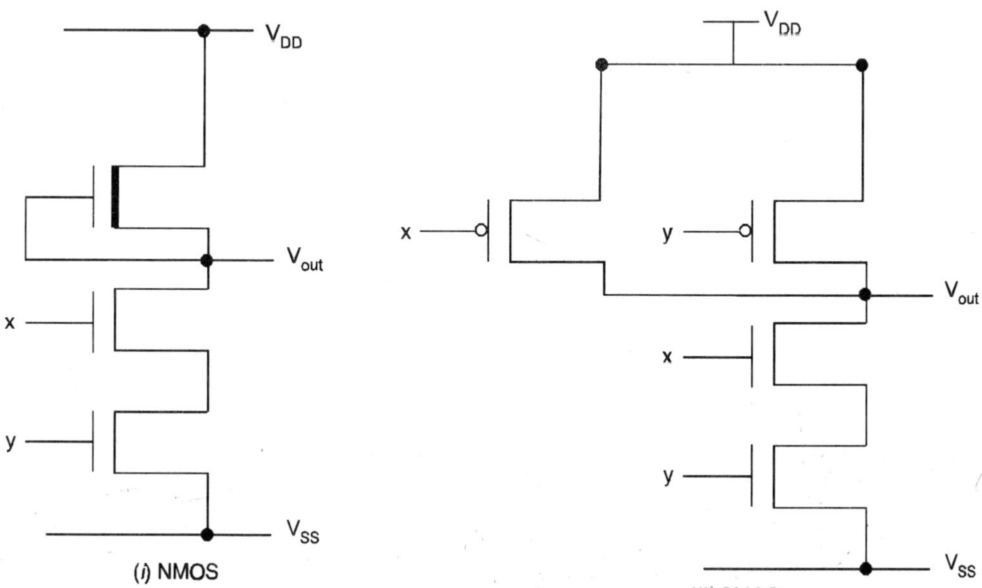

(*i*) NMOS (*ii*) CMOS

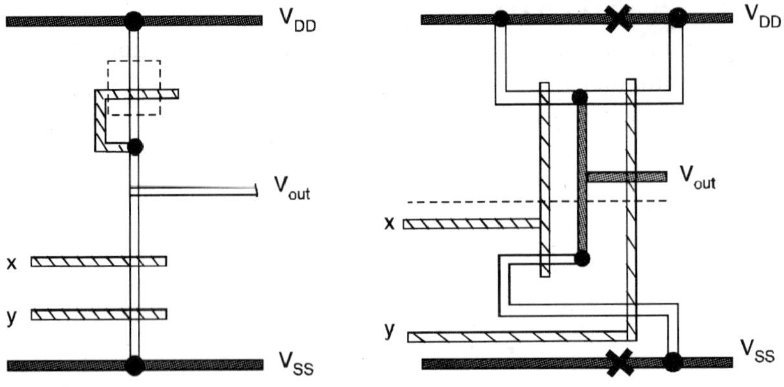

Fig. 4.2.6. Stick diagrams for NAND.

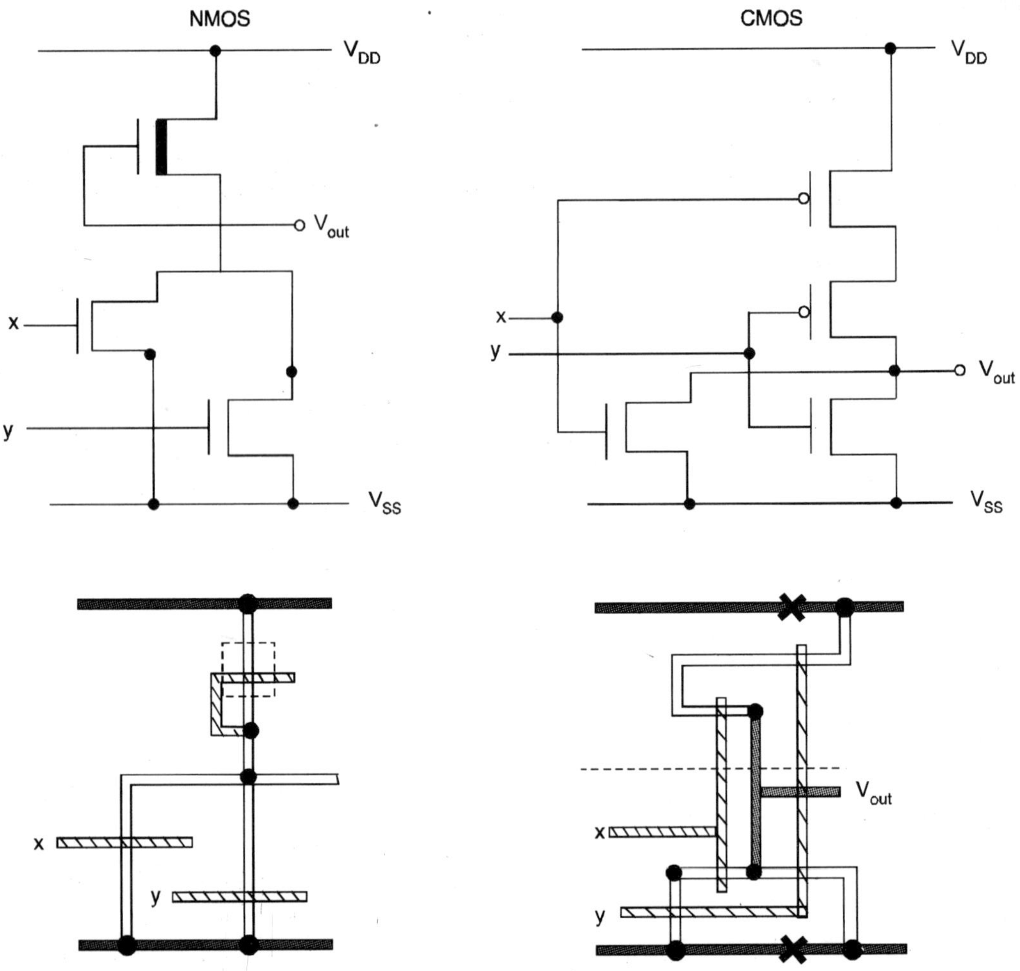

Fig. 4.2.7. Stick diagram for NOR.

4.3. DESIGN RULES AND LAYOUT

The object of a set of design rules is to allow a ready translation of circuit design concepts, usually in stick diagram (or) symbolic form into actual geometry in silicon. The design rules are the effective interface between the circuit designer and the fabrication engineer. The simple 'lambda (λ) based' design rules are set out first. The design rules are based on a single parameter 'λ' which leads to simple set of rules for the designer and wide acceptance of the rules by a large cross section of the fabrication houses and silicon brokers and allows for scaling of the designs to a limited extent. This latter feature may help to give designs a longer life time. The simplicity of lambda based rules also provides a simple introduction to design rules and to mask layout design in general and helps to set the scene for the micron – based rule sets which follow.

4.3.1. Lambda Based Design Rules

All paths in all layers will be dimensioned in ' λ' units and subsequently λ can be allocated an appropriate value compatible with the feature size of the fabrication process. This concept mean that the actual mask layout design takes little account of the value allocated to feature size, but the design rules are such that, if correctly obeyed, the mask layout will produce working circuits for a range of values allocated to λ. For example, λ can be allocated a value of 1.0 mm so that minimum feature size on chip will be 2 μm (2λ). Design rules specify line widths, separations and extensions in terms of λ. Design rules can be set out in diagrammatic form for the widths and separation of conducting paths and also for extensions and separations associated with transistor layouts. For example, from figure, label 'a' indicates the minimum separation between two different layers.

Where no seperation is specified, wires may over lap or corss (*e.g.,* metal is not constrained by any other layer). For *p*-well CMOS, note that *n*-diffusion wires can only exist inside and *p*-diffusion wires outside the *p*-well).

• **Note :** Many fabrication houses now accept 2λ diffusion to diffusion separation and 2λ metal 1 width and separation.

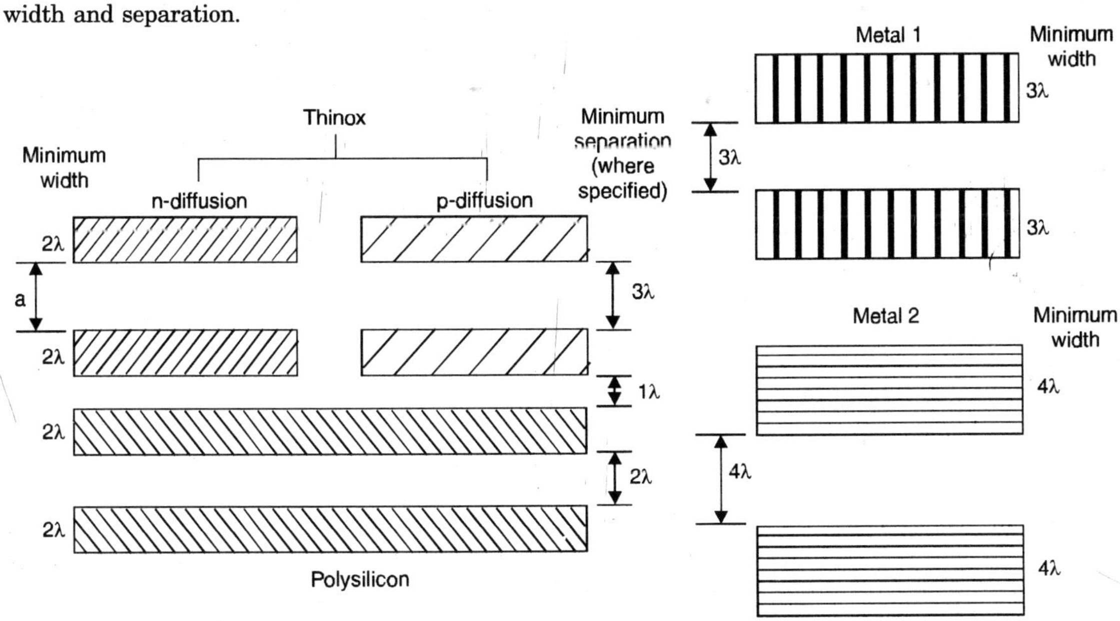

Fig. 4.3.1. Design rules for wires (NMOS and CMOS).

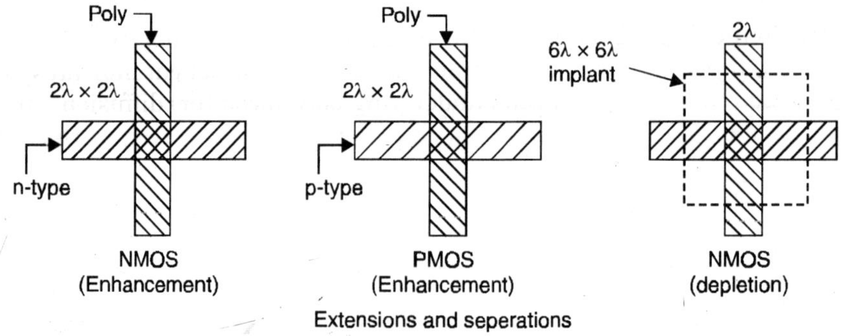

Fig. 4.3.2. Transistor design rules.

4.3.2. Contact Cuts

When making contacts between polysilicon and diffusion in NMOS circuits, it should be recognised that there are 3 possible approaches.

1. Poly to metal then metal to diffusion
2. buried contact poly to diffusion
3. butting contact (poly to diffusion using metal)

Of the latter two, the buried contact is the most widely used, giving economy in space and a reliable contact. Butting contacts were widely used at one time but have been mostly superseded by buried contacts. In CMOS designs, poly to diffusion contacts are almost always made via-metal.

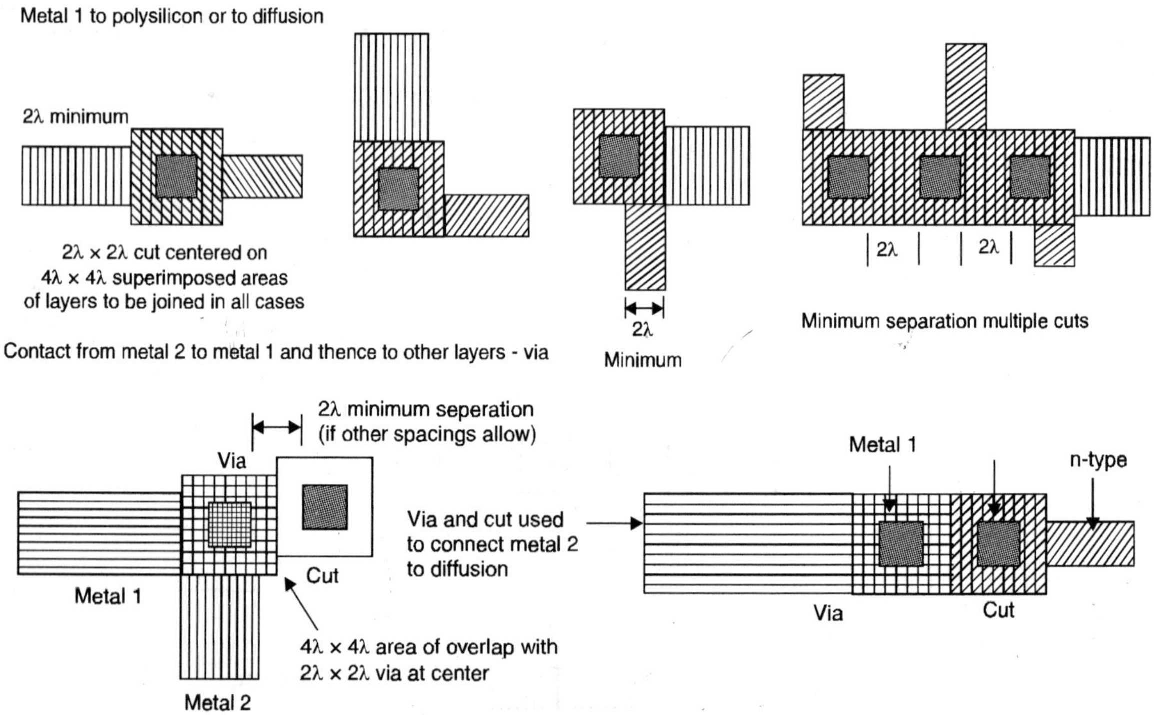

Fig. 4.3.3. Contacts.

When making connections between metal and either of the other two layers (either polysilicon or diffusion) the process is quite simple. The 2λ × 2λ contact cut indicates an area in which the oxide is to be removed down to the underlying polysilicon (or) diffusion surface. The metal is deposited through the contact cut areas on to the underlying area so that contact is made between the layers.

When connecting diffusion to polysilicon using the butting contact approach, the process is rather more complex. In effect, a 2λ × 2λ contact cut is made down to each of the layers to be joined. The layers are butted together in such a way that these two contact cuts become contiguous. Since the polysilicon and diffusion outlines overlap and thinoxide under polysilicon acts as a mask in the diffusion process, the polysilicon and diffusion layers are also butted together. The contact between the two butting layers is then made by a metal overlay as shown in figure.

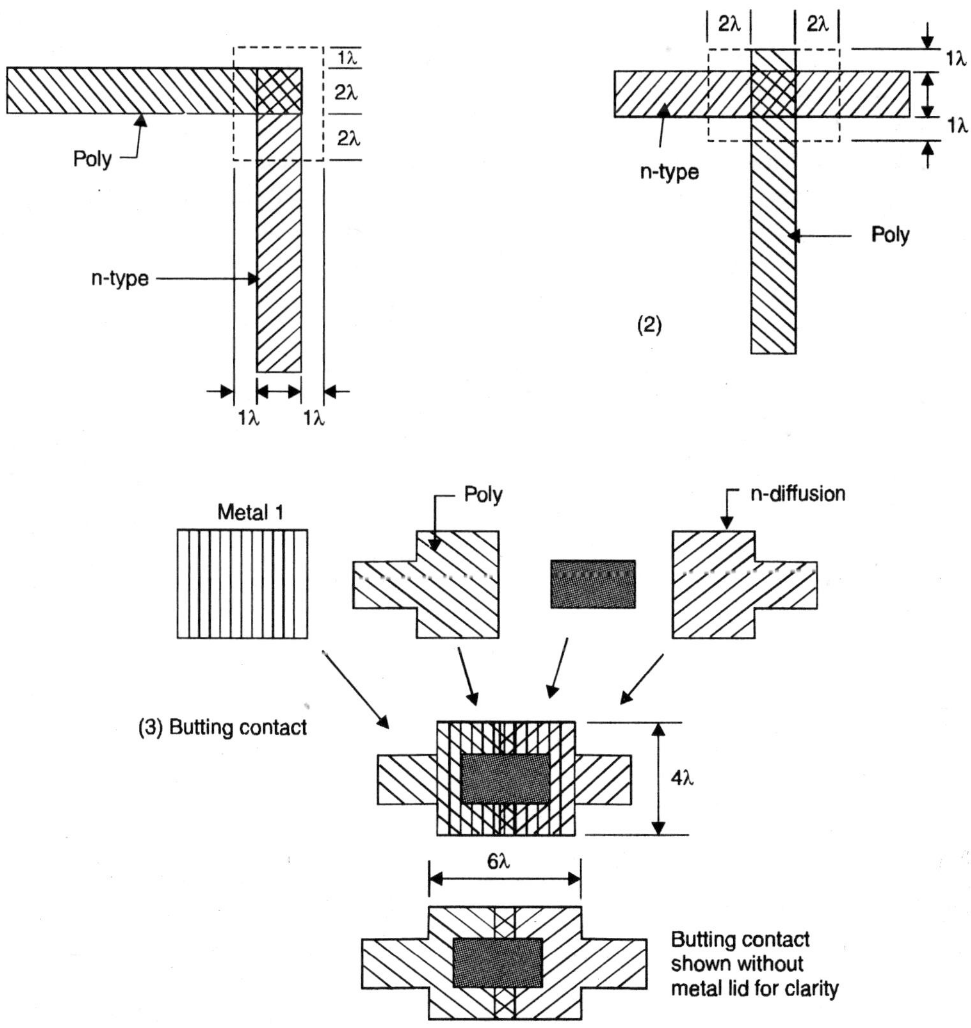

Fig. 4.3.4. Contacts-Polysilicon to diffusion.

Buried contacts can be smaller in area than their butting contact counterparts and since they use no metal layer, they are subject to fewer design rule restrictions in a layout. The design rules are such that they also avoid the formation of unwanted diffusion to polysilicon contacts and protect the gate oxide of any transistors in the vicinity of the buried contact cut area.

In buried contacts basically, layers are joined over $2\lambda \times 2\lambda$ area with the buried contact cut extending by 1λ in all directions around the contact area except that the contact cut extensions increased to 2λ in diffusion paths leaving the contact area. This is to avoid forming unwanted transistors. This is illustrated in Fig. 4.3.4.

4.3.3. Double Metal MOS Process Rules

A powerful extension to the process so far described is provided by a second metal layer. This gives a much greater degree of freedom, for example, in distributing global V_{DD} and V_{SS} (V_{ND}) rails in a system. From the overall chip interconnection aspect, the second metal layer in particular is important and although the use of such a layer is readily envisaged, its disposition relative to its connection to other layers using metal 1 to metal 2 contacts called "Vias" can be easily established.

Usually, second level metal layers are coarser than the first layer and the isolation layer between the layers may also be of relatively greater thickness. To distinguish contacts between the layers may also be of relatively greater thickness. To distinguish contacts between first and second metal layers, they are known as Vias rather than contact cuts.

For the sake of completeness, the process steps for a two metal layer process are briefly outlined as follows. The oxide below the first metal layer is deposited by atmospheric chemical vapor deposition (CVD) and the oxide layer between the metal layers is applied in a similar manner. Depending on the process, removal of selected areas of the oxide is accomplished by plasma etching, which is designed to have a high level of vertical ion bombardment to allow for high and uniform etch rates.

To revert to the double metal process it is convenient at this point to consider the layout strategy commonly used with this process: The approach taken may be summarised as follows :

1. Use the second level metal for the global distribution of power buses, that is V_{DD} and G_{ND} (V_{SS}) and for clocks lines.

2. Use the first level metal for local distribution of power and for signal lines.

3. Layout the two metal layers so that the conductors are mutually orthogonal wherever possible.

4.4. LAYOUT DIAGRAMS

Once a circuit design is complete, it becomes necessary to provide an area efficient layout of the circuit to generate the masks necessary for fabrication. One of the first considerations in the layout is sizing the devices as well as interconnections. But to simplify layouts, 'Unit' sized transistors will generally be shown in diagrams.

Mask layout diagrams may be hand-drawn on, say 5 mm squared paper. In the case of lamda-based rules, the side of each square is taken to represent λ and for micron based rules, it will be taken to represent the least common factor associated with the rules (for *e.g.,* 0.25 μm/side for the 2 μm process and 0.2 μm/side for the 1.2 μm orbit process layout).

General CMOS Logic-gate Layout Guidelines

1. Complete the electrical gate design
2. Run V_{DD} and V_{SS} in metal at the top and bottom of the cell
3. Run a Vertical Polysilicon line for each gate input.
4. Order the polysilicon gate signal to allow the maximal connection between transistors. (Optimize Horizontal and Vertical transistor connectivity)
5. Place n-gate segments close to V_{SS} and P-gate segments close to V_{DD}.
6. Connections to complete the logic gate should be made in polysilicon, metal or where appropriate in diffusion.

If required, user defined conventions can also be used.

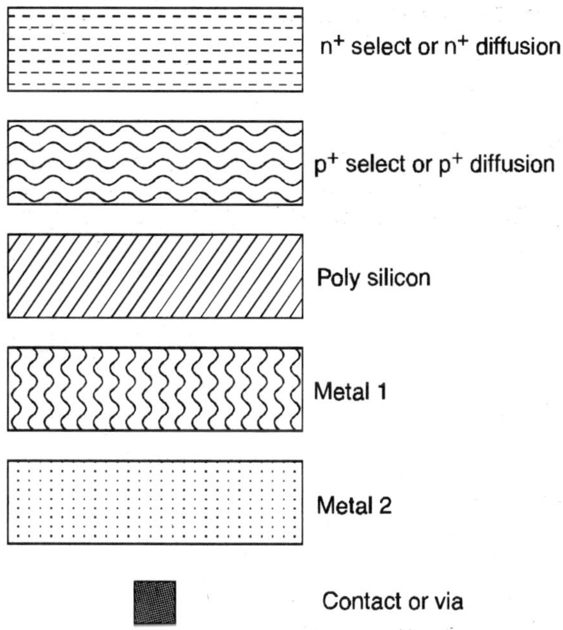

Fig. 4.4.1. User Defined Layout drawing conventions.

The Inverter. The figure shows the schematic of an inverter. The physical layout can be drawn by substituting layout symbols for the schematic symbols. In a schematic, lines drawn between device terminals represent connections. However, in a physical layout, we have to concern ourselves with the interaction of physically different interconnection layers.

We know, the source and drain of the n-transistor are the n-diffusion regions, while the p-transistor uses p-diffusion regions. Also, in a CMOS Process, we cannot make a direct connection from n-diffusion to p-diffusion. Thus we have to use two contacts. And this connection has to be in metal. Similarly, the simple connections to power, V_{DD} and ground V_{SS} could be made using metal wires and contacts. The common gate connection may be a simple polysilicon wire. The resulting Layout diagram is shown in Fig. 4.3.3.

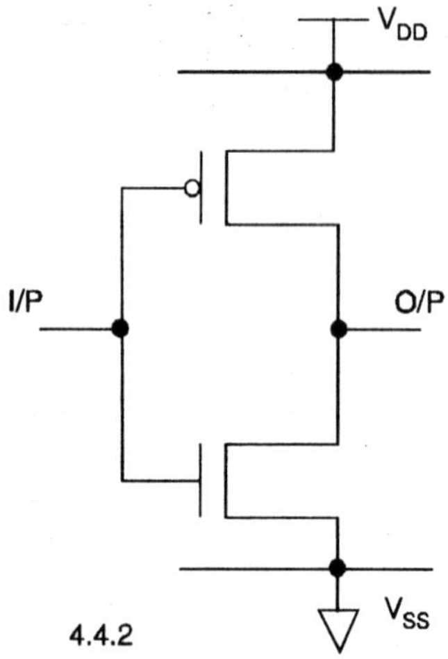

Fig. 4.4.2. Inverter.

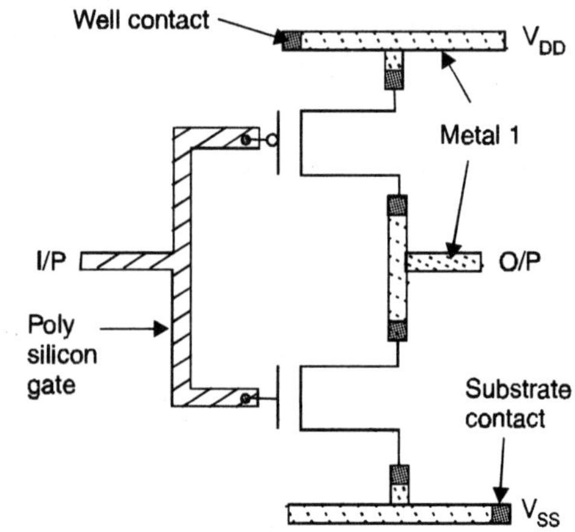

Fig 4.4.3. Layout diagram of an inverter.

Now substituting layout symbols for n transistor and p-transistor the final Layout diagram of inverter is shown in Fig. 4.4.4.

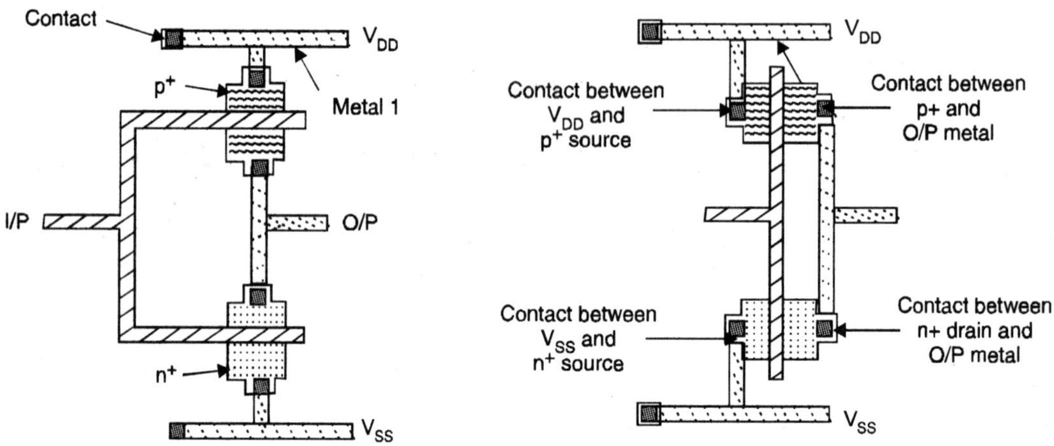

(*a*) Transistors aligned vertically. (*b*) Transistors aligned horizontally.

Fig. 4.4.4. Symbolic layout for the CMOS Inverter.

NAND and NOR Gates Layout. For multiple input gates, Transistors are oriented horizontally and polysilicon gate run vertically.

Steps for Drawing NAND Gate Layout (User defined conventions are used). Fig. 4.2.6. shows the CMOS implementation of NAND gate. To obtain the layout diagram the following steps are used. The steps are illustrated in Fig. 4.4.5.

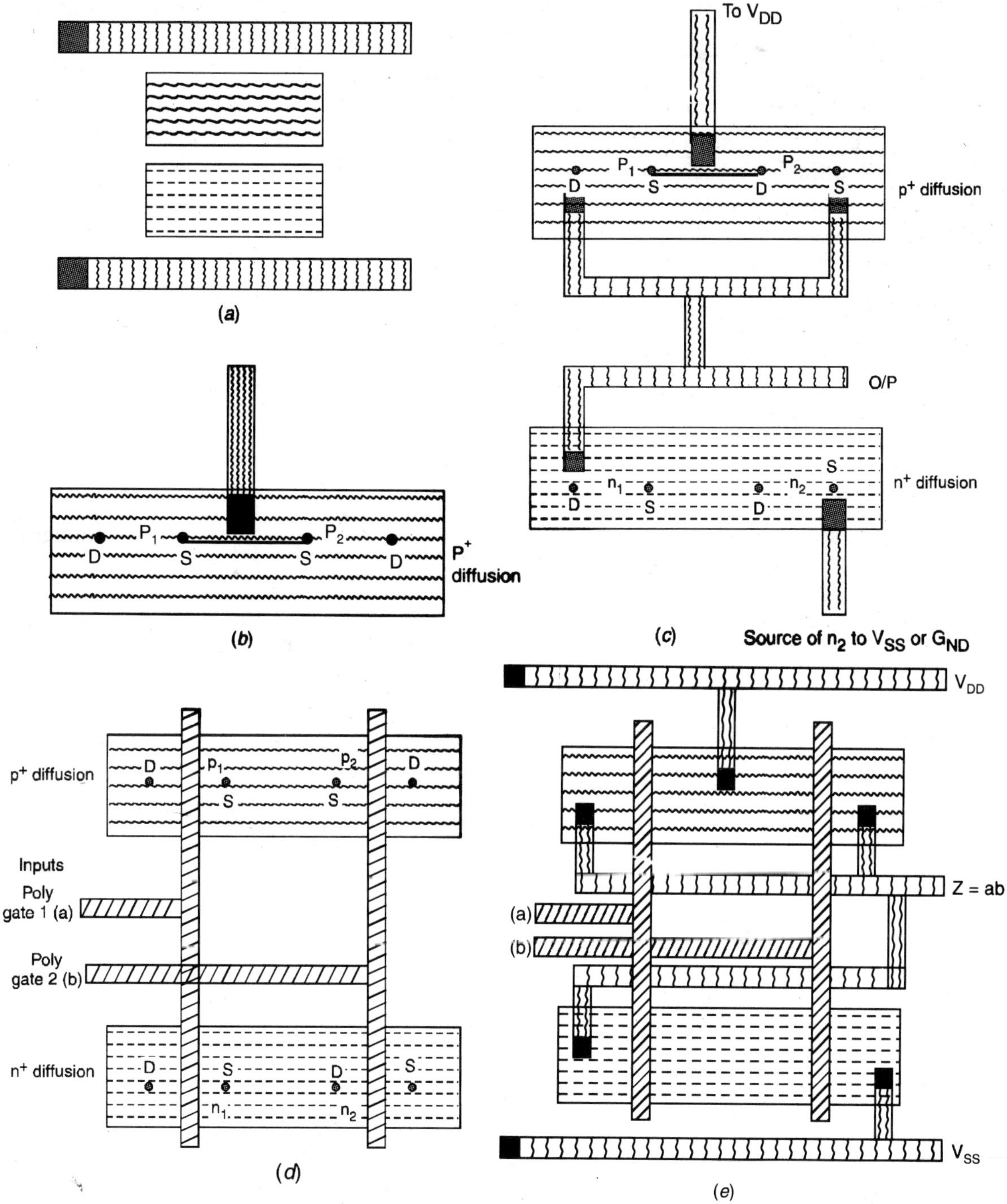

Fig. 4.4.5. Steps for drawing NAND gate layout. **Complete layout diagram of NAND.**

1. Draw V_{DD} and V_{SS} lines using metal 1 parallely leaving enough separation between th two lines.
2. Draw P+ diffusion near to V_{DD} and n+ diffusion near GND. Fig. (a).
3. From schematic of NAND gate we can understand that there are two p-type transistors and two n-type transistors.

(*i*) Sources of P_1 and P_2 are connected together and connected to V_{DD} using metal 1. Fig. (*b*).

(*ii*) Drains of P_1 and P_2 are connected together and connected with drain of n_1 and from there OUTPUT connection is drawn using metal 1.

(*iii*) Source of n_1 is connected to drain of n_2 and source of n_2 is connected to GND. Fig. (c)

(*iv*) Draw polysilicon gate for input '*a*' between source and drain of n_1 and P_1 and polysilicon gate 2 for input '*b*' between source and drain of n_2 and p_2 as shown in Fig. (*d*).

Note : The selection of drain and source points in *p*-diffusion and *n*-diffusion are arbitrary and can be done to our convenience. The final Layout diagram of two input NAND gate is shown after incorporating all the features which have been explained so far Fig. (*e*).

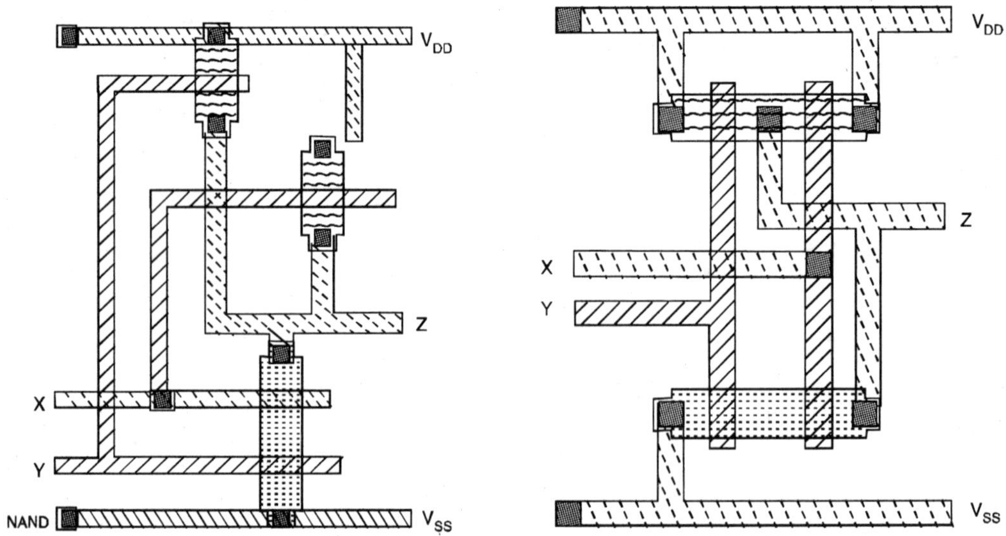

Fig. 4.4.6. Layout Diagram of NAND with a different convention.

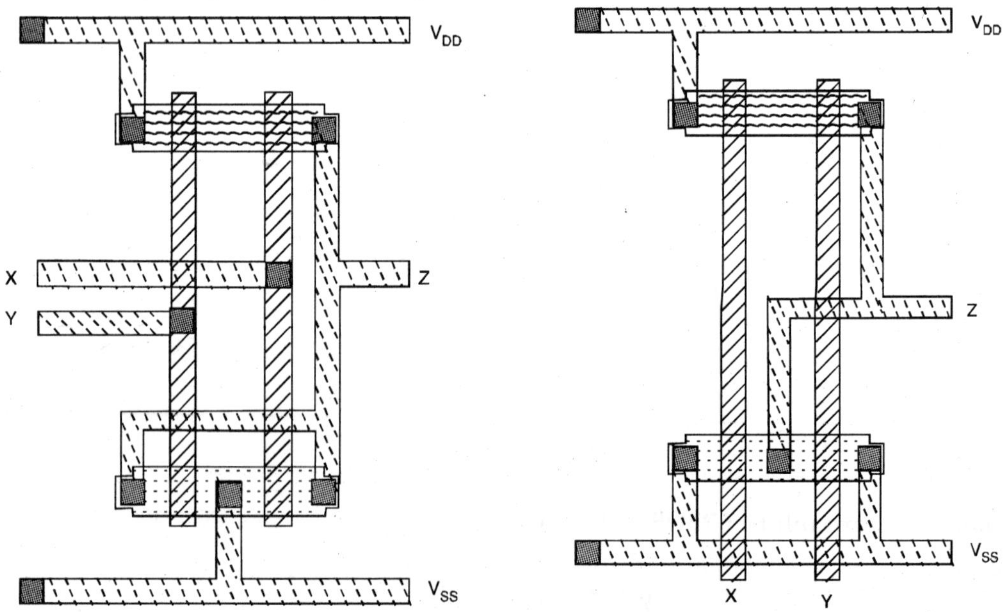

Fig. 4.4.7. Typical NOR gate symbolic layouts.

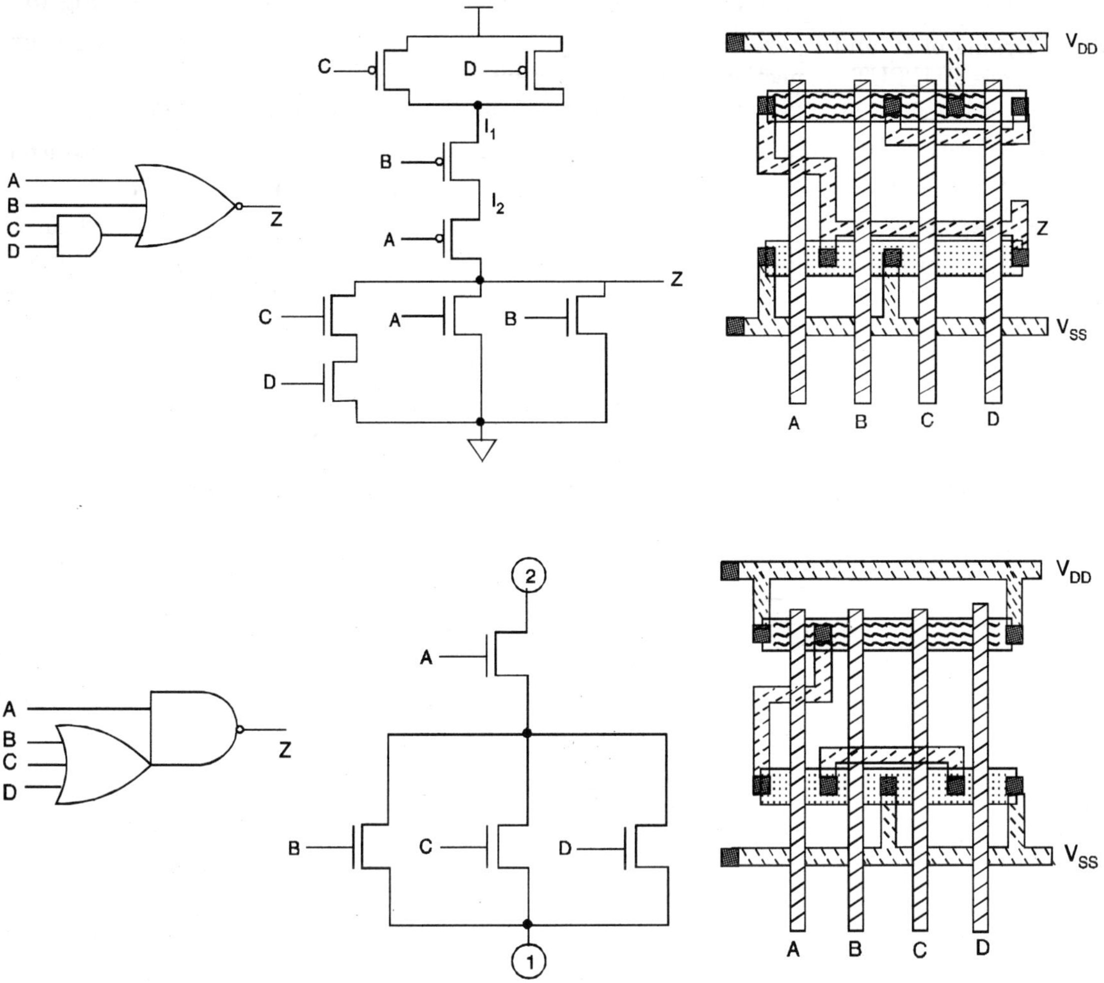

Fig. 4.4.8. Example layout diagram of compound gates.

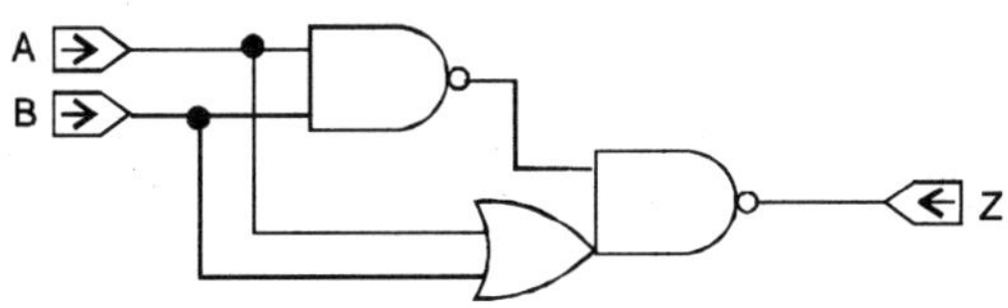

Fig. 4.4.9. CMOS XNOR gate alternative layout styles.

Automated approach to CMOS-gate layout

In this style also transistors are arranged horizontally with polysilicon vertically. Power and ground are run at the top and bottom of the cell.

The remaining steps are as follows:

1. Group Transistor in strips to allow maximum source/drain connection by abutment.
2. For better grouping, the position of polysilicon (gate i/ps) can be interchanged.
3. The resultant groups are then placed in rows with groups maximally connected to the V_{SS} and V_{DD} rails placed toward these signals. Row placement is then based on the density of other connections.

Finally routing is achieved by metal.

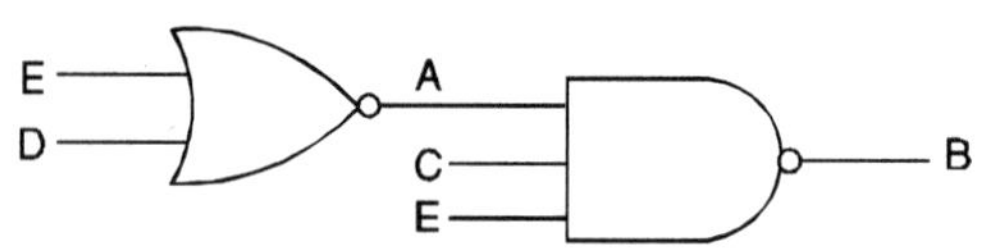

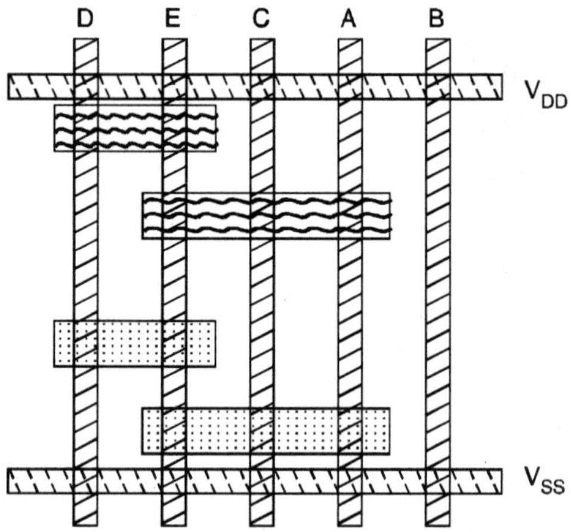

(*b*) Order gate adn outputs to optimize horizontal transistor connectivity.

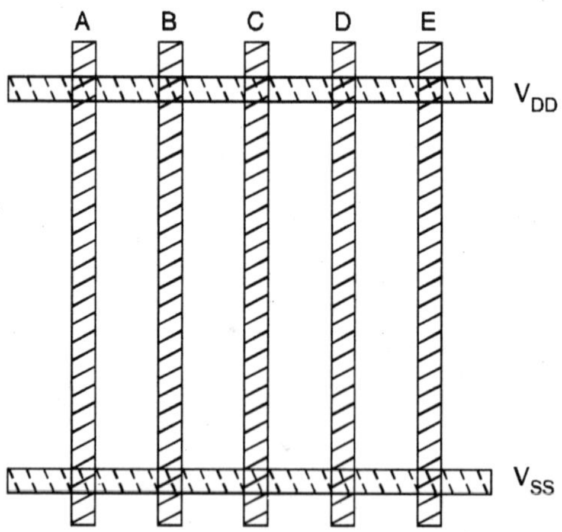

(*a*) Route power, ground, gate and output signals.

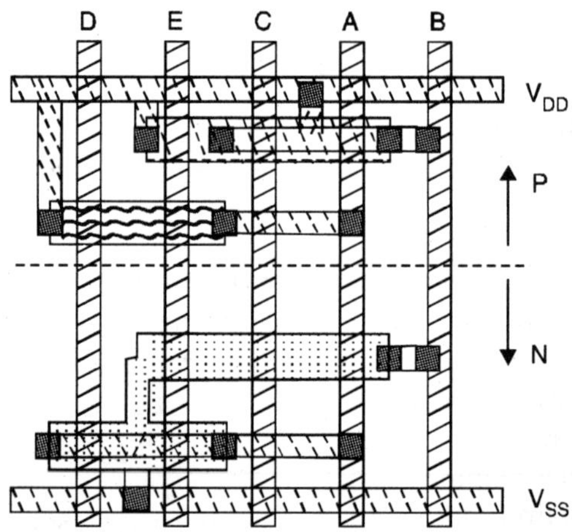

(c) Rearrange vertical strip ordering to optimize power routing internal gate connection and output connections.

Fig. 4.4.10. Outline of automated approach to CMOS-gate layout.

4.5. SCALING OF MOS CIRCUITS

VLSI fabrication technology is still in the process of evolution which is leading to smaller line width and feature size and to higher packing density of circuitry on a chip. The scaling down of feature size generally leads to improved performance and it is important therefore to understand the effects of scaling.

Micro electronic technology may be characterized in terms of several figures of merit. They are

Minimum feature size

Number of gates on one chip

Power dissipation

Maximum operational frequency

Die size

Production cost

The above figure of merit can be improved by

(*i*) Shrinking the dimensions of transistors

(*ii*) Interconnections and the separation between features

(*iii*) Adjusting the doping levels and supply voltages.

4.5.1. Scaling Models and Scaling Factors

The most commonly used models are the

(*i*) Constant electric field scaling

(*ii*) Constant voltage scaling

(*iii*) Combined voltage and dimensions scaling or lateral scaling.

MOS scaling theory indicates that the characteristics of a MOS device can be maintained and the basic operational characteristics preserved if the critical parameters of a device are scaled in accordance to a given criterion.

Constant field scaling. In this the scaled device is obtained by applying a dimension less factors α to

- all dimensions, including those vertical to the surface
- device voltages
- the concentration densities

In this, if device dimensions channel length, L; channel width, W, oxide thickness, t_{ox} junction depth, x_j ; applied voltages; and substrate concentration density, N; are scaled by the constant parameter α, then the depletion layer thickness, d, the threshold voltage, V_t, and the drain-to-source current, Ids are also scaled. Since the voltage is scaled, electric field, E, in the device remains constant. This has the desirable effect that many nonlinear factors essentially remain unaffected.

Lateral scaling. In this only the gate length is scaled. This is commonly called a gate shrink.

Constant voltage scaling. Here V_{DD} voltage is kept constant, while the process is scaled. With constant voltage scaling, E increases, which has led to process development to reduce the deleterious effects of high fields.

The depletion regions associated with the pn junction of the source and drain determine how small we can make the channel. As a rule, the source drain distant must by greater than the sum of the width of the depletion layers to ensure that the gate is able to exercise control over the conductance of the channel. Thus is order to reduce the length of the channel you need to reduce the width of the depletion layers. This is accomplished by increasing the doping level of the substrate silicon. In constant field scaling as we scale device dimensions by $1/\alpha$, the drain-to-source current. Ids, per transistor decreases to $1/\alpha$, the number of transistors per unit area; that is, circuit density scales up by α^2, which subsequently results in the current density scaling linearly with α. Constant-voltage scaling exacerbates this problem (increasing the current density by α^3), while a gate shrink increases the current density by α^2. Thus proportionally wider metal power conductors are necessary for more densely packed structures. This is usually solved by adding metal layers that are used solely for power and ground.

Table 4.5. Shows the scaling effects on various parameters

Parameters		*Combined V and D*	*Constant E*	*Constant V*
V_{DD}	Supply voltage	$1/\beta$	$1/\alpha$	1
L	Channel length	$1/\alpha$	$1/\alpha$	$1/\alpha$
W	Channel width	$1/\alpha$	$1/\alpha$	$1/\alpha$
D	Gate oxide thickness	$1/\beta$	$1/\alpha$	1
A_g	Gate area	$1/\alpha^2$	$1/\alpha^2$	$1/\alpha^2$
C_o(or C_{ox})	Gate C Per unit area	β	α	1
C_g	Gate capacitance	β/α^2	$1/\alpha$	$1/\alpha^2$
C_x	Parasitic capacitance	$1/\alpha$	$1/\alpha$	$1/\alpha$
Q_{on}	Carrier density	1	1	1
R_{on}	Channel resistance	1	1	1
I_{dss}	Saturation current	$1/\beta$	$1/\alpha$	1
A_c	ConductorX-section area	$1/\alpha^2$	$1/\alpha^2$	$1/\alpha^2$
I	Current density	α^2/β	α	α^2
V_g	Logic I level	$1/\beta$	$1/\alpha$	1
E_g	Switching energy	$1/\alpha^2\cdot\beta$	$1/\alpha^3$	$1/\alpha^2$
P_g	Power dissipation per gate	$1/\beta^2$	$1/\alpha^2$	1
N	Gates per unit area	α^2	α^2	α^2
P_a	Power dissipation per unit area dissipation	α^2/β^2	1	α^2
T_d	Gate delay	β/α^2	$1/\alpha$	$1/\alpha^2$
F_o	Max.operating frequency	α^2/β	α	α^2
P_T	Power-speed product	$1/\alpha^2.\beta$	$1/\alpha^3$	$1/\alpha^2$

For constant E : $\beta = \alpha$; Constant V : $\beta = 1$

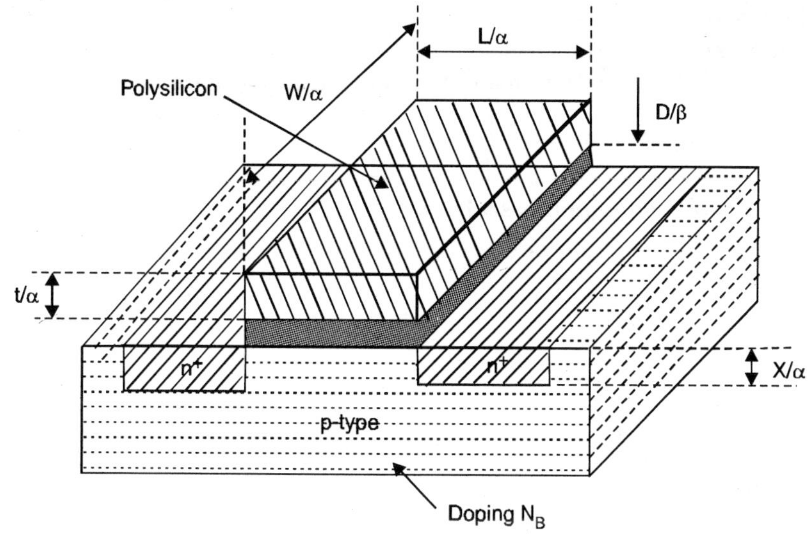

Fig. 4.5.1. Scaled NMOS Transistor.

The Fig. 4.5.1. indicates device dimension associated with the scaling of a transistor. The scaling factors used are $1/\alpha$ and $1/\beta$. For the constant field to constant voltage scaling, $\beta = \alpha$.

Scaling factors for device parameters

(*a*) Gate area – A_g

$$A_g = LW$$

where L – channel length

W – channel width

L and W are scaled by $\frac{1}{\alpha}$

Hence A_g is scaled by $1/\alpha^2$

(*b*) Gate capacitance/unit area – C_o

$$C_o = \varepsilon_{ox}/D$$

where ε_{ox} – permittivity of gate oxide

D – gate oxide thickness

D is scaled by $\frac{1}{\beta}$, hence C_o is scaled by $\frac{1}{\frac{1}{\beta}} = \beta$

(*c*) Gate capacitance C_g

$$C_g = C_o \text{ L.W.}$$

C_g is scaled by $\beta . \frac{1}{\alpha^2} = \frac{\beta}{\alpha^2}$

(*d*) Parasitic capacitance C_x

$$C_x \propto \frac{A_x}{d}$$

where d – depletion width is scaled by $\frac{1}{\alpha}$

A_x – area of depletion region is scaled by $\frac{1}{\alpha^2}$

Hence C_x is scaled by $\frac{1}{\alpha^2} \cdot \frac{1}{\frac{1}{2}} = \frac{1}{\alpha}$

(*e*) Carrier density in channel Q_{on}

$$Q_{on} = C_o.V_{gs}$$

where Q_{on} – average charge/unit area in the channel in 'ON' state

As C_o is scaled by β and V_{gs} is scaled by 1/β.

Q_{on} is scaled by 1.

(*f*) Channel Resistance R_{on}

$$R_{on} = \frac{L}{W} \cdot \frac{1}{Q_{on}\mu}$$

where μ – carrier mobility and is assumed constant

Hence R_{on} is scaled by $\frac{1}{\alpha} \cdot \frac{1}{1/\alpha} 1 = 1$

(*g*) Gate delay T_d

T_d is proportional to R_{on} Lg

T_d is scaled by $\frac{1.\beta}{\alpha^2} \cdot \frac{\beta}{\alpha^2}$

(*h*) Maximum operating frequency f_o

$$f_0 = \frac{L}{W} \cdot \frac{\mu C_o V_{dd}}{C_g}$$

(or) $$f_o \propto \frac{1}{T_\delta}$$

Hence f_o is scaled by α^2/β

(*i*) Saturation current I_{dss}

$$I_{dss} = \frac{C_{o\mu}}{2} \cdot \frac{W}{L} (V_{gs} - V_t)^2$$

V_{ds} and V_t are scaled by 1/β, & hence I_{dss} is scaled by 1/β (*j*) Current density $J = I_{dss}/A$

where A – cross sectional area of the channel in 'on' state. It is scaled by $1/\alpha^2$

Hence J is scaled by $= \dfrac{1/\beta}{1/\alpha^2} \cdot \dfrac{\alpha^2}{\beta}$

(*k*) Switching energy/gate E_g

$$E_g = \frac{C_g}{2} (V_{DD})^2$$

Hence, E_g is scaled by $\dfrac{\beta}{\alpha^2} \cdot \dfrac{1}{\beta^2} = \dfrac{1}{\alpha^2 \beta}$

(*l*) Power dissipation/gate P_g

$$P_g = P_{gs} + P_{g\delta}$$

P_{gs} – static component $= \dfrac{V_{DD}^2}{R_{on}}$

P_{gd} – dynamic component $= E_g f_o$

Both P_{gs} and P_{gd} are scaled by $1/\beta^2$

$\therefore$ P_g is scaled by $1/\beta^2$

(*m*) Power dissipation/unit area P_a

$$P_a = P_g/A_g$$

Hence P_a is scaled by $\dfrac{1}{\beta^2} \Big/ \dfrac{1}{\alpha^2} = \dfrac{\alpha^2}{\beta^2}$

(*n*) Power speed product P_T

$$P_T = P_g T_\delta.$$

Now P_T is scaled by $\dfrac{1}{\beta^2} \cdot \dfrac{\beta}{\alpha^2} = \dfrac{1}{\alpha^2 \beta}$

Sometimes, the scaling effect may cause problems, which eventually become severe enough to prevent further miniaturization. The parameters which influence the above are substrate doping, depletion width, interconnect and contact resistance, sub threshold currents, logic levels and supply voltage due to noise, current density.

4.6. PROPAGATION DELAYS

4.6.1. Pass Transistors

We know a pass transistors can be used as series or parallel switches in logic arrays. The Fig. 4.6.1 shows four pass transistors connected in series. The gates of all transistors are tied to V_{DD} (logic 1) in order to propagate the signal to the OUTPUT. The delay through the network is calculated by using the Elmore delay model which is shown in Fig. 4.6.1.

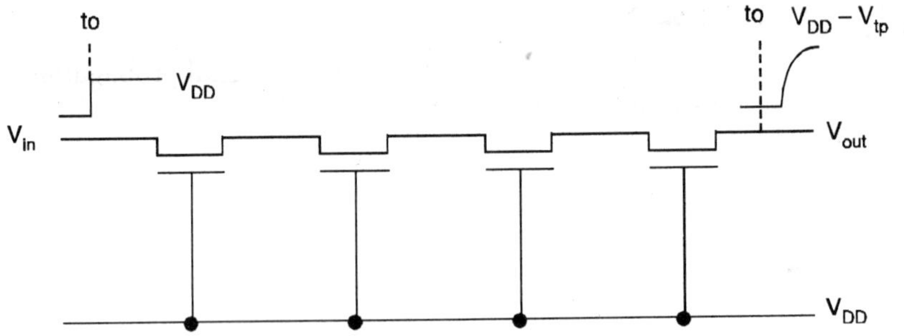

(*a*) Pass transistors connected in series.

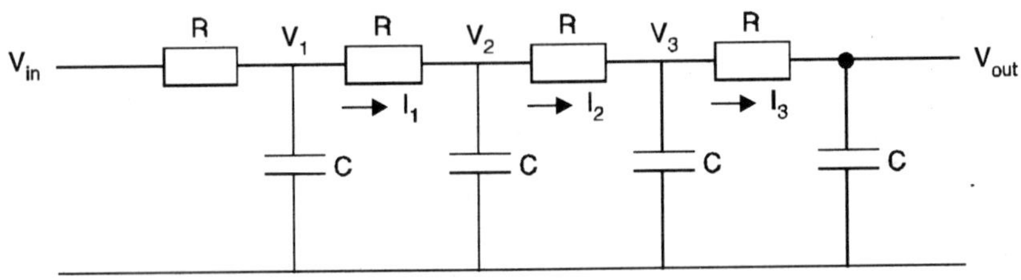

(*b*) Corresponding delay model.

Fig. 4.6.1.

At node V_2 we have

$$C\frac{dV_2}{dt} = I_1 - I_2 = \frac{V_1 - V_2}{R} - \frac{V_2 - V_3}{R}$$

$$RC = \frac{dV_2}{dt}(V_1 - V_2 - V_2 + V_3) = V_1 - V_3$$

If the number of pass transistors in the chain becomes large, we can write

$$RC\frac{dV}{dt} = \frac{d^2V}{dd^2}$$

where R → Resistance / unit length

C → Capacitance / unit length

d – distance along the network from input

Now the time taken by a signal to propagate a distance d along the network be denoted as τ_p *i.e.*, τ_p – propagation time is proportional to d^2.

If all Resistors and Capacitors in the network are lumped together with $R_{total} = n\, rR_s$ and $C_{total} = ncC_g$

where r – relative resistance/section in terms of R_s

c – relative capacitance/section in terms of C_s

n – number of sections

Now over all delay τ_d for n sections is

$$\tau_d = n^2 rc(\tau)$$

4.6.2. Long Polysilicon Wires

The R and C of long polysilicon wires also influence in signal propagation. Particularly for wires in diffusion, the value of "C" will be very high. Therefore delay in network is increased. So it is advisable to run signal in diffusion only for very short distances. For long polysilicon runs, we can opt for buffers. It has two advantages.

1. Signal propagation is not delayed much
2. Sensitivity to noise is less.

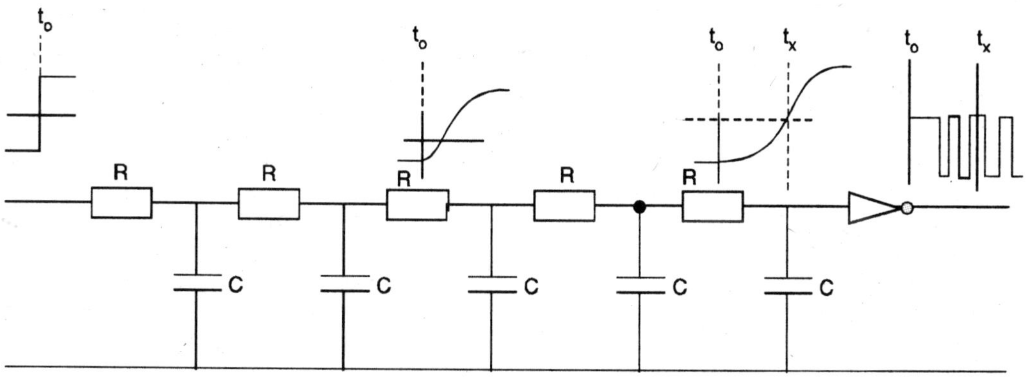

Fig. 4.6.2. Effects of delays in long polysilicon wires.

In figure The effect of delay introduced by the long polysilicon wire is explained. At the end of poly silicon wire i.e. at output node Vo, we can see that the signal takes a significant amount of time to change from one level to next level (0 →1 in this figure). Now a small disturbance due to noise may change the output. Therefore it is essential that long polysilicon wires be driven by suitable buffers to protect against noise and to speed up the rise time of propagated signal edges.

5

Building Blocks of Digital Systems

INTRODUCTION

Combinational logic (or) non-regenerative circuits forms the core of most digital integrated circuits such as fast arithmetic units and controllers. The output of the circuit is related directly to its input signals by some Boolean expression (ignoring the short propagation delay of the composing gates). No intentional connection between outputs and inputs is present.

In another class of circuits, known as sequential (or) regenerative circuits, the output is not only a function of a current input data but also of previous values of the input signals. A common characteristics of sequential circuits is that one (or) more outputs are intentionally connected back to inputs. The difference between combinational and sequential circuits is shown below :

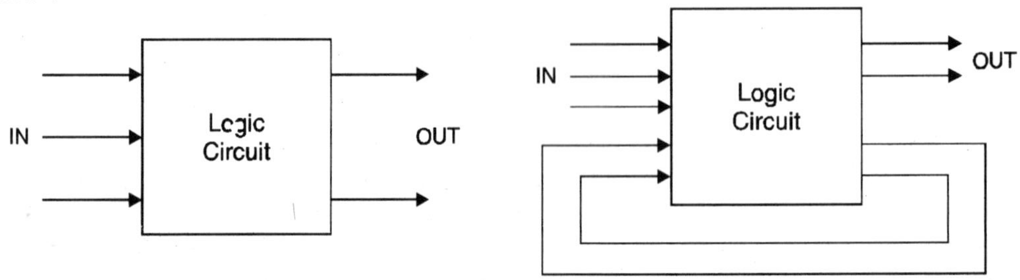

(*a*) Combinational. (*b*) Sequential.

Classification of logic circuits.

This chapter discusses on more popular combinational circuits, sequential circuits and data path circuits.

5.1. COMBINATIONAL LOGIC CIRCUITS

5.1.1. Multiplexers

A multiplexer circuit has a number of data inputs, one (or) more select inputs and one output. The data input is selected by the values of the select inputs.

For 2^n – data inputs, the select inputs $= n$

Fig. 5.1.1 (*a*) shows the graphical symbol of 2 to 1 Multiplexer. The select inputs choose the output of multiplexer. Based on the select input, the output of the multiplexer will be either x_0 or x_1.

Truth Table of Mux

s	f
0	x_0
1	x_1

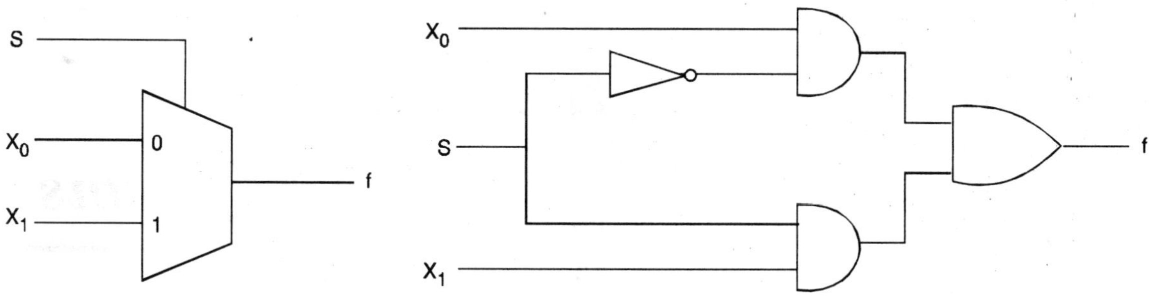

(*a*) Graphical symbol.

(*b*) SOP implementation of 2 to 1 mux.

Fig. 5.1.1.

The truth table describes the functionalities of the multiplexer.

Sum of products circuit

$$f = \overline{S} . X_0 + S . X_1$$

Fig. 5.1.1 (*b*) shows the sum-of-products implementation of the 2-to-1 multiplexer, and Fig. 5.1.1 (*c*) shows how it can be constructed with transmission gates

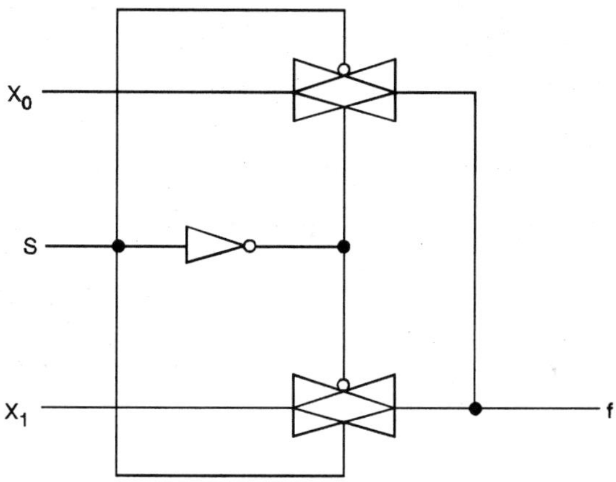

Fig. 5.1.1 (*c*) Transmission gates implementation of Mux.

4 to 1 Multiplexer

It has four input lines, one output line and two select lines S_1 and S_0. Fig. 5.1.1 (*d*) depicts the graphical symbol of 4 to 1 MUX and truth table describes the functionality.

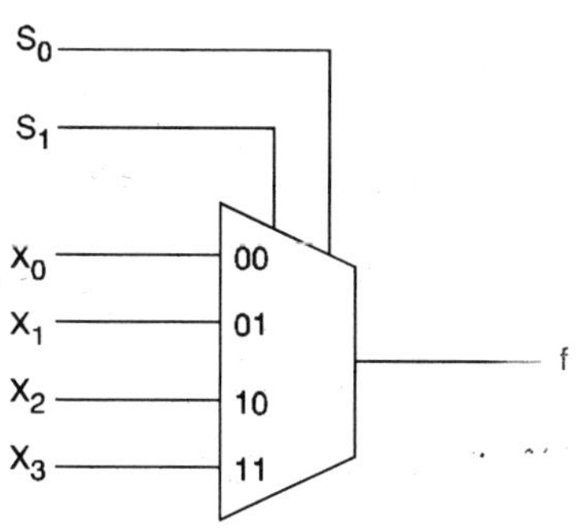

Fig. 5.1.1 (*d*) Graphical Symbol of 4-to-1 Mux.

Truth table for 4 to 1 Mux

s_0	s_1	f
0	0	x_0
0	1	x_1
1	0	x_2
1	1	x_3

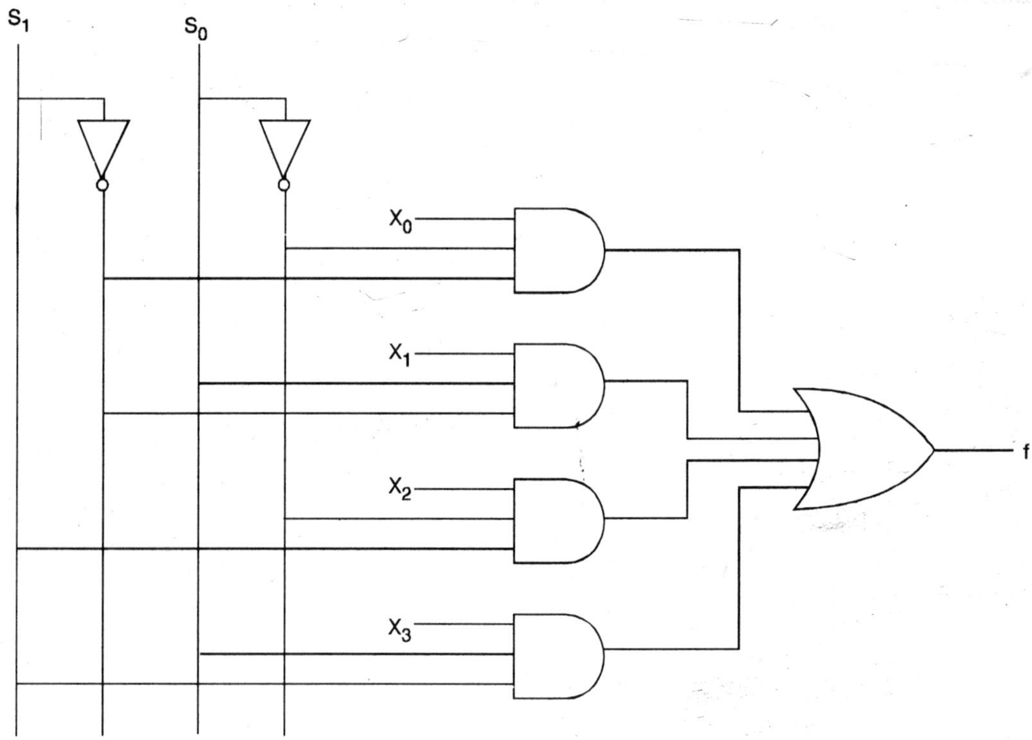

Fig. 5.1.1 (*e*) SOP implementation of the 4-to-1 multiplexer.

Fig. 5.1.1 (*e*) Represents the SOP implementation of the 4 to 1 mux. The realization function is given by $f = \overline{S_1} \cdot \overline{S_1} \cdot X_0 + S_1 \cdot \overline{S_1} \cdot X_1 + \overline{S_1} \cdot S_0 \cdot X_2 + S_1 \cdot S_0 \cdot X_3$. Larger multiplexers can also be constructed from smaller multiplexers. For example the 4-to-1 multiplexer can be built using three 2-to-1 multiplexers as shown in Fig. 5.1.1 (*f*) and Fig. 5.1.1 (*g*) shows how a 16-to-1 multiplexer can be constructed with five 4 to 1 Mux.

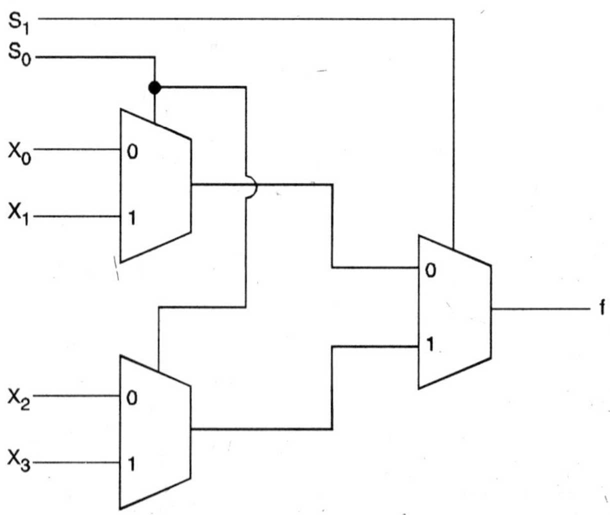

Fig. 5.1.1.(*f*) Implementation of 4 to 1 Mux using 2 to 1 Mux.

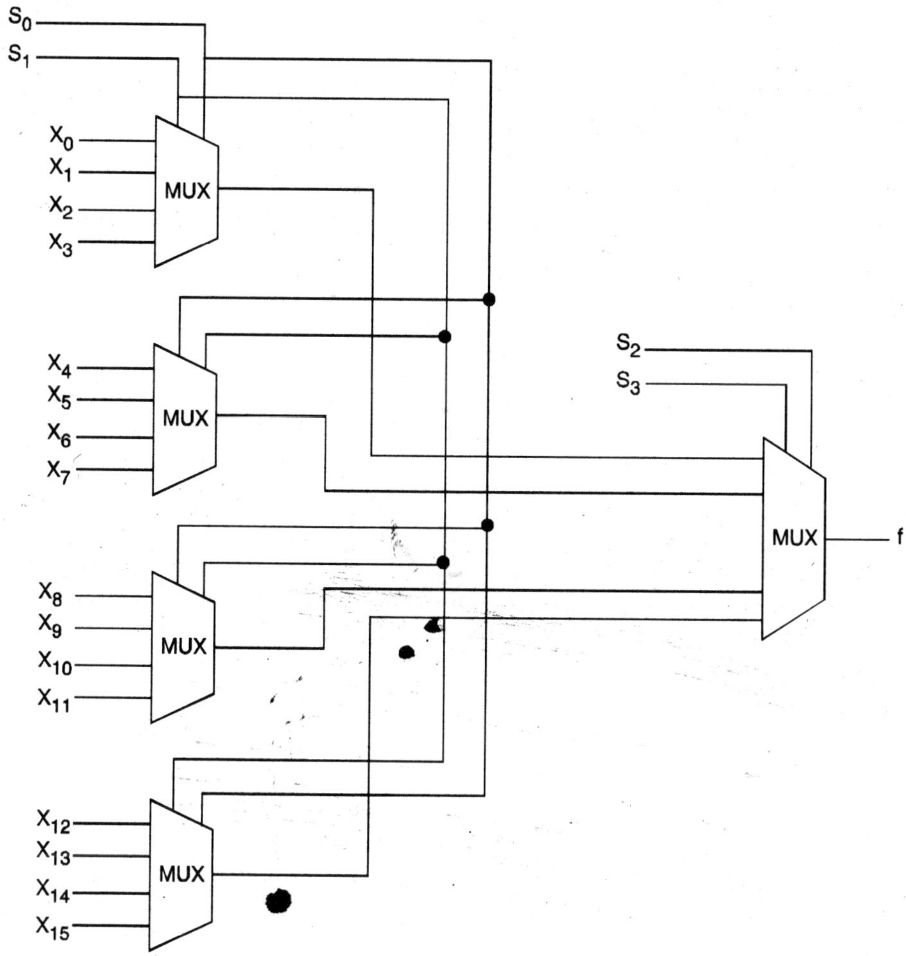

Fig. 5.1.1 (*g*) 16 to 1 Mux using 4 to 1 Mux.

5.1.2. Synthesis of Logic Functions using Multiplexers

Multiplexer are useful in many applications. They can also be used to synthesize logic functions. Consider the following example which represents an Ex-OR gate.

Truth table of EX-OR gate

S_1	S_2	y
0	0	0
0	1	1
1	0	1
1	1	0

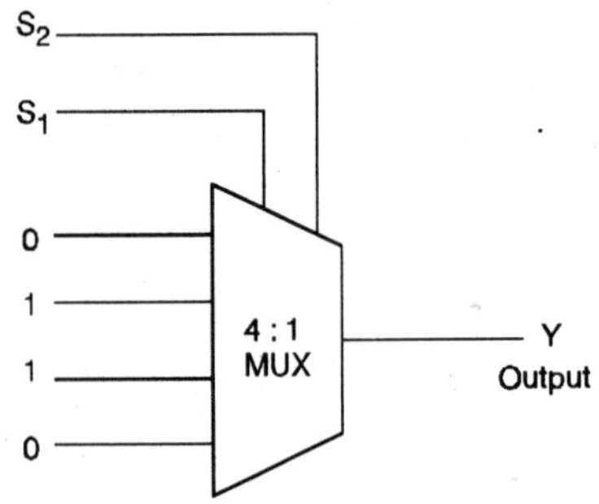

Fig. 5.1.2. (*a*) Implementation using 4-to-1 Mux.

The truth table defines the function y = $S_1 \oplus S_2$. A 4-to-1 multiplexer can implement this function. The values of y in each row are connected to the data inputs of multiplexer. The multiplexer's select inputs are driven by S_1 and S_2. The above implementation is straightforward but it is not very efficient.

A better implementation is shown in Fig. 5.1.2 (*b*) which allows y to be implemented using single 2-to-1 multiplexer. One of the S_1 and S_2 can be taken as a select input to the 2-to-1 multiplexer. Here S_1 is taken as the select input. Based on that the values for y are given as shown in the truth table. The circuit is implemented with respect to truth table as given in Fig. 5.1.2 (*c*). This procedure can be applied to synthesize a circuit that implements any logic.

S_1	S_2	y
0	0	0
0	1	1
1	0	1
1	1	0

S_1	Y
0	S_2
1	$\bar{S}_2$

Fig. 5.1.2. (*b*) Modified truth table.

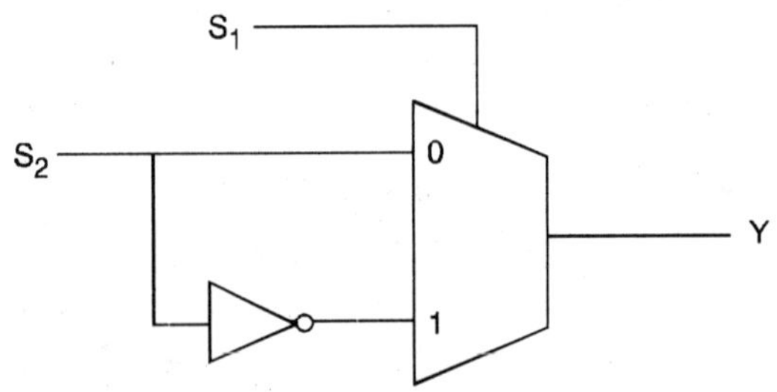

Fig. 5.1.2. (*c*) Circuit implementation for the modified truth table.

Example 1. The following Fig. 5.1.2 (*d*) shows the truth table for the three input majority function and it shows how the truth table can be modified to implement the function using a 4-to-1 multiplexer.

S_1	S_2	S_3	y
0	0	0	0
0	0	1	1
0	1	0	0
0	1	1	1
1	0	0	0
1	0	1	1
1	1	0	1
1	1	1	1

→

S_1	S_2	Y
0	0	0
0	1	S_3
1	0	$\bar{S}_3$
1	1	1

Fig. 5.1.2. (*d*) Modified truth table for the 3-input majority function.

Any two of the 3-inputs may be considered as the multiplexer select inputs. We have chosen S_1 and S_2 as the select inputs for the multiplexer. The circuit implementation for the three-input majority function using 4-to-1 multiplexer is shown in Fig. 5.1.2 (*e*).

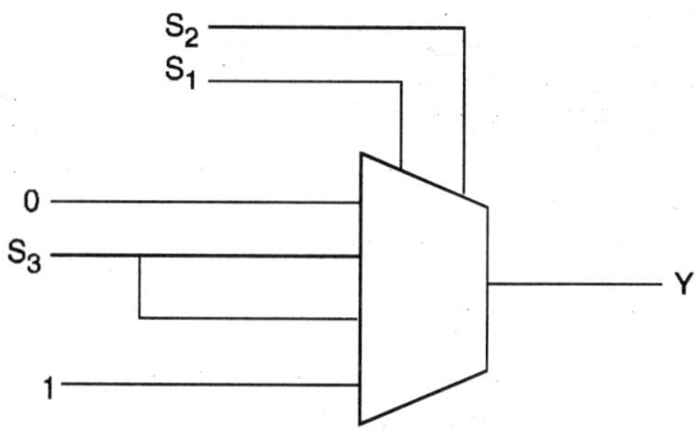

Fig. 5.1.2. (*e*) Circuit implementation of 3-input majority function using 4-to-1 Mux.

Example 2. This example shows that how the function $y = S_1 \oplus S_2 \oplus S_3$ can be implemented using 2-to-1 multiplexers.

Table for 3-input XOR

S_1	S_2	S_3	y	
0	0	0	0	$S_2 \oplus S_3$
0	0	1	1	
0	1	0	1	
0	1	1	0	
1	0	0	1	$\overline{S_2 \oplus S_3}$
1	0	1	0	
1	1	0	0	
1	1	1	1	

Fig. 5.1.2. (*f*) Truth table for 3-input XOR and XNOR function.

When $S_1 = 0$, y is equal to the XOR of S_2 and S_3 and when $S_1 = 1$, y is the XNOR of S_2 and S_3.

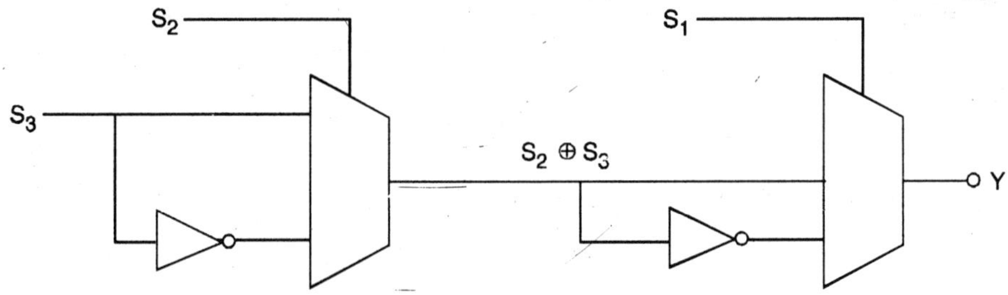

Fig. 5.1.2 (*g*). Circuit implementation for 3-input XOR function using 2 to 1 MUX.

5.1.3. Demultiplexer

The purpose of multiplexer circuit is to multiplex the n-data inputs onto the single data output under control of the select inputs. A circuit that performs the opposite function namely placing the value of a single data input onto multiple data outputs is called a demultiplexer. Truth table shows the functionality of Demux.

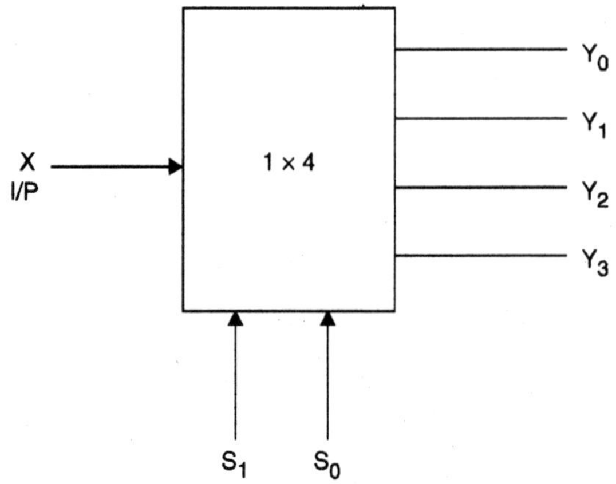

Fig. 5.1.3. (*a*) Demultiplexer.

Truth Table

S_1	S_0	x	y_0	y_1	y_2	y_3
0	0	1	1	0	0	0
0	1	1	0	1	0	0
1	0	1	0	0	1	0
1	1	1	0	0	0	1

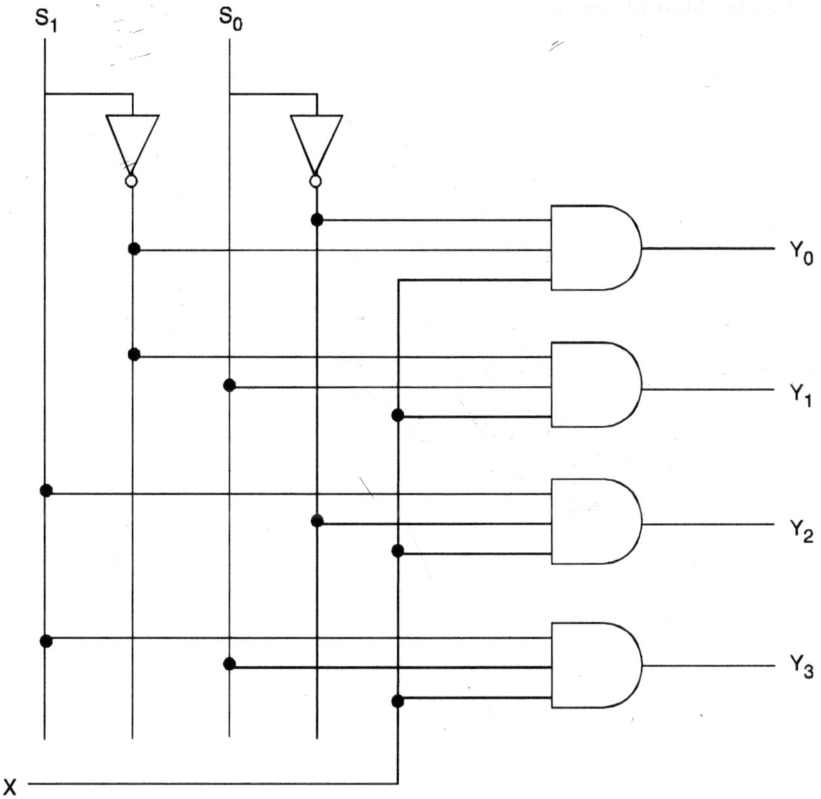

Fig. 5.1.3. (*b*) Logic Circuit.

Table 5.1.3. 3-to-8 Demultiplexer

S_2	S_1	S_0	x	y_0	y_1	y_2	y_3	y_4	y_5	y_6	y_7
0	0	0	1	1	0	0	0	0	0	0	0
0	0	1	1	0	1	0	0	0	0	0	0
0	1	0	1	0	0	1	0	0	0	0	0
0	1	1	1	0	0	0	1	0	0	0	0
1	0	0	1	0	0	0	0	1	0	0	0
1	0	1	1	0	0	0	0	0	1	0	0
1	1	0	1	0	0	0	0	0	0	1	0
1	1	1	1	0	0	0	0	0	0	0	1

Logic circuit can be constructed in similar manner as in Fig. 5.1.3 (*b*). In general, an n-to 2^n decoder circuit can be used as a 1-to-n demultiplexer. However in practice decoder circuits are used much more often as decoders rather than as demultiplexers. Table 5.1.3 shows the functionality of the 3 to 8 Demux.

5.1.4. Encoders and Decoders

A binary encoder encodes information from 2^n-inputs into an n-bit code. Exactly one of the input signals should have a value of 1 and the outputs present the binary number that identifies which input is equal to 1.

Encoders are used to reduce the number of bits needed to represent given information. A practical use of encoders is for transmitting information in a digital system.

Encoding the information allows the transmission link to be built using fewer wires. Encoding is also useful if information is to be stored for later use because fewer bits are needed to be stored.

1. Binary Encoders. An encoder encodes the given information into a more compact form. It performs the opposite function of a decoder. It encodes 2^n inputs into *n*-bit code. The truth table for 4-to-2 encoder is shown below, which describes the functionality of the 4 to 2 encoder.

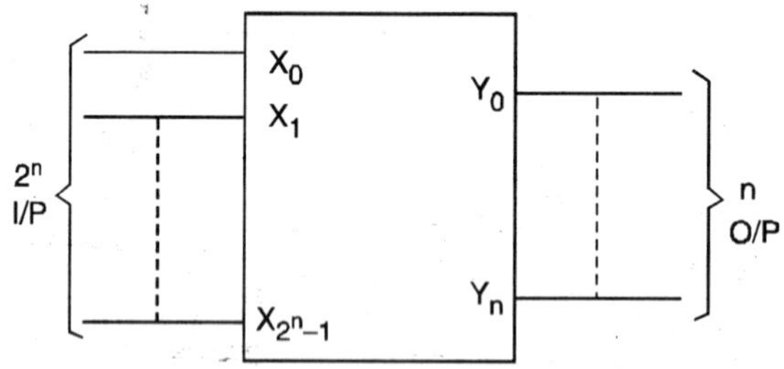

Fig. 5.1.4. (*a*) Graphical Symbol of binary encoder.

Truth table

x_3	x_2	x_1	x_0	y_1	y_0
0	0	0	1	0	0
0	0	1	0	0	1
0	1	0	0	1	0
1	0	0	0	1	1

The outputs present the binary numbers that identifies which input is equal to 1. From the truth table, it is observed that the output y_0 is 1 when either x_1 (or) x_3 is 1, and output y_1 is 1 when input x_2 (or) x_3 is 1.

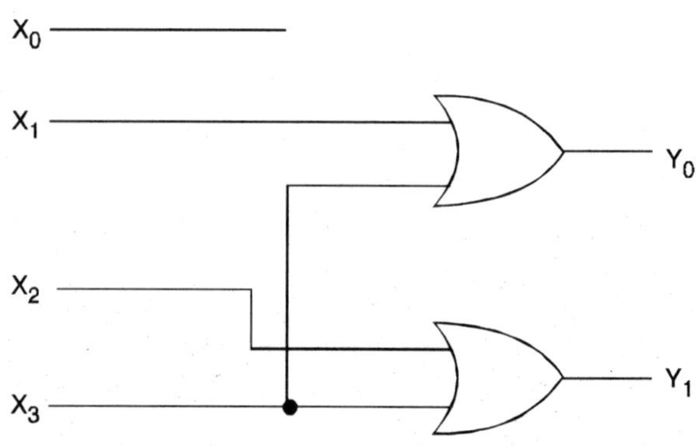

Fig. 5.1.4. (*b*) 4-to-2 binary encoder.

Encoders are used to reduce the number of bits needed to represent the given information and thus it allows the transmission link to be built using fewer wires in practical usage. The logic circuit for 4-to-2 encoder is shown in Fig. 5.1.4 (*b*).

2. Priority Encoders. It is based on the priority of input signals. Each input has a priority level. When an input with a high priority is asserted, the other inputs with lower priority are ignored. The truth table describes the functionality of a 4-to-2 priority encoder.

Truth Table

x_3	x_2	x_1	x_0	y_1	y_0	Z
0	0	0	0	x	x	0
0	0	0	1	0	0	1
0	0	1	x	0	1	1
0	1	x	x	1	0	1
1	x	x	x	1	1	1

From the truth table, it is observed that the output indicates the active input that has the highest priority. It assumes that x_3 has the highest priority and x_0 has the lowest priority.

When all the inputs are zero *i.e.*, no priority is encountered, the outputs y_0 and y_1 are not meaningful in this case and hence the y_0 and y_1 are treated as don't care condition and to indicate this condition, z is provided for all the other inputs, z is set to 1.

When an input with highest priority is asserted, then the other input are not considered and so they are treated as don't care condition as given in the truth table. For example, if an input $x_3 = 1$, then x_2, x_1 and x_0 do not matter and they are noted as x. Also, the outputs are set to 1 ($y_1y_0 = 1\ 1$) for $x_3 = 1$, similarly all the cases are considered.

To implement the truth table the intermediate signals are assumed as,

$$i_0 = \overline{X_3}\ \overline{X_2}\ \overline{X_1}\ X_0$$

$$i_1 = \overline{X_3}\ \overline{X_2}\ X_1$$

$$i_2 = \overline{X_3}\ X_2$$

$$i_3 = X_3$$

The outputs y_0 and y_1 are considered as,

$$y_0 = i_1 + i_3$$

$$y_1 = i_2 + i_3$$

The output z is considered as,

$$Z = i_0 + i_1 + i_2 + i_3$$

3. Decoders. Decoders are used to decode encoded information. It is a logic circuit with n inputs and 2^n outputs. The graphical symbol of decoder is shown in Fig. 5.1.4 (*c*). At any time only one output is asserted. The decoder has an enable input E, that is used to enable the outputs. When E = 0, the decoder output is disabled.

Truth Table

E	I_1	I_0	y_0	y_1	y_2	y_3
0	X	X	0	0	0	0
1	0	0	1	0	0	0
1	0	1	0	1	0	0
1	1	0	0	0	1	0
1	1	1	0	0	0	1

Fig. 5.1.4. (*d*) shows a 2-to-4 decoder. The two data inputs are I_1 and I_0. When input I_1, I_0 is set to '00' the output y_0 will be set to 1. Similarly setting if I_1I_0 is set '00' the output y_0 will be set to 1. Similarly setting of I_1I_0 as '01', '10', '11' causes the outputs y_1, y_2 and y_3 to bet set to 1, respectively.

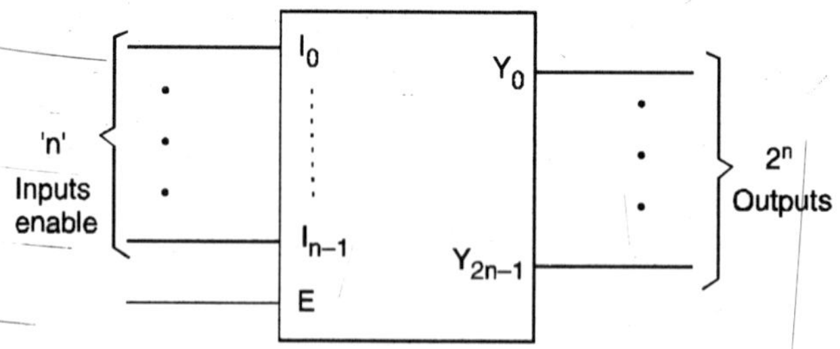

Fig. 5.1.4. (*c*) Graphical symbol of decoder.

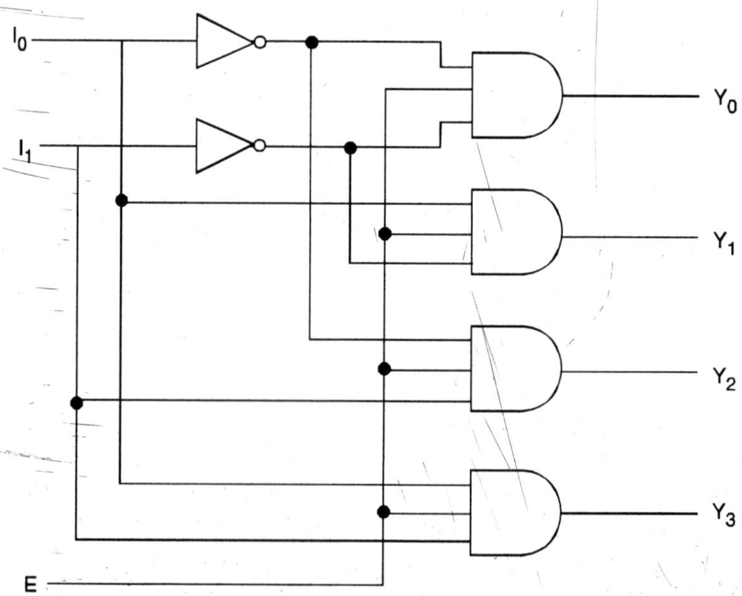

Fig. 5.1.4. (*d*) Logical circuit of 2-to-4 decoder.

Large decoders can be constructed from small decoders. Fig. 5.1.4 (*e*) shows how an 3 to 8 decoder can be built from two 2 to 4 decoders. The decoder D_1 is enabled when $I_2 = 0$, and the decoder D_2 is enabled when $I_2 = 1$.

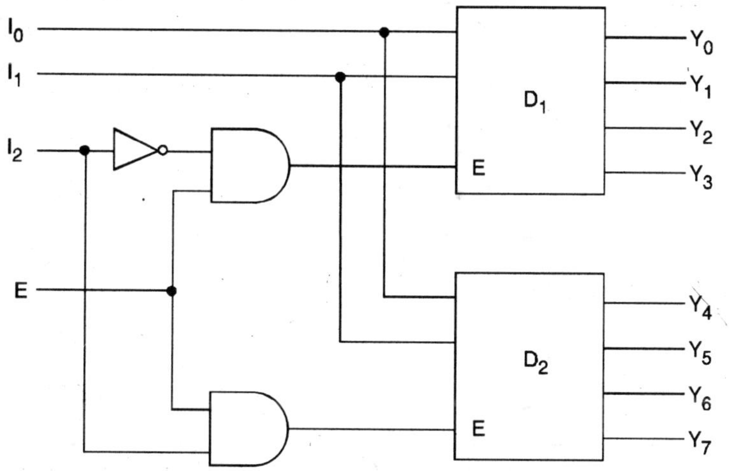

Fig. 5.1.4. (*e*) 3-to-8 decoder using 2-to-4 decoders.

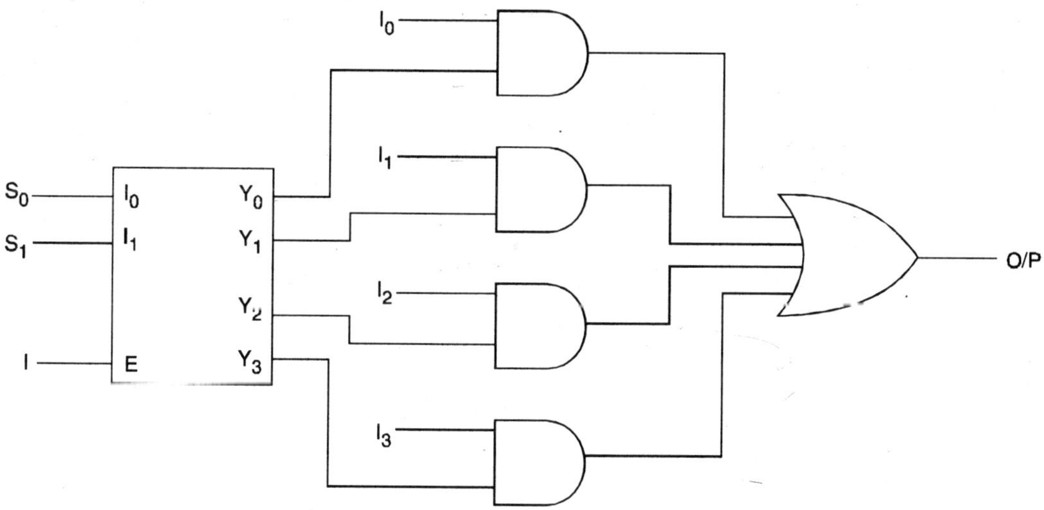

Fig. 5.1.4. (*f*) A 4-to-1 Mux using a decoder.

Decoders are useful for many practical purposes. As an example the Fig. 5.1.4 (*f*) shows how a multiplexer can be constructed using decoder. Here the enable input of decoders is set to '1'. The four outputs of the decoder represent the four values of the select inputs.

5.1.5. Code Converters

***(i)* BCD to 7 segment display.** Code converters are mainly used to convert one form of coding to another form. For example, a BCD-to-7-segment decoder is one, which converts binary coded decimal (BCD) digit into information suitable for display.

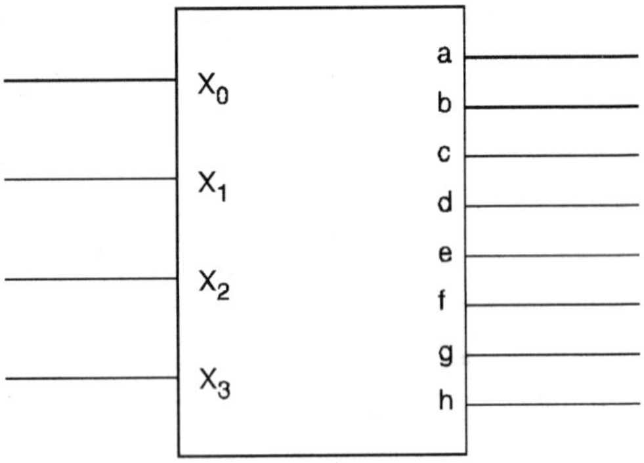

Fig. 5.1.5. (*a*) Graphical symbolb of code converter.

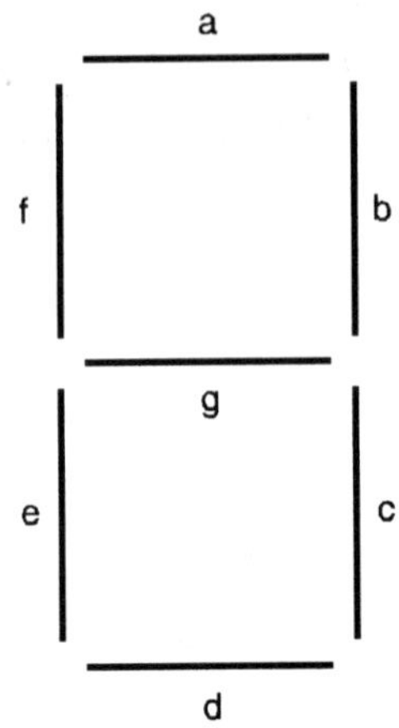

Fig. 5.1.5. (*b*) 7-Segment display.

Truth Table for BCD-to-7 Segment Display

x_3	x_2	x_1	x_0	a	B	c	d	e	f	g
0	0	0	0	1	1	1	1	1	1	0
0	0	0	1	0	1	1	0	0	0	0
0	0	1	0	1	1	0	1	1	0	1
0	0	1	1	1	1	1	1	0	0	1
0	1	0	0	0	1	1	0	0	1	1
0	1	0	1	1	0	1	1	0	1	1
0	1	1	0	1	0	1	1	1	1	1
0	1	1	1	1	1	1	0	0	0	0
1	0	0	0	1	1	1	1	1	1	1
1	0	0	1	1	1	1	1	0	1	1

Each segment in the 7-segment display is a small light emitting diode (LED), which glows when driven by an electrical signal. The segments are labelled from a to g as in Fig. 5.1.5 (*b*). Actually, the circuit converts the BCD digit into seven signals that are used to drive the segments in the display. The truth table for BCD-to-7 segment display code converter is shown which describes the functionality of BCD to 7-segment display. For each combination of inputs, the 7-segments are set to display the appropriated BCD digit.

***(ii)* Gray code to binary code converter.** This takes Gray code as input and converts it to a binary code. The table below gives the Gray to Binary code conversion.

	Gray				Binary			
	G_3	G_2	G_1	G_0	B_3	B_2	B_1	B_0
0	0	0	0	0	0	0	0	0
1	0	0	0	1	0	0	0	1
2	0	0	1	1	0	0	1	0
3	0	0	1	0	0	0	1	1
4	0	1	1	0	0	1	0	0
5	0	1	1	1	0	1	0	1
6	0	1	0	1	0	1	1	0
7	0	1	0	0	0	1	1	1
8	1	1	0	0	1	0	0	0
9	1	1	0	1	1	0	0	1
10	1	1	1	1	1	0	1	0
11	1	1	1	0	1	0	1	1
12	1	0	1	0	1	1	0	0
13	1	0	1	1	1	1	0	1
14	1	0	0	1	1	1	1	0
15	1	0	0	0	1	1	1	1

The figure below shows Gray to Binary code converter. It employs two input Ex-OR-gate. Many arrangements are possible for implementing this code conversion. But here Ex-OR gate is used which is constructed using AND and OR gates. The figure itself is self-explanatory.

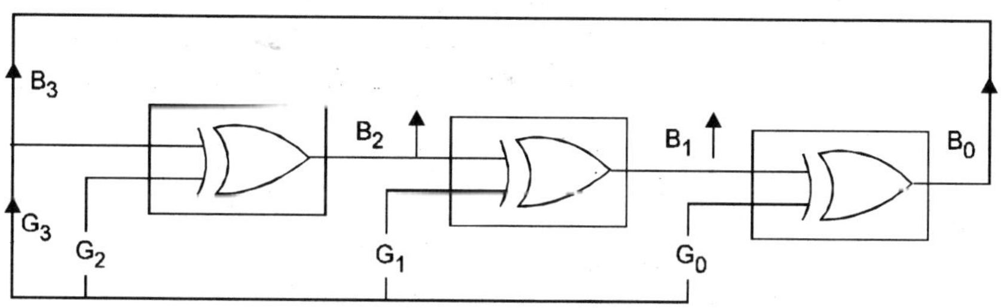

Fig. 5.1.5. (*c*) Gray to binary code converter.

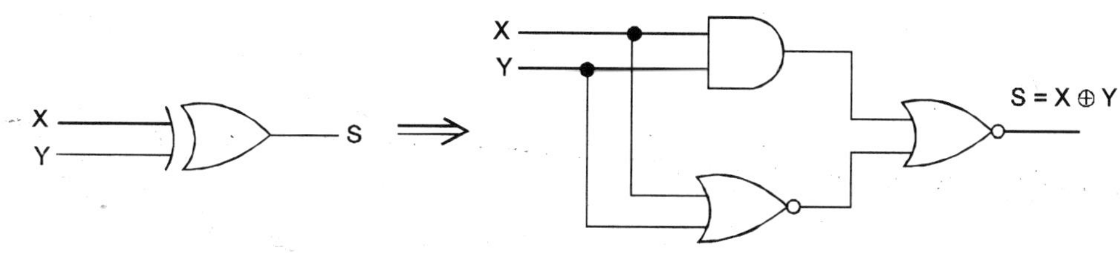

Fig. 5.1.5. (*d*) One possible realization of Ex-OR gate.

Truth Table

Input		*Output* (*S*)
X	*Y*	
0	0	0
0	1	1
1	0	1
1	1	0

5.1.6. Arithmetic Comparison Circuits

1. 1-Bit Comparator

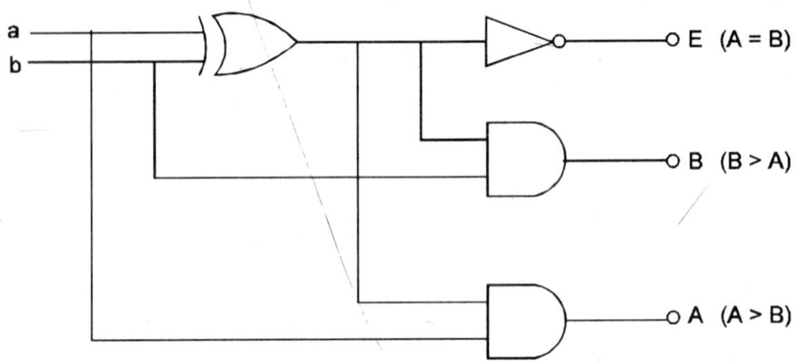

Fig. 5.1.6. (*a*) 1-bit comparator.

A comparator circuit compares 2 binary datas and produces 3 outputs called A eq B, A lt B and A gt B. *i.e.*, the comparator circuit presents whether one binary data is greater, equal (or) lesser than another binary data.

The output A eqB is set to 1, if A = B, otherwise it is set to Zero.

The output A gtB is set to 1, if A > B, otherwise it is set to Zero.

The output A ltB is set to 1, if A < B, otherwise it is set to Zero.

The logic circuit for the 1-bit comparator is shown in Fig. 5.1.6 (*a*).

2. A four bit comparator circuit

It compares two 4 bit datas and produces 3 outputs namely A eq B, A gt B, A lt B.

Let the datas be $A = a_3 a_2 a_1 a_0$

$$B = b_3 b_2 b_1 b_0$$

Initially most significant bit of each data namely a_3 and b_3 are compared and then it is continued upto least significant bit by considering the status of previous comparison of each bits.

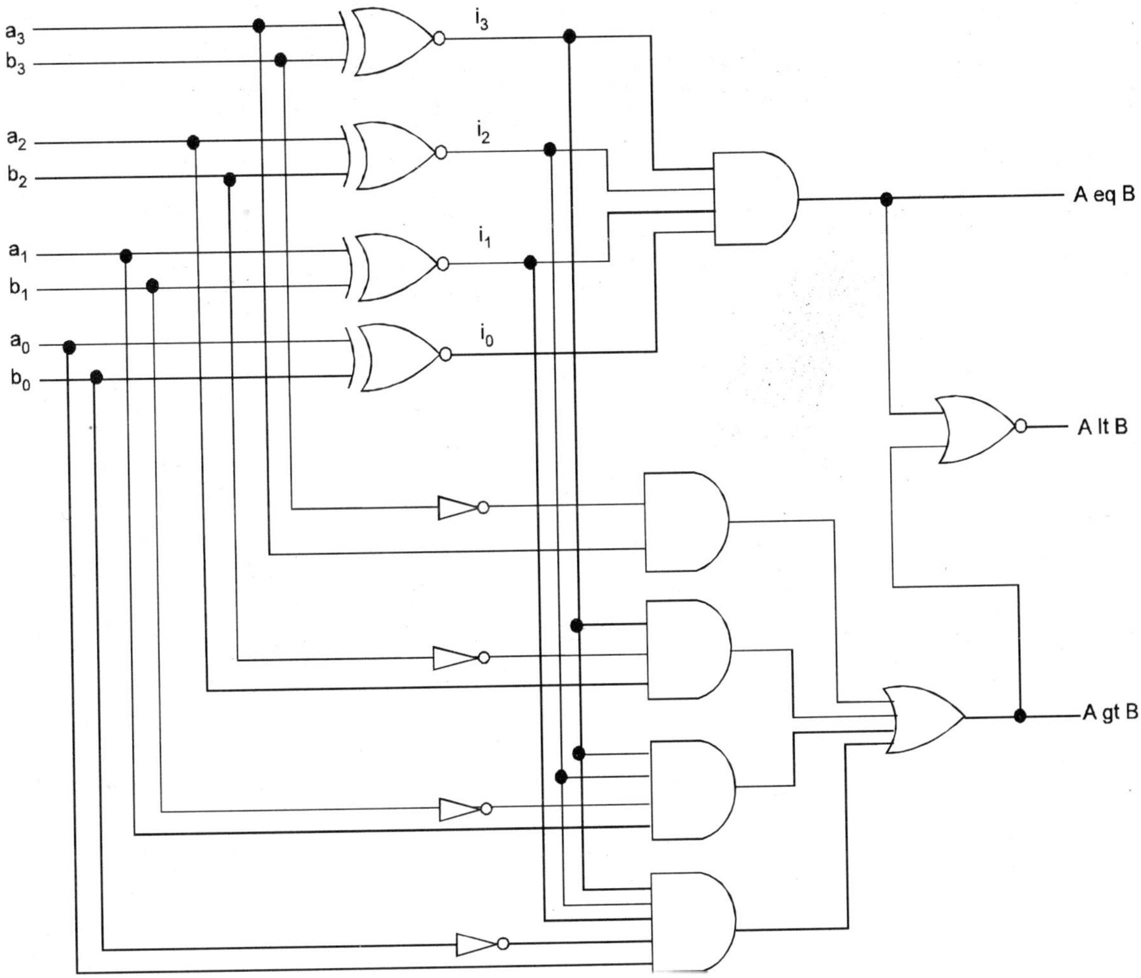

Fig. 5.1.6. (*b*) A 4-Bit comparator circuit.

In this, the comparison is done as follows :

(*i*) MSB of each data is compared for equivalence and if they are equal it is continued upto LSB. If all bits are equal then the output A eq B is set to 1.

(*ii*) In step 1 if any unequivalence is detected then a check for A gt B is done bit by bit.

(*iii*) If both the above conditions are not found to be true, then A lt B is set to 1.

This can be well understood from the figure shown.

5.1.7. Parity Generator

The parity generator will indicate the parity of a binary word. The basic block diagram of a parity generator is shown in Fig. 5.1.7 (*a*).

P = 1; When number of 1^s at the input is Even

P = 0; When number of 0^s at the input is odd

A n bit parity generator may be formed from a standard or basic one bit cell.

Bi denotes *i*th Bit

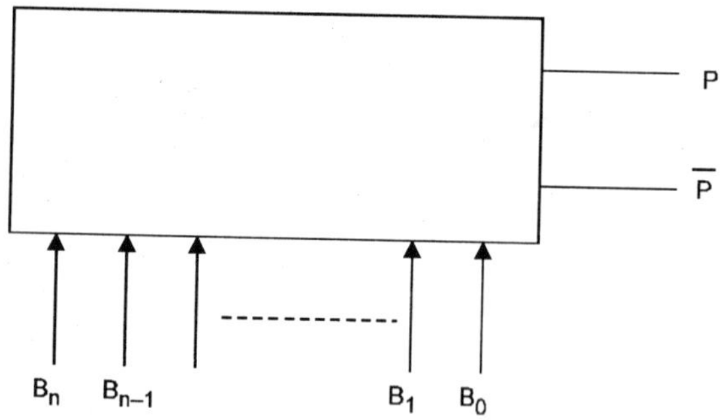

Fig. 5.1.7. (*a*) Basic block diagram of a parity generator.

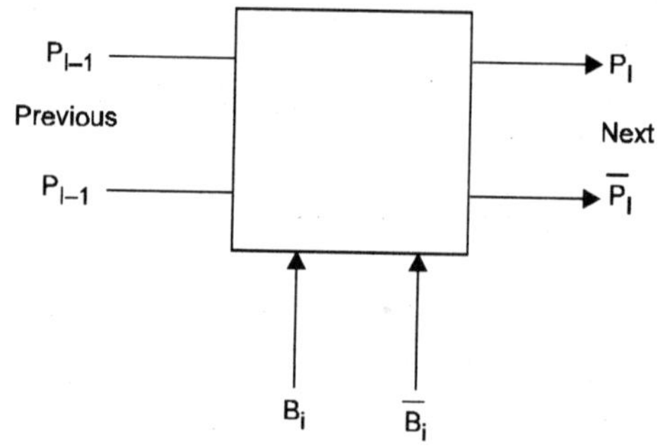

Fig. 5.1.7 (*b*) Parity generator basic one bit cell.

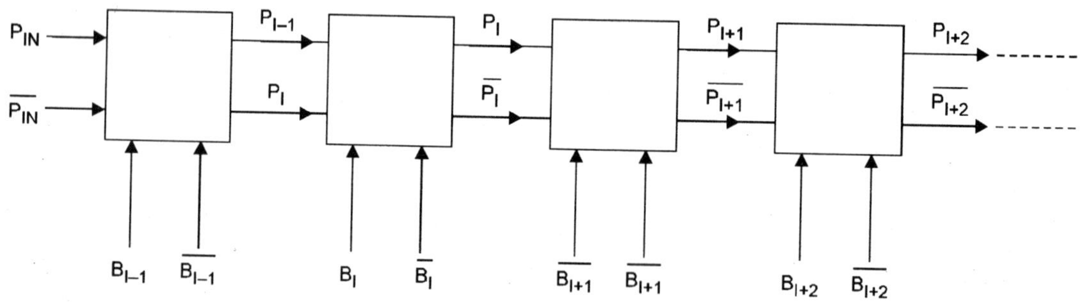

Fig. 5.1.7. (*c*) Parity generator for *n*-bit cell.

Parity requirements are set at the left-most cell. When $P_{in} = 1$, even parity is set and when $P_{in} = 0$, odd parity is set. The parity information is passed from one cell to the next. Depending on the state of input lines Bi and $\overline{B}$i, the parity is changed

If $B_i = 1$ parity is changed, $P_i = \overline{P}_i - 1$

$B_i = 0$ parity is not changed, $P_i = \overline{P}_i - 1$

5.2. SEQUENTIAL LOGIC CIRCUITS

5.2.1. Shift Registers

1. Serial-in and Serial Out: [SISO]. A register that has the ability to shift its contents is called a shift register. A register which has the ability to receive data serially and shift its content in a serial fashion is called serial in serial out register. The data bits are loaded into the shift register in a serial fashion. The contents of each flip-flop are transferred to the next flip-flop at each positive edge of clock. As an example, the below figures shows how data A = 1010 is loaded and are shifted serially.

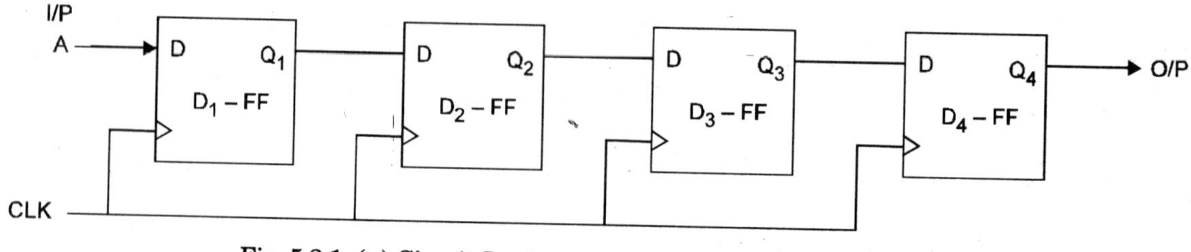

Fig. 5.2.1. (*a*) Circuit Implementation for SISO Shift register.

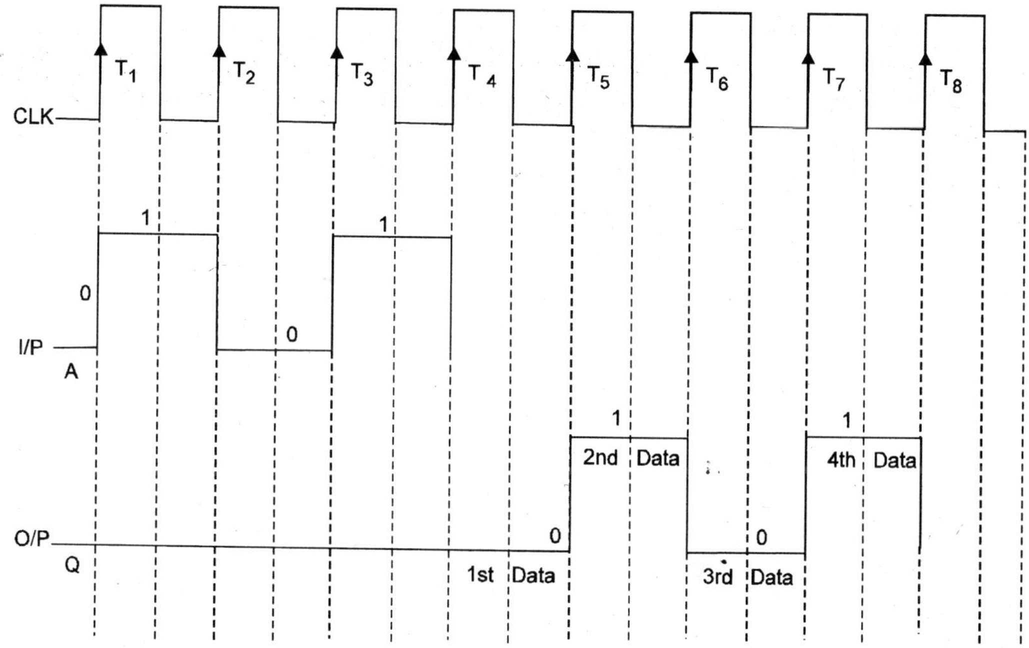

Fig. 5.2.1. (*b*) Timing Diagram for SISO.

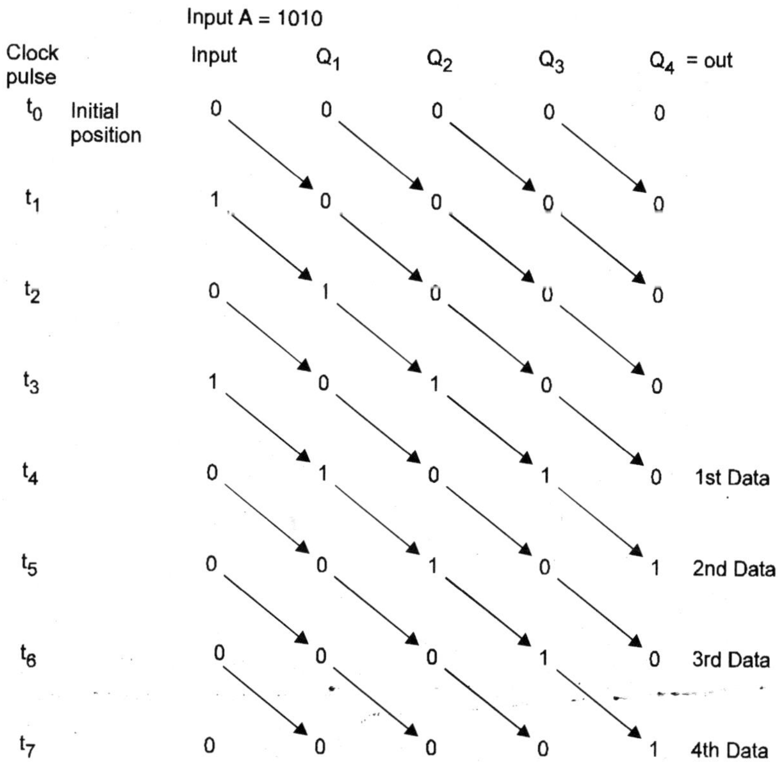

Fig. 5.2.1. (*c*) A sample sequence.

2. Serial-in and Parallel out [SIPO]. A register which has the ability to receive its data serially and shift its contents parallely is called serial in and parallel out shift register. Here the output datas are taken parallely. As an example, the Fig. illustrates how a data A = 1010 is loaded serially and the output are taken parallely.

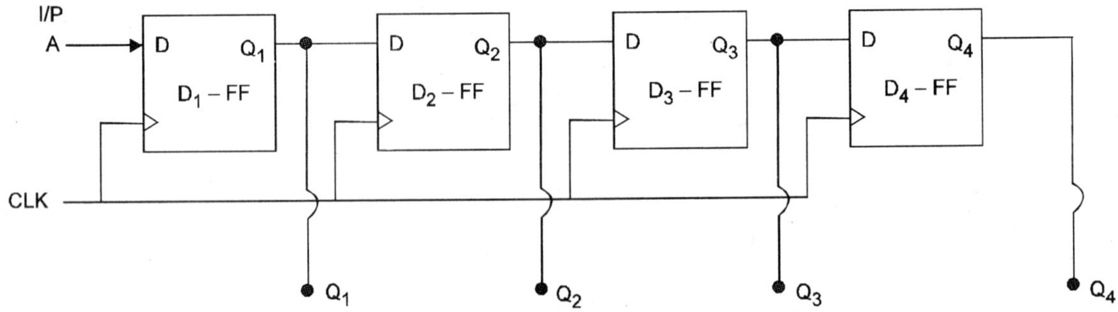

Fig. 5.2.1. (*d*) Circuit implementation for SIPO.

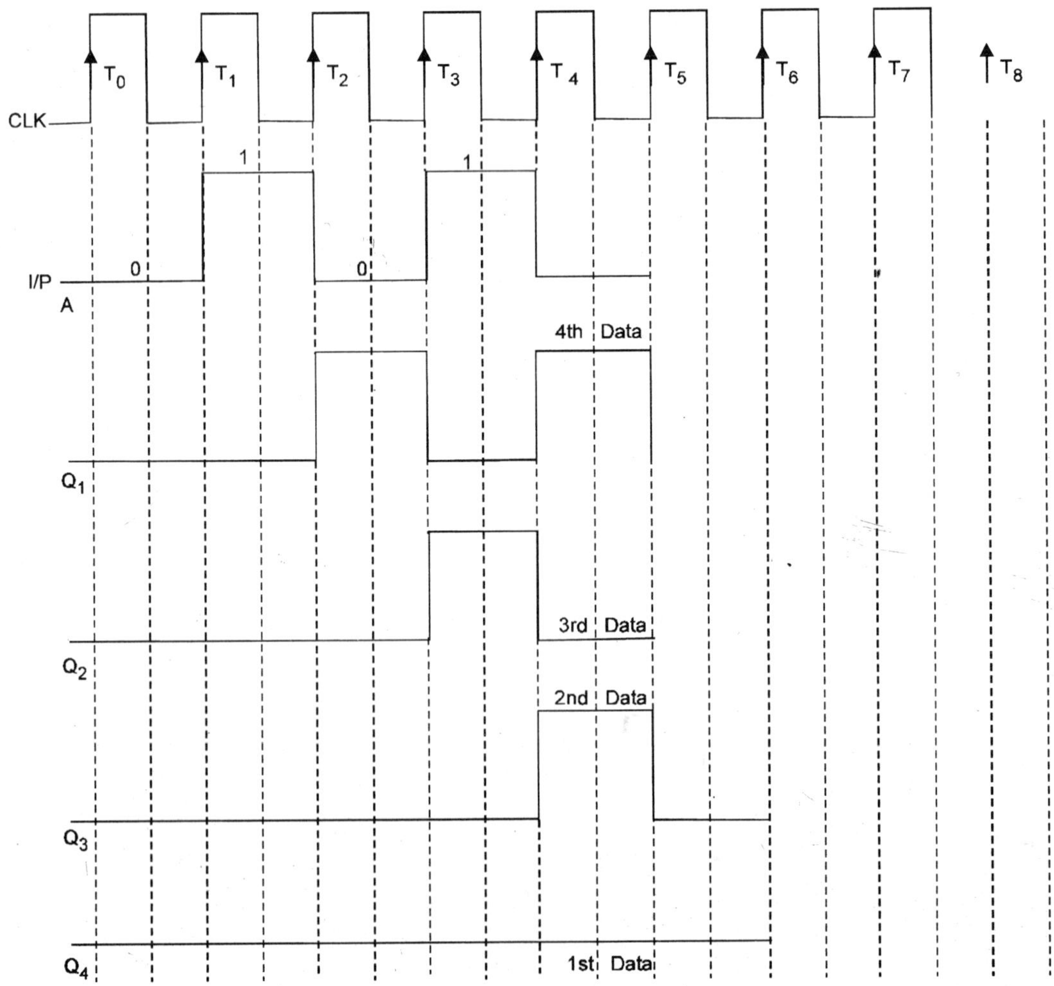

Fig. 5.2.1. (*e*) Timing diagram for SIPO.

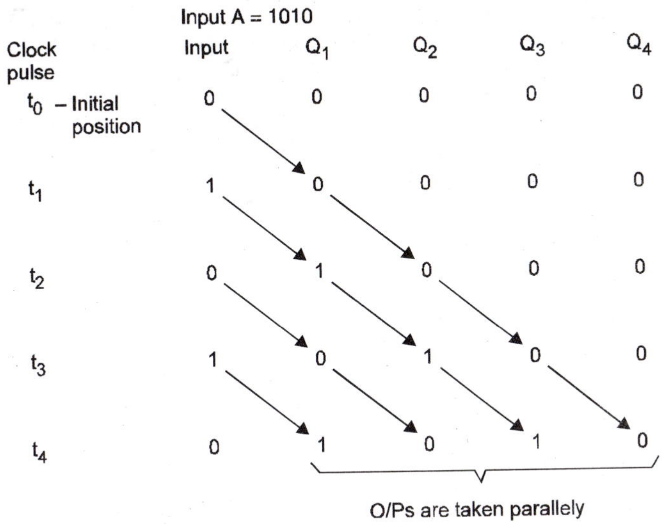

Fig. 5.2.1. (*f*) A sample sequence.

At t_1, the clock pulse is applied and after that the datas are shifted to the output of each flip-flop.

3. (*i*) Parallel in serial out–PISO

(*ii*) parallel in parallel out–PIPO

When H = 0, input datas K, L, M and N are loaded into respective D-flipflop.

When H = 1, the input datas are passed serially.

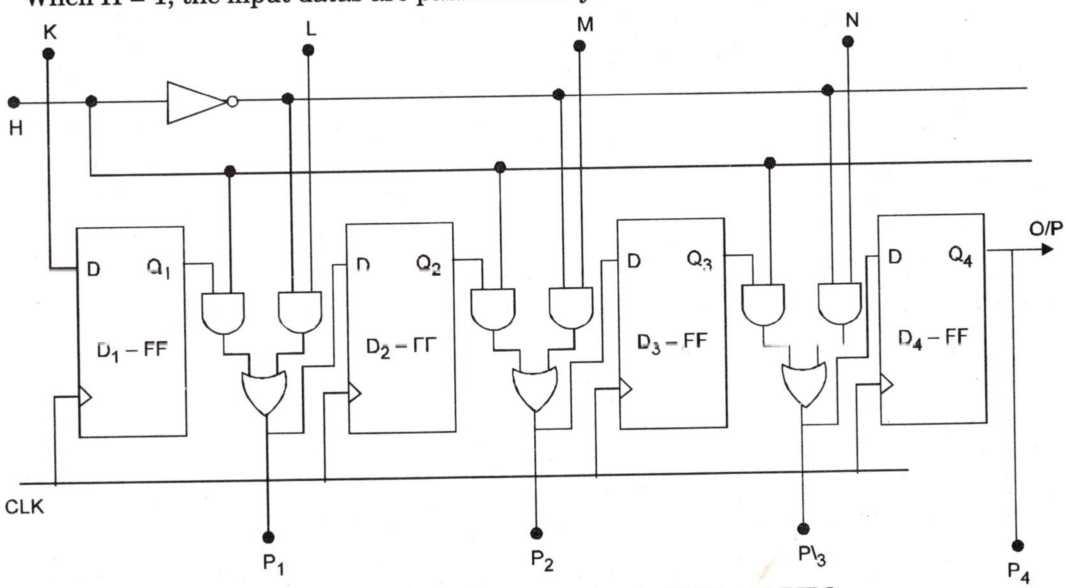

Fig. 5.2.1. (*g*) Circuit implementation for PISO and PIPO.

Parallel IN and Serial OUT. After loading the data parallely by applying H = 0, we are making H = 1 for the data to be shifted and the applied datas are taken serially. For example, a data 0101 is loaded parallely through K, L, M and N and the datas are taken at the output P_4 serially as shown in Fig. 5.2.1 (*g*).

Parallel IN and Parallel OUT. On making H = 0, datas are loaded parallelly through K, L, M and N. On making H = 1, the datas are allowed to shift through flip flop and finally output datas are taken parallely through P_1 to P_4 as shown in Fig. 5.2.1 (*g*).

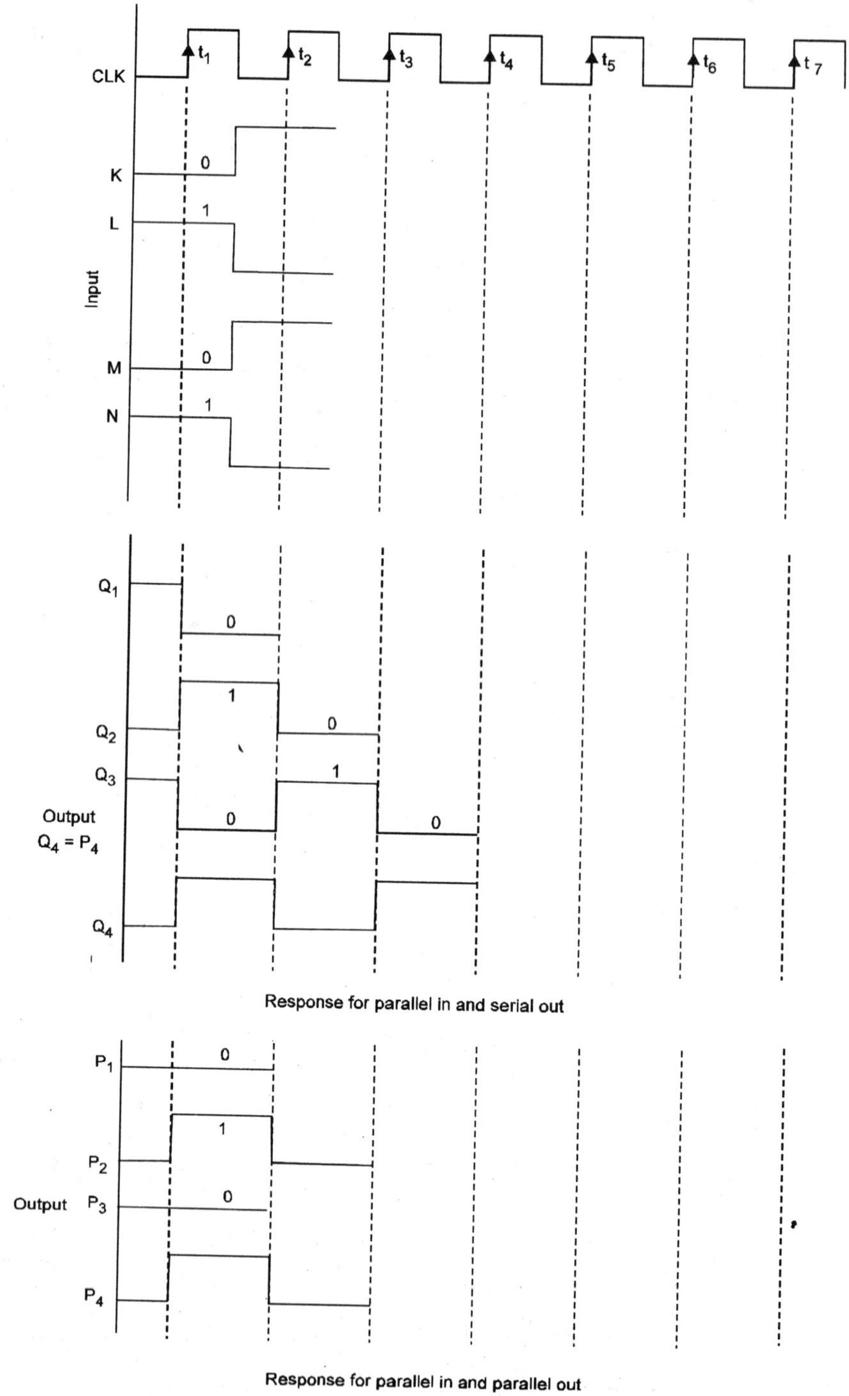

Fig. 5.2.1. (*h*) Response for Parallel in and Serial out and Response for Parallel in and Parallel out:

5.2.2. Counters

A counter is probably one of the most useful and versatile subsystems in a digital system. A counter driven by a clock can be used to count the number of clock cycles. Since the clock pulses occur at known intervals, the counter can be used as an instrument for measuring time and therefore period or frequency. There are basically two types of counters.

1. Asynchronous or ripple counter. This is simple and straightforward in operation and requires a minimum of Hardware. But the speed is less as each flip-flop is triggered by the previous flip-flop.

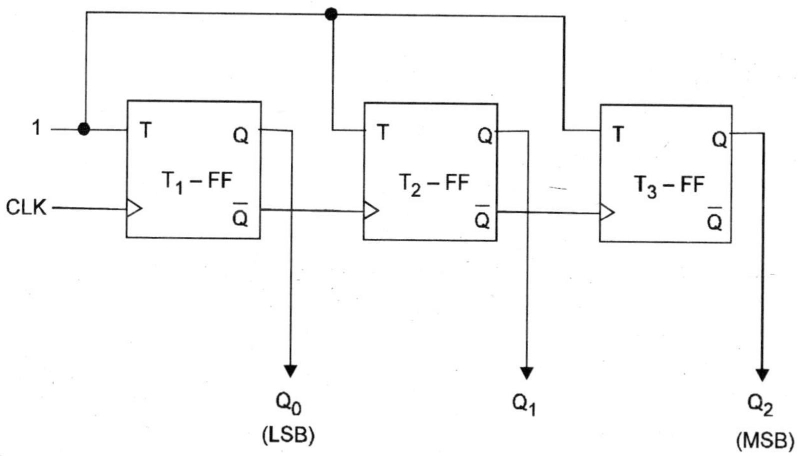

Fig. 5.2.2. (*a*) 3 Bit up counter with T flip-flops.

Clock pulse	Q_0 (*LSB*)	Q_1	Q_2 (*MSB*)
	0	0	0
1ST	1	0	0
2ND	0	1	0
3RD	1	1	0
4TH	0	0	1
5TH	1	0	1
6TH	0	1	1
7TH	1	1	1

Fig. 5.2.2. (*b*) Response is Tabulated.

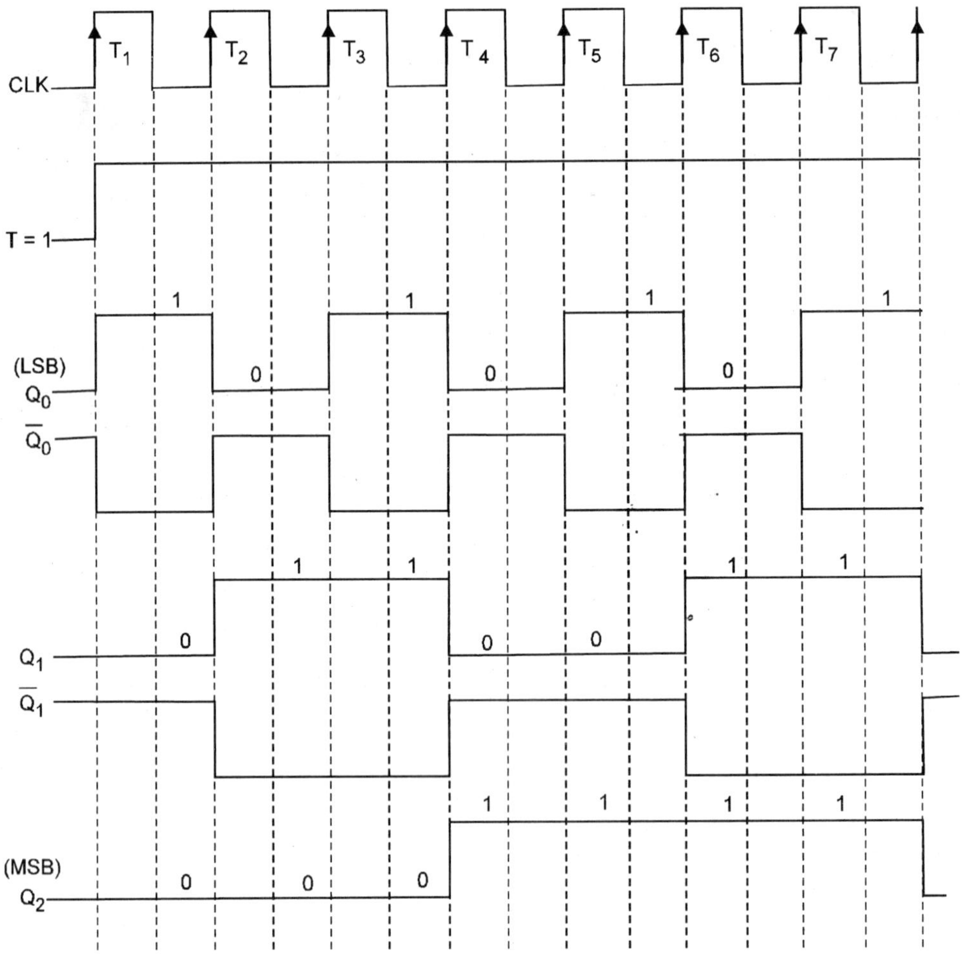

Fig. 5.2.2. (*c*) Timing Diagram for 3 bit counter.

3-Bit up-counters with T-flipflop. A 3 bit upcounter is capable of counting from 0 to 7. The T input is connected to a constant 1 and so that the sate of the flip-flop will be toggled (reversed) at each positive edge of its clock. Clock-input of the first flip-flop is connected to the clock line. Clock inputs of the successive flip-flops are connected to the $\overline{Q}$ output of the preceding flip-flops. Therefore they toggle their state whenever the proceeding flip-flop changes its state from Q = 1 to Q = 0 *i.e.*, during the positive edge of the $\overline{Q}$ signal.

The output Q_0 changes its state at every positive clock-pulse. The output Q_1 changes its state at every positive clock pulse of $\overline{Q}_0$. Finally, the output Q_2 changes its state at every positive clock pulse of $\overline{Q}_1$. As a result, if we look at the values of $Q_2Q_1Q_0$ from the timing diagram, it indicates that the counting sequence is 0, 1, 2, 3, 4, 5, 6, 7, 0, 1 and so on. Since this circuit counts in the upward direction, we call it an up counter.

3-Bit Down Counter with T Flip-flop. The clock input of the first flip-flop is connected to the clock line. The clock input of the second and third flip-flops are driven by the Q output of the preceding stages, rather than by the $\overline{Q}$ outputs..

The timing diagram shows that this circuit counts in the sequence 0, 7, 6, 5, 4, 3, 2, 1, 0, 7 and so on. Since this circuit counts in the downward direction, it is called as a down-counter.

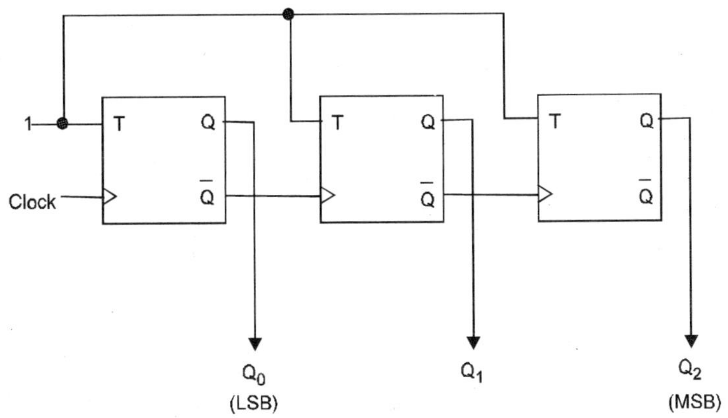

Fig. 5.2.2. (*d*) 3-bit down counter using T-FF.

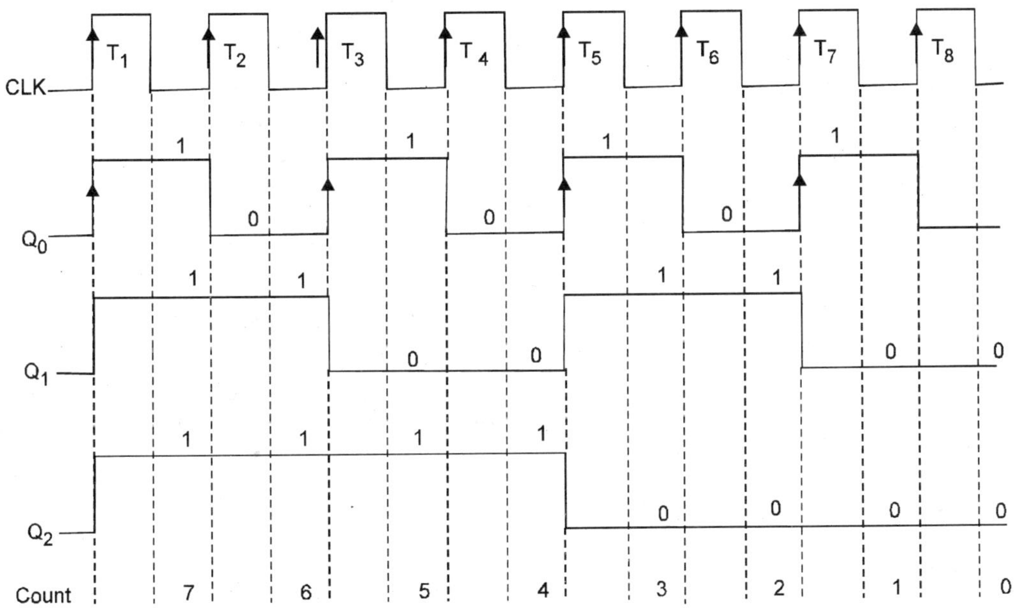

Fig. 5.2.2. (*e*) Timing diagram for 3-bit down counter.

Note : While drawing timing diagram the propagation delays have been considered negligible.

2. Synchronous Counters. In the asynchronous counters, the clock pulses are not given at the same time to all the flip-flops. If a counter with a large number of bits is constructed in this manner then the delays may become too large to meet the desired performance.

We can construct a faster counter by clocking all the flip-flops at the same time. These type of counters are named as synchronous counters.

2. (*a*) Synchronous Counter with T flip-flop

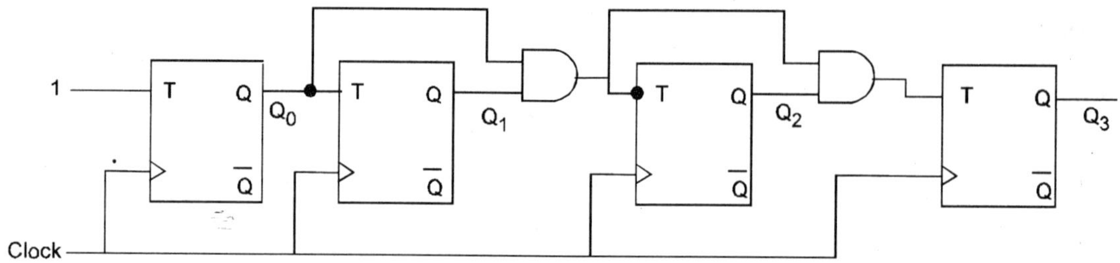

Fig. 5.2.2. (*f*) 4-bit synchronous up counter.

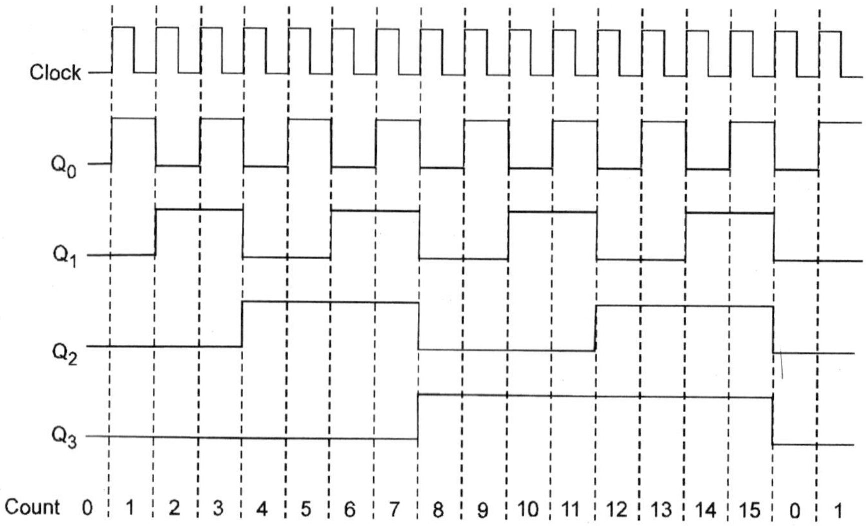

Fig. 5.2.2. (*g*) Timing diagram of 4-bit synchronous up counter.

The clock line drives the clock inputs of all the flip-flops at the same time. For example, a four bit synchronous up-counter is given for the discussion. The T-input of the first flip-flop is connected to a constant "1". So that, Q_0 changes its state on each clock cycle. Q_0 is connected as an input to the T-input of 2nd Flip-flop. Thus Q_1 changes its state with respect to Q_0. Thus Q_2 and Q_3 change its state according to the output of the preceding flip-flops.

Then the T-inputs are defined as,

$$T_0 = 1$$

$$T_1 = Q_0$$

$$T_2 = Q_0 \cdot Q_1$$

$$T_2 = Q_0 \cdot Q_1 \cdot Q_2$$

•

•

•

•

$$T_n = Q_0 Q_1 \;\ldots..\; Q_{n-1}$$

Enable and Clear Capability. In general, the counters change their contents in response to each clock pulse. But it is desirable to be able to inhibit counting so that the count remains in its present state. It may be accomplished by including an enable control signal, which controls directly the T-input of the first flip-flop. Since the enable signal is also connected to first AND gate, if Enable = 0, then all T-inputs will be equal to 0. If enable = 1, then the counter operates normally as discussed previously.

In many applications, it is necessary to start with the count equal to zero. This is achieved using clear control signal. The clear inputs of all the flip-flops are connected to clear control signal.

2. (*b*) Synchronous Counter with D-flip-flops

Counters can be constructed using other types of flip-flops. If the J and K flip-flops are tied together, a JK flip-flop becomes a T-flip flop and thus JK flip-flops are used to build counters. We now consider the construction of counter using D-flip-flops.

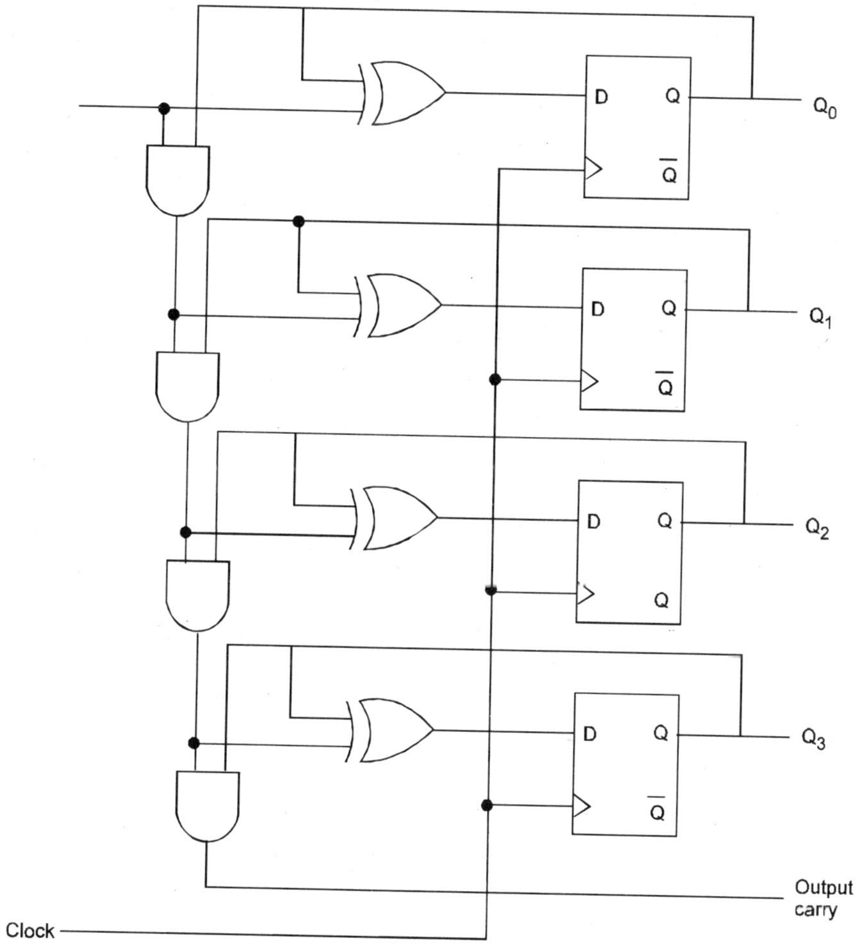

Fig. 5.2.2. (*h*) A four bit counter with D-flip flop.

The figures show a four bit up counter that counts in the sequence 0, 1, 2, ... 14, 15, 0, 1 and so on. This count is indicated by the flip-flop outputs Q_3, Q_2, Q_1, Q_0. If the Enable signal is one, then the D-inputs of the flip flops are defined by the expressions.

$$D_0 = 1 \oplus Q_0$$
$$D_1 = Q_1 \oplus Q_0$$
$$D_2 = Q_2 \oplus Q_1\, Q_0$$
$$D_3 = Q_3 \oplus Q_2 Q_1\, Q_0$$

Since the Enable control signal is included, the counter counts the clock pulses only when Enable = 1. So that, the above equations modified as shown below.

$$D_0 = Q_0 \oplus .\ \text{Enable}$$
$$D_1 = Q_1 \oplus Q_0 \cdot \text{Enable}$$
$$D_2 = Q_2 \oplus Q_1 \cdot Q_0 \cdot \text{Enable}$$
$$D_3 = Q_3 \oplus Q_2 \cdot Q_1 \cdot Q_0 \cdot \text{Enable}$$

Operation : Assume that the outputs of all the flip-flops are zero initially. After making the Enable = 1, and when the 1st clock pulse is applied to the circuit, the inputs of D_0 D_1 D_2 D_3 are 1000 respectively. So that, the outputs of Q_3 Q_2 Q_1 Q_0 are 0010. When the 2nd clock pulse is applied to the circuit, the inputs of D_0 D_1 D_2 D_3 are 0100. So, the outputs of Q_3 Q_2 Q_1 Q_0 are 0001. Thus the count sequence becomes 0, 1, 2..... and so on. Even though the circuit is only a four-bit counter, we have included an extra AND that produces the "output carry". This signal makes it easy to concatenate two such four-bit counter to create an eight bit counter.

Counter with Parallel Load

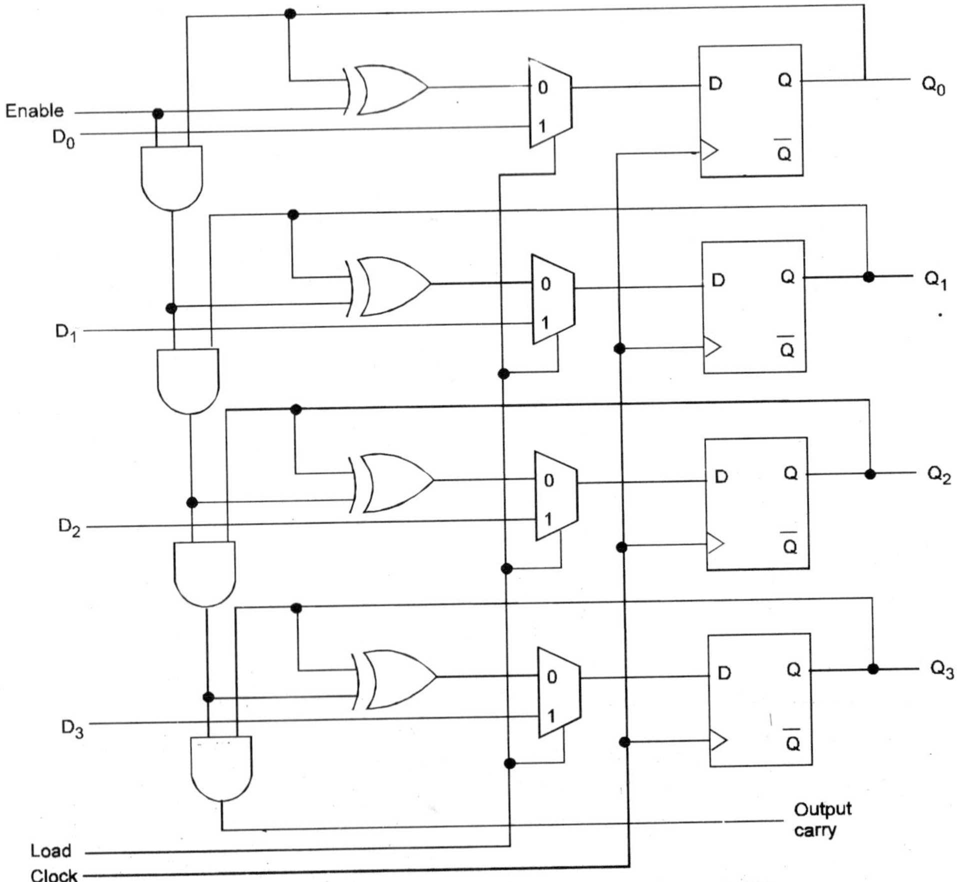

Fig. 5.2.2. (*i*) A counter with parallel load capability.

Generally, we use to start the counting from zero. But sometimes it is necessary to start with different count. To achieve this kind of operation, the counter circuit must have some inputs through which the required initial count can be loaded.

The Fig. 5.2.2. (*i*) shows the counter with parallel-load capability. Here, a two–input multiplexer is inserted before each D-input. One input to the multiplexes is used to provide the normal counting operation. The other input is a data bit that can be loaded directly into the flip-flop. A control input, Load, is used to choose the mode of operation. The circuit counts when load = 0. A new initial value, D_3 D_2 D_1 D_0 is loaded into the counter when load = 1.

Reset synchronisation. Before to start the counting operation, the contents of a counter should be cleared. This can be done using clear control signal as discussed previously. But, sometimes, it is necessary to reset the count to 0 during normal counting process. An n-bit up counter functions naturally as a modulo 2^n counter. For example, we may require to design a modulo – 6 counter, for which the counting sequence is 0, 1, 2, 3, 4, 5, 0, 1 and so on.

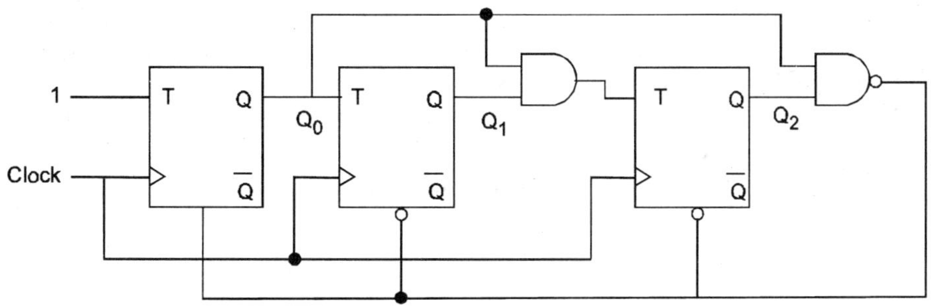

Fig. 5.2.2. (*j*) A MOD-6 counter with asynchronous reset.

Since the clear inputs are active when low, a NAND gate is used to detect the occurrence of the count of 5 and cause the clearing of all three flip-flops. As soon as the count reaches 5, the NAND gate triggers the resetting action. The flip-flops are cleared to 0-a short time after the NAND gate has detected the count of 5. This time delay depends on the gate delays in the circuit, but not on the clock.

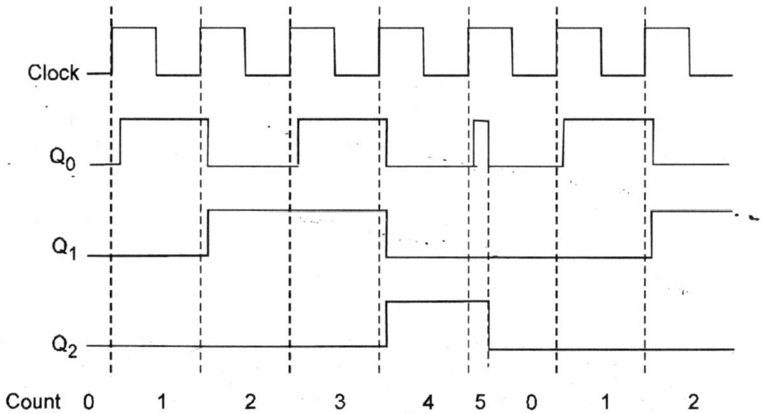

Fig. 5.2.2. (*k*) Timing diagram of mod-6 counter.

3. Ring Counter. A ring counter is a circular shift register with only one flip-flop being set at any particular time and all others are cleared. The single bit is shifted from one flip-flop to the other to produce the sequence of timing signals. Fig. 5.2.2 (*l*) shows a 4 bit shift register connected as a ring counter. The initial value of the register is 1000, which produces the variable To. The single bit is shifted right with every clock pulse and circulates back from T3 to To. Each flip-flop is in the 1 state once every four clock pulses and produces one of the four timing signals shown in Fig 5.2.2 (*n*). Each output becomes a 1 after the negative-edge transistion of a clock pulse and remains 1 during the next clock pulse. The timing signals can be generated by continuously enabling a 2 bit counter that goes through four distinct states. The decoder shown in Fig 5.2.2 (*m*) decodes the four states of the counter and generates the required sequence of timing signals.

The timing signals, when enabled by the clock pulses, will provide multiple-phase clock pulses. For example if T_0 is ANDed with CLK, the output of the AND gate will generate clock pulses at one-fourth the frequency of the master-clock pulses. Multiple-phase clock pulses can be used for controlling different registers with different time scales.

To generate 2^n timing signals, we need either a shift register with 2^n flip-flops or an n-bit counter together with an *n* to 2^n-line decoder. For example, 16 timing signals can be generated with a 16-bit shift register connected as ring counter or with a 4-bit counter and a 4 to 16 line decoder. In the first case, we need 16 flip-flops. In the second case, we need four flip-flops and 16 4-input AND gates for the decoder. It is also possible to generate the timing signals with a combination of a shift register and a decoder. In this way, the number of flip-flops required is less than a ring counter, and the decoder requires only 2-input gates. This combination is sometimes called a Johnson counter.

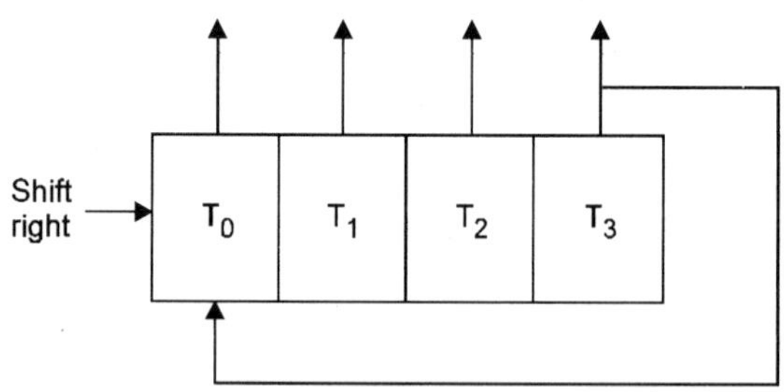

Fig. 5.2.2. (*l*) Shift register as ring counter.

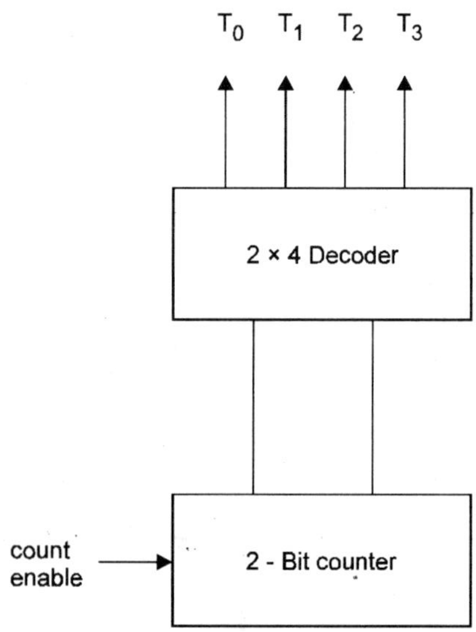

Fig. 5.2.2. (*m*) Counter and decoder.

Fig. 5.2.2. (*n*) Sequence of four timing signals.

4. Johnson Counter. A k-bit ring counter circulates a single bit among the flip-flops to provide k distinguishable states. The number of states can be doubled if the shift register is connected as a switch-tail ring counter. A switch-tail ring counter is a circular shift with the complement output of the last flip-flop connected to the input of the first flip-flop. Fig. 5.2.2 (*o*) shows such a shift register. The circular connection is made from the complement output of the rightmost flip-flop to the input of the leftmost flip-flop. The register shifts its contents once to the right with every clock pulse, and at the same time, the complement value of E flip-flop is transferred into the A flip-flop. Starting from a cleared state, the switch-tail ring counter goes through a sequence of eight states, as listed in Fig. 5.2.2 (*p*). In general, a kit-bit switch-tail ring counter will go through a sequence of 2k states. Starting from all 0's, each shift operation insert 1's from the left until the register is filled with all 1's. In the following sequences, 0's are inserted from the left until the register is again filled with all 0's.

A Johnson counter is a k-bit switch-tail ring counter with 2k decoding gates to provide outputs for 2k timing signals. The decoding gates are not shown in Fig. but are specified in the last column of the table. The eight AND gates listed in the table, when connected to the circuit, will complete the construction of the Johnson counter. Since each gate is enabled during one particular state sequence, the outputs of the gates generate eight timing sequences is succession.

The decoding of a k-bit switch-tail ring counter to obtain 2k timing sequence follows a regular pattern. The all-0's state is decoded by taking the complement of the two extreme flip-flop outputs. The all-1's state is decoded by taking the normal outputs of the two extreme flip-flops. All other states are decoded from an adjacent 1, 0 or 0, 1 pattern in the sequence. For example, sequence 7 has an adjacent 0, 1 pattern in flip flops B and C. The decoded output is then obtained by taking the complement of B and the normal output of C, or B' C.

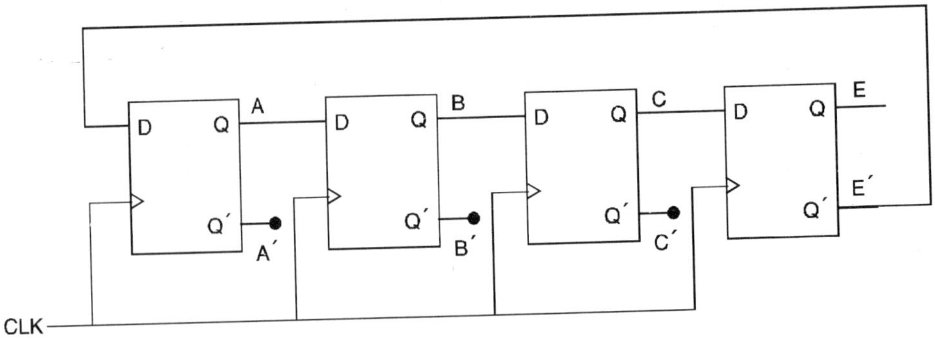

Fig. 5.2.2 (*o*). Four stage switch-rail ring counter.

Sequence number	*Flip-flop outputs*				*AND gate required for output*
	A	B	C	D	
1	0	0	0	0	A′E′
2	1	1	0	0	AB′
3	1	1	0	0	BC′
4	1	1	1	0	CE′
5	1	1	1	1	AE
6	0	1	1	1	A′B
7	0	0	1	1	B′C
8	0	0	0	1	C′E

Fig. 5.2.2 (*p*). Count sequence and required decoding.

One disadvantage of the circuit in Fig. 5.2.2 (*o*) is that if it finds itself in an unused state, it will persist in moving from one invalid state to another and never find its way to a valid state. This difficulty can be corrected by modifying the circuit to avoid this undesirable condition. One correcting procedure is to disconnect the output from flip-flop B that goes to the D input of flip-flop C, and instead enable the input of flip-flop C by the function: DC = (A + C) B Where DC is the flip-flop input function for the D input of flip-flop C.

Johnson counters can be constructed for any number of timing sequences. The number of flip-flops needed is one-half the number of timing signals. The number of decoding gates is equal to the number of timing signals and only 2 input gates are employed.

5.3. DATA PATH CIRCUITS

Fig. 5.3.1 (*a*) shows Basic digital processor structure. Relating to the digital processor architectures, the data path is the core of the processor where all computations are performed The other blocks in the processor are support units that store either the results produced by the data path (or) help to determine what will happen in the next cycle. The data path has been separated out in the Fig. 5.3.1 (*b*) and it will be seen that the structure comprises a unit which processes data applied at one port and presents its output at a second port.

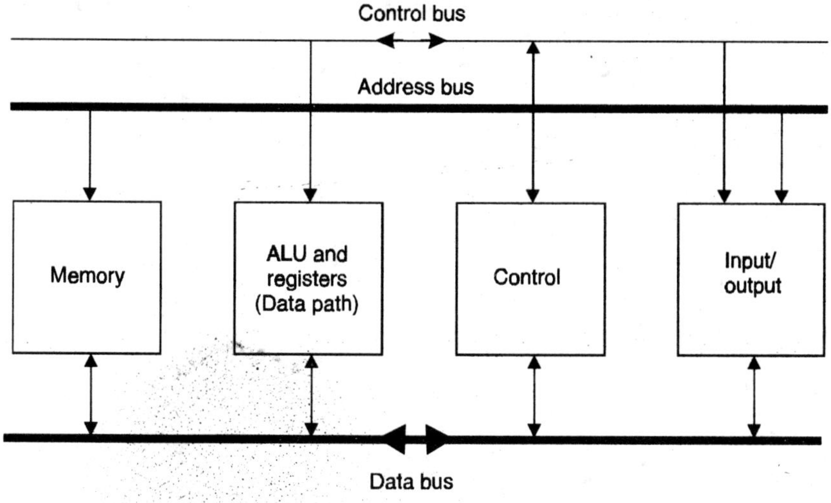

Fig. 5.3.1 (*a*). Basic digital processor structure.

A typical data path consists of an interconnection of basic combinational function such a logic (AND, OR, EX-OR) or arithmetic operators (addition, multiplication, comparison, shift).

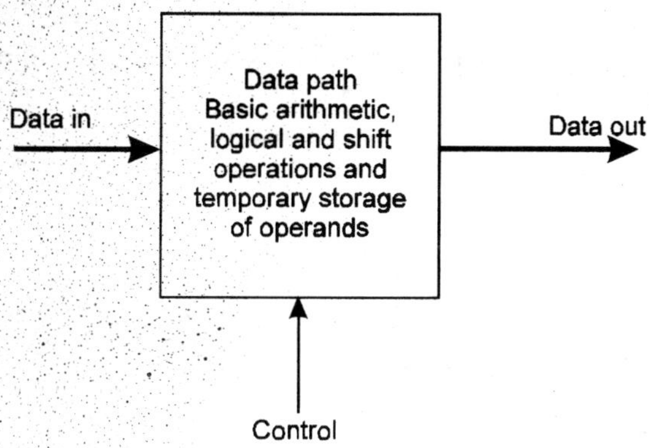

Fig. 5.3.1 (*b*). Data path communication strategy.

The data path can be decomposed into a block diagram showing the main sub units. As in Fig. below.

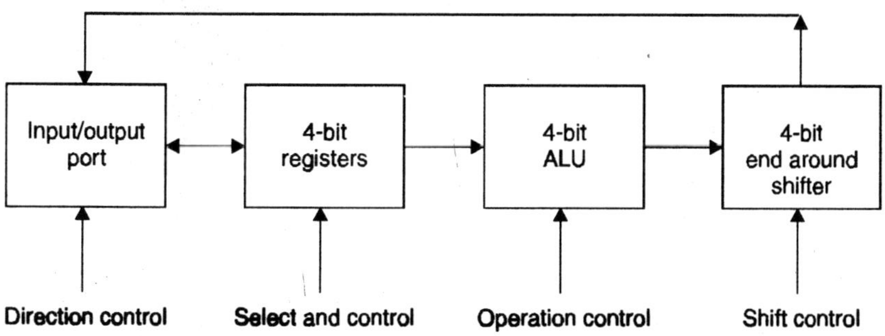

Fig. 5.3.2. Subunits and basic interconnections for data path.

From the above, we can understand that the data path mainly consists of adders, multipliers, shifters, registers, comparators and ALU. As enough explanations have already been given on shifters, registers and comparators now we will concentrate our discussion towards adders and multipliers.

5.4. ADDER/SUBTRACTOR

Addition forms the basis for many processing operations for counters, multipliers, filters etc. As a result, adder circuits that add two binary numbers are very essential. A wide variety of adder implementations are available to serve different speed/density requirements.

Types of adders

1. Ripple carry adder or carry propagate adder.
2. Carry save adder.
3. Carry select adder.
4. Carry look ahead adder.

5.4.1. Ripple Carry Adder

Fig. 5.4.1 (*a*) below shows the block diagram of Ripple carry adder, where *ai* and *bi* form input bits and S_i and C_i are sum and carry respectively.

The carryout from first full adder is given as carry input for the second full adder. In a similar way the carry is propagated to the subsequent stages. In this adder circuit, the delay is high because of the rippling effects of the carry bits. For example in a 4 bit adder the carry bit has to propagate through all the four stages to produce a final carry at the output as shown in Fig. 5.4.1 (*a*). In this processing time is proportional to word width 'n'. With little modification, i.e. by introducing X-OR at one of the inputs and by using a control bit 'add-sub', the same fig. can be used to perform both addition and subtraction by selection.

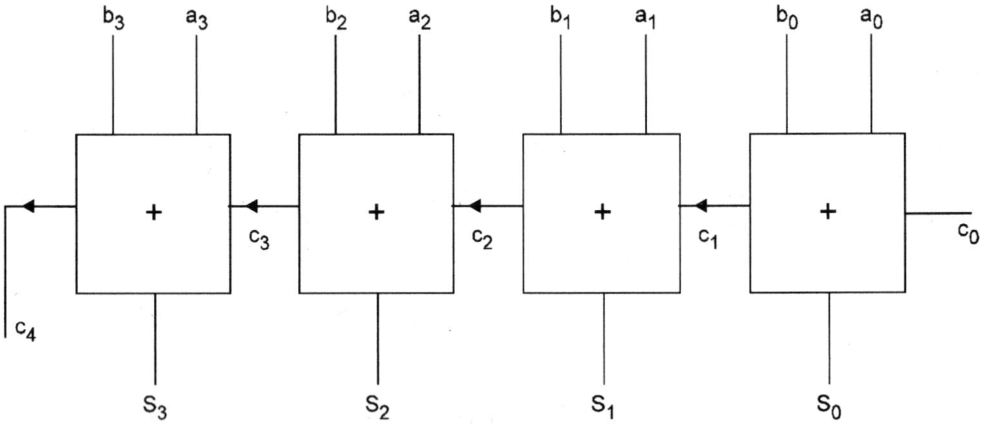

Fig 5.4.1. (*a*) Ripple Carry Adder.

For example when the control bit 'add-sub' is at level '1', then input bit bi will be complemented. The C_0 should be forced to zero in order to get 2s complement of 'bi'. Now 2s complement of bi is obtained which is added with ai. So here subtraction is implemented through 2's complement addition. When control bit 'add-sub' is at low level '0', the circuit will function as a normal adder. The Fig. 5.4.1 (*b*) below shows the 4 bit adder-subtractor circuit.

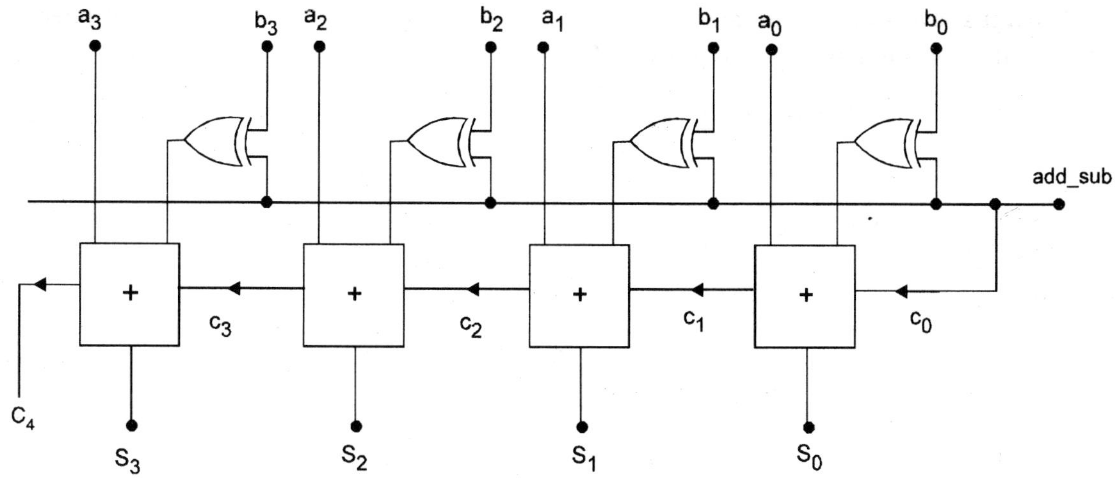

Fig 5.4.1. (*b*) 4-bit adder/Subtractor circuit.

5.4.2. Carry Look Ahead Adder

Carry look-ahead (CLA) adders are designed to overcome the latency introduced by the rippling effect of the carry bits. The CLA algorithm is based on the origin of the carry-out bit in the equation

$$C_{i+1} = a_i\, b_i + c_i\, (a_i \oplus b_i)$$

for the cases that give $c_{i+1} = 1$. Since either term may cause this output, we treat the two separately. First, if $a_i\, b_i = 1$, then $c_{i+1}=1$. We call

$$g_i = a_i\, b_i$$

the **generate** term, since the inputs are viewed as "generating" the carry out bit. Note that if $g_i = 1$, then we must have $a_i = b_i = 1$. The second term represents the case where an input carry $c_i = 1$ may be "propagated" through the full-adder. This will happen if the **propagate** term

$$p_i = a_i \oplus b_i$$

is equal to 1; if $p_i = 1$ then $g_i = 0$ since the XOR operation produces a 1 if the inputs are not equal. Table 5.4.2 shows the behaviour of the generate and proparate terms.

Table 5.4.2. Basis of the carry look-ahead algorithm

		g_i	p_i
Ai	*bi*	*ai bi*	*ai* ⊕ *bi*
0	0	0	0
0	1	0	1
1	0	0	1
1	1	1	0

From the above we can understand that,

(*i*) if $a_i = b_i = 0$ $g_i = p_i = 0$

(*ii*) if $a_i = b_i = 1$ $g_i = 1$ and $p_i = 0$

(*iii*) if $a_i \neq b_i$ $g_i = 0$ and $p_i = 1$

With these definitions, the equation for the carry out bit is

$$c_{i+1} = g_i + p_i\, c_i$$

The main idea of the CLA is to first calculate the values of p_i and g_i for every bit, then use them to find the carry bits c_{i+1} Once these are found, the sum bits are given by

$$s_i = p_i \oplus c_i \qquad \text{... }(a)$$

for every i. This avoids the need to ripple the carry bits serially down the chain.

Let us analyze the 4-bit CLA equations. With c_0 assumed known, we have

$$c_1 = g_0 + p_0\, c_0$$

The expressions for c_2 c_3 and c_4 have the same form with

$$c_2 = g_1 + p_1\, c_1$$

$$c_3 = g_2 + p_2\, c_2$$

$$c_4 = g_3 + p_3\, c_3$$

These can be expressed using primitive generate and propagate terms by noting that c_i can be substituted into c_{i+1} in succession. The first reduction is obtained by substituting c_1 into the c_2 equation to arrive at

$$c_2 = g_1 + p_1\,(g_0 + p_0\, c_0)$$

Expanding, $$c_2 = g_1 + p_1\, g_0 + p_1\, p_0 c_0$$

Similarly, substituting c_2 into c_3 gives

$$c_3 = g_2 + p_2\,(g_1 + p_1 \cdot g_0 + p_1\, p_0\, c_0$$

$$= g_2 + p_2\, g_1 + p_2\; p_1\, g_0 + p_2\, p_1\, p_0\, c_0$$

Finally, the carry out bit is

$$c_4 = g_3 + p_3\,(g_2 + p_2 \cdot g_1 + p_2\, p_1\, g_0 + p_2\, p_1\, p_0\, c_0)$$

$$= g_3 + p_3\, g_2 + p_3 \cdot p_2\;\, g_1 + p_3\, p_2\, p_1\, g_0 + p_3\, p_2\;\, p_1\, p_0\, c_0$$

These equations show that every carry bit can be found from the generate and propagate terms. Moreover, the algorithm yields nested SOP expressions. The logic diagram for the 4 bit network is shown in Fig. 5.4.2 (*a*) using the expanded expressions. Note the structured nature of the gate arrangement. Once the carry-out bits have been calculated, the sums are found using the simple XOR in equation (*a*). The complete adder circuit is shown in Fig. 5.4.2 (*b*) where the 'CLA Network" box represents the carry bit logic in Fig. 5.4.2 (*a*). In this also as with ripple carry addition processing time is proportional to word width '*n*'.

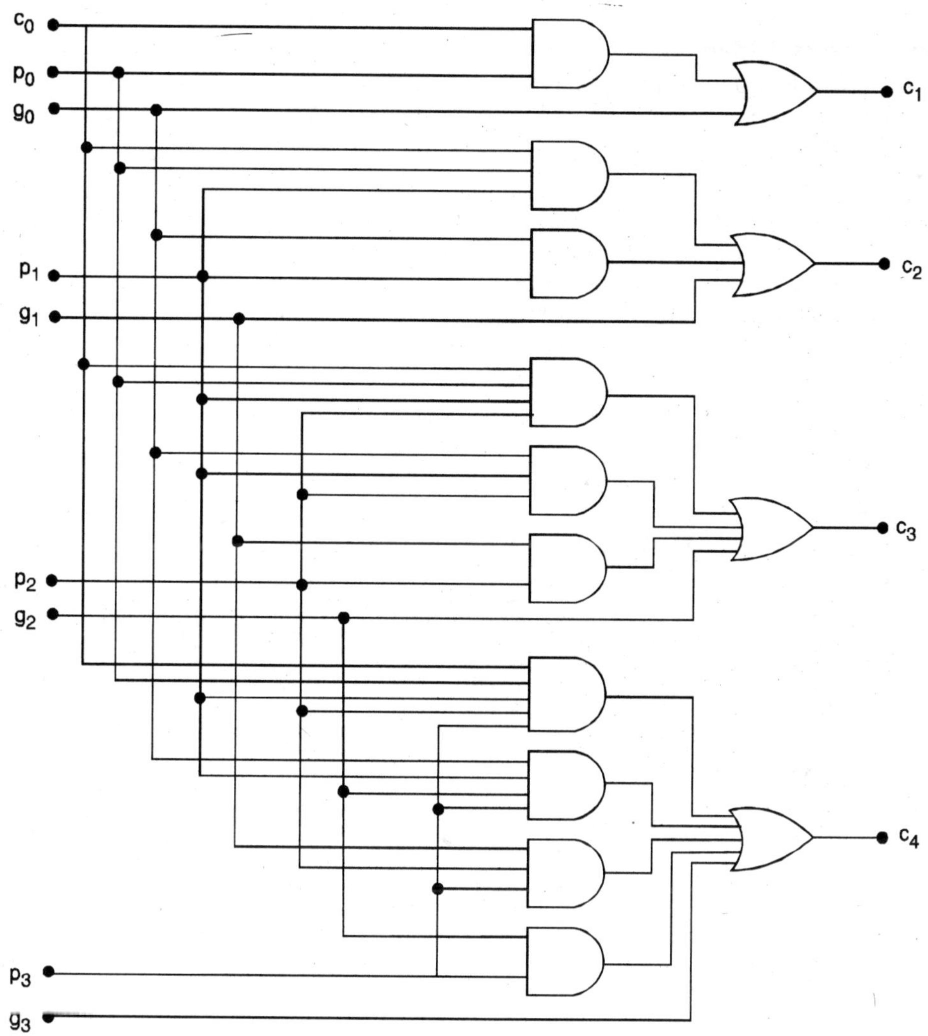

Fig. 5.4.2. (*a*) Logic network for 4-bit CLA carry bits.

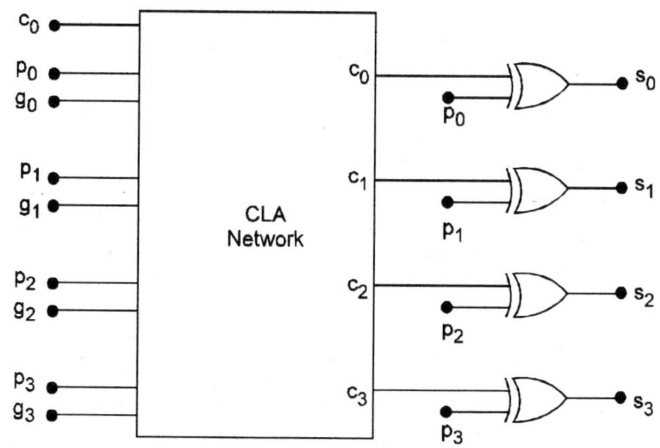

Fig. 5.4.2. (*b*) Sum calculation using the CLA network.

5.4.3. Carry-select Adder

Carry-select adders use multiple narrow adders to create fast wide adders. Consider the addition of two n-bit numbers with $a = a_{n-1} \dots a_0$ and $b = b_{n-1} \dots b_0$. At the bit level, the adder delay increases from the least significant 0th position upward, with the (n-1)th requiring the most complex logic. A carry-select adder breaks the addition problem into smaller groups. For example, the n-bit problem is split into two (n/2) bit sections. Then special attention is given to the higher order group that adds the word segments $a_{n-1} \dots a_{n/2}$ and $b_{n-1} \dots b_{n/2}$. The carry delay will then center around the carry-out bit $c_{n/2}$ produced by the sum of lower order word segments $a_{(n/2)} \dots a_0$ and $b_{(n/2)-1} \dots b_0$. There are only two possibilities for the carry bit :

$$c_{n/2} = 0 \text{ or } c_{n/2} = 1$$

A carry-select adder provides two separate adders for the upper words one for each possibility. A MUX is then used to select the valid result.

As an example, consider an 8-bit adder that is split into two 4 bit groups. The lower-order bits $a_3a_2a_1a_0$ and $b_3b_2b_1b_0$ are fed into the 4 bit adder L to produce the sum bits $s_3s_2s_1s_0$ and a carry-out bit c_4 as shown in Fig. 5.4.3. The higher order bits $a_7a_6a_5a_4$ and $b_7b_6b_5b_4$ are used as inputs to two 4-bit adders. Adder U_0 calculates the sum with a carry-in of c = 0., while U1 does the same only it has a carry-in value of c = 1. Both sets of results are used as inputs to an array of 2 : 1 MUXes. The carry bit c_4 from the adder L is used as the MUX select signal. If c_4 = 0. then the results of U_0 are sent to the output, while a value of c_4 = 1 selects the results of U1 for $s_7s_6s_5s_4$. The carry out bit c_8 is also selected by the MUX array.

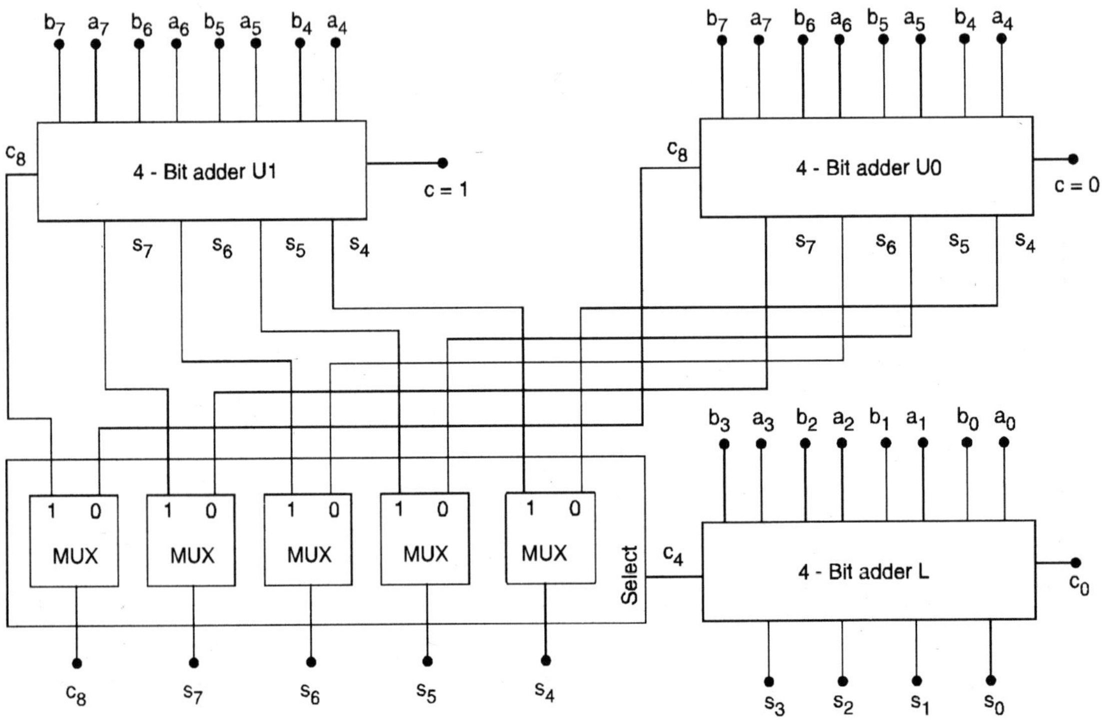

Fig. 5.4.3. 8-bit carry select adder.

This design speeds up addition of the word by allowing the upper and lower portions of the sum to be calculated simultaneously. The price paid is that it requires an additional word adder, a set of multiplexers, and the associated interconnect wiring. The design becomes viable if speed is more important than area consumption. Carry-select adders can be created using multiple levels, but the hardware costs increase accordingly.

5.4.4. Carry Save Adders (CSA)

The bit parallel adders handled upto this point are carry propagate adders (CPA) (where carry signal is evaluated to determine the sum.) For summation of large number of operands Carry Save Adders are used. The name 'carry save' arises from the fact that we save the carry-out instead of using it immediately to calculate a final sum. Here carry signals are not used for current addition, but used for its successor.

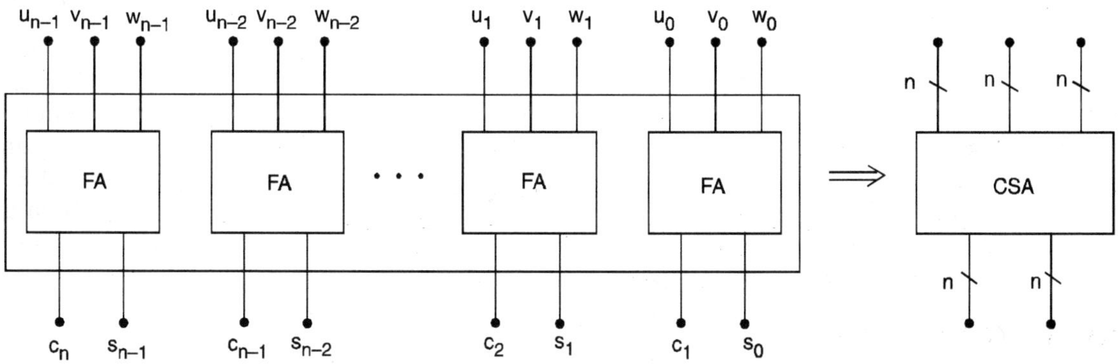

Fig 5.4.4. (*a*) n-bit carry save adder.

Fig. Shows 'n' bit CSA with 'n' FA to merge three operands U, V, W produce 'n' sum bits 's_i', 'n' carry bits 'c_{i+1}'.

S_i constitute sum vector = $\{s_{n-1} \ldots s_1, s_0\}$ S

C_{i+1} constitute carry vector $\{c_n \ldots c_2, c_1\}$ C [shifted by one position)

Then S and C are again added in a CPA to have the complete result. [n+2 bits are necessary to represent the sum of three operands with 'n' bits]. Such an adder is called Vector merging adder.

$$U + V + W = S + 2C$$

A multi operand adder, CSA are used as long as possible until the final step when just a sum and carry vector remain. Then, these two are merged in vector merging adder.

The particular shift by one bit of intermediate result vectors must be observed at each stage. This is signified by arrows. When using 2's complement. negative numbers, the sign digits must be expanded accordingly to achieve correct results.

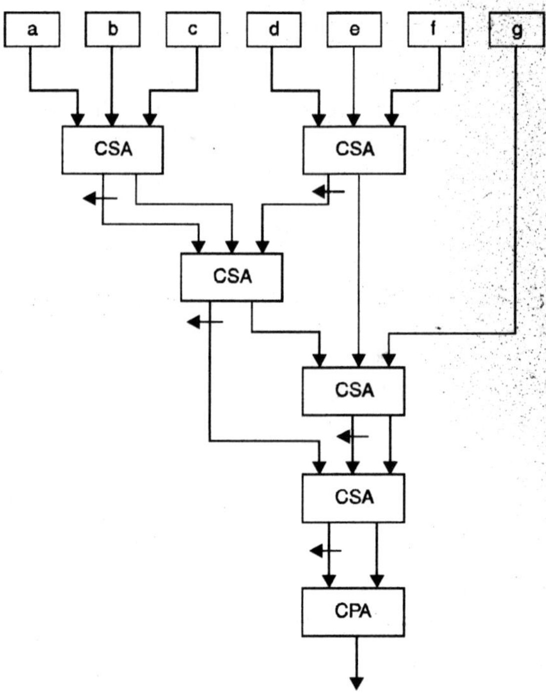

Fig. 5.4.4. (*b*) A 7-to-2 reduction using carry save adder.

5.5. ACCUMULATORS

While performing addition, for space saving, accumulators can be used. In this operands are read in sequentially and added to previous results.

$$S_A = \sum_{k=1}^{N} A_k$$

Accumulator based on CPA

(*i*) Register is cleared.

(*ii*) In N clock cycles, the operands A_k are read in and added to intermediate sum and finally stored in the REG.

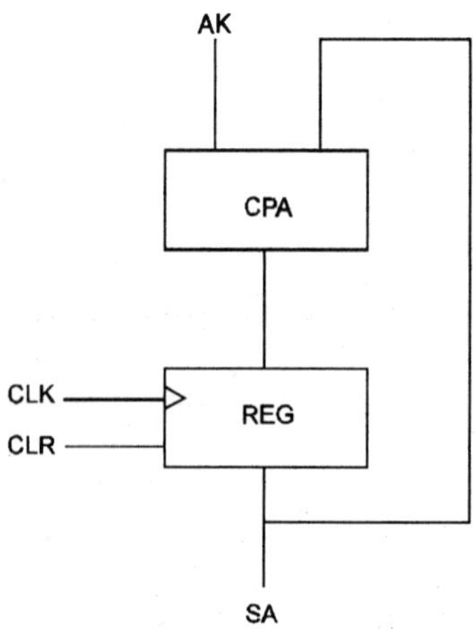

Fig. 5.5. (*a*) Accumulator based on CPA.

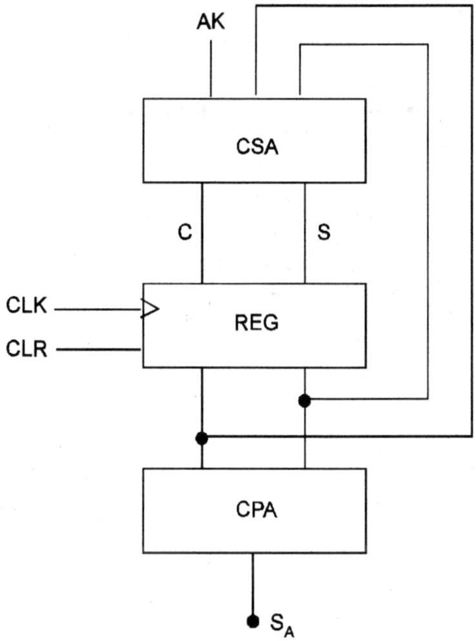

Fig. 5.5. (*b*) Accumulator based on CSA..

Accumulator according to CSA. Here intermediate results are 'S' and 'C' vector. Both should be stored in register and fed back for following addition. After N clock cycles the two vectors are merged in CPA.

Comparison

Accumulator based on CPA	*Accumulator based on CSA*
Low CLK frequency for accumulation delay is given by n bit CPA & D-FF.	High CLK frequency for accumulation because the delay that decides CLK frequency is given by one Full adder and D-FF
Requires lesser transistors	But due to additional CPA at the output and higher number of D-FF for registers, CSA requires more transistors.

5.6. MULTIPLIERS

Multiplication of two unsigned numbers A and B creates the product P = A.B.

A - Multiplicand (bit +ve whole number)

B - Multiplier (n bit +ve whole number)

Then numeric representation of P requires (m+n) bits

Value of product P $V(P) = V(A)\, V(B) = V(A) \sum_{j=0}^{n-1} b_j 2^j$

(*i*) partial products are formed by the product between bits a_j of A and bits b_j of B

(*ii*) Shifted versions of partial products are added to get product P. If $b_j = 0$ partial product $V_A b_j = 0$. otherwise it is equal to V_A.

$$V(P) = \sum_{i=0}^{m-1} \sum_{j=0}^{n-1} (a_j b_j) 2^{i+j}$$

If m = n = 4 partial product and final product is given by

			a_3	a_2	a_1	a_0		
			a_3b_0	a_2b_0	a_1b_0	a_0b_0	b_0	
		a_3b_1	a_2b_1	a_1b_1	a_0b_1		b_1	Partial products
	a_3b_2	a_2b_2	a_1b_2	a_0b_2			b_2	
a_3b_3	a_2b_3	a_1b_3	a_0b_3				b_3	
P_7	P_6	P_5	P_4	P_3	P_2	P_1	P_0	
Final Product								

5.6.1 Ripple Carry Array Multipliers

Partial product is implemented by AND. Then using adder cells computation of product can be implemented. Fig. Shows array multiplier based on Ripple carry adders. The function of each multiplier cell includes both the evaluation of partial products and their summation. Their function is given by

$S = ab \oplus S' \oplus C'$ $\quad\quad C = abs' + abc' + s'c'$

ab → bits of operand A and B

S′ → sum bit of preceding cell

C′ → carry bit of preceding cell

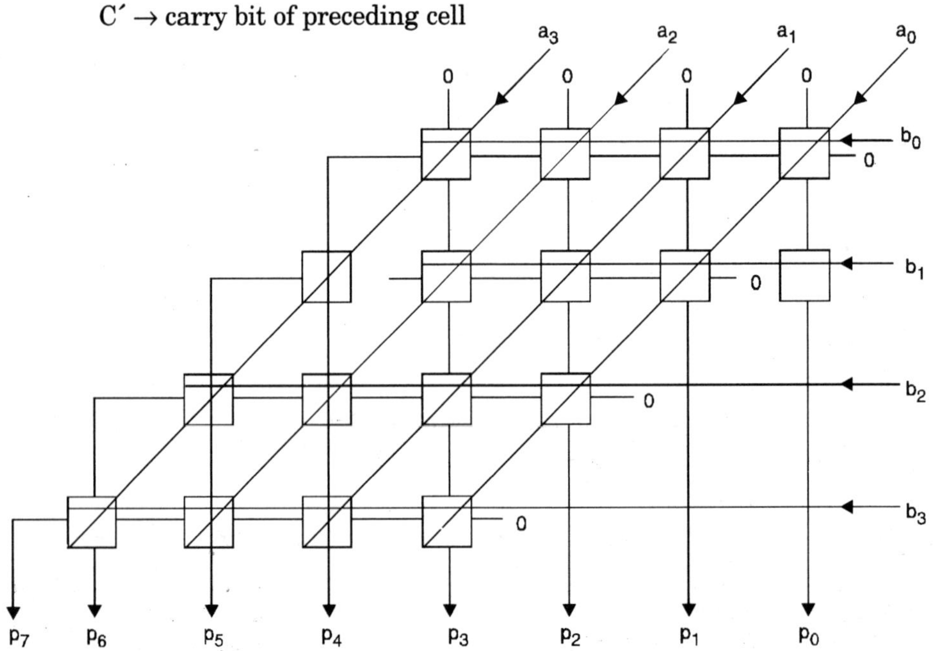

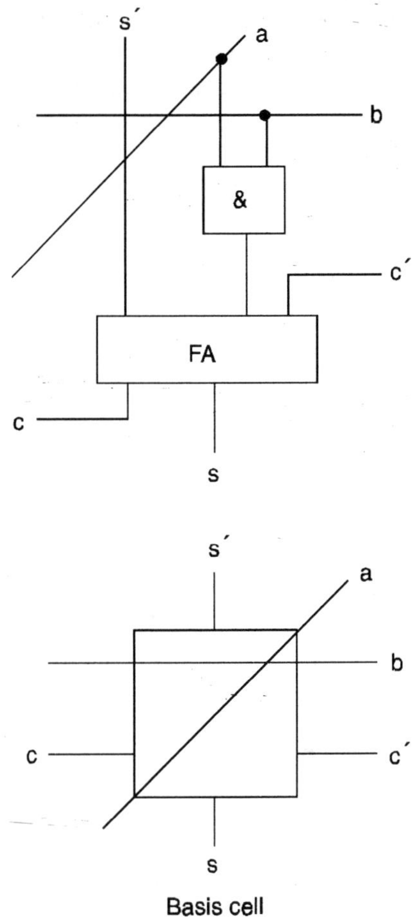

Fig 5.6.1. Ripple carry array multiplier.

5.6.2. Array Multipliers with Bit-pair Evaluation (Low Hardware Cost and Improved Delay Characteristic)

(*i*) Here two consecutive multiplier bits are simultaneously evaluated for creation of partial products.

$$V(P) = \sum_{j=0}^{n/2-1} V(A)(2b_{2j+1} + b_{2j})2^{2j}$$

For each bit pair $b_{2j+1}\, b_{2j}$ the following values can occur

$$2_{b2j+1} + b_{2j} \in \{0, 1, 2, 3\}$$

in addition to 0 and A , 2A and 3A also be available

2A is simple A shifted by one position

3A is computed in separate adder from (A + 2A)

partial products are created using bit-pair controlled MUX. Digit shifting corresponding to 2^{2j} is achieved by hardwire shifting of the bit levels. Partial products are combined to final product in CSA adder tree. Due to the operands that have been shifted by two bits relative to one another in the CSA adder, not all Full Adders, have three operand bits. Those with two operands reduce to a HA. Those with one operand are simply hard wired. There is no significant improvement in delay characteristic. because of lengthy computation of 3A. But the use of hierarchical CPA adders and special drivers for A and 3A enables a decrease in the delay.

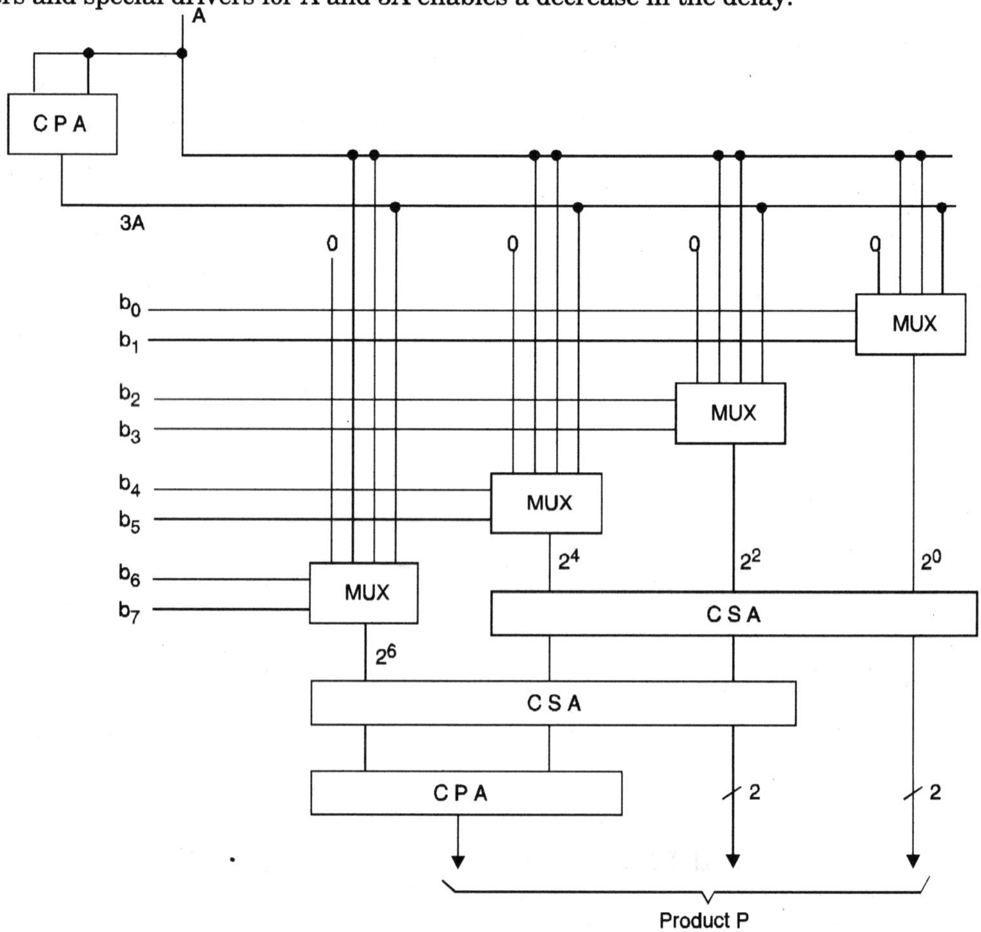

Fig. 5.6.2. Array multiplier with bit-pair evaluation.

5.6.3. Tree Based Multipliers

The speed of operation of multipliers is improved in tree-based multipliers. Also the power dissipation is reduced in tree-based structure. There are two types of multipliers based on this.

1. Wallace tree multiplier (working forward)
2. Dadda multiplier (working backward)

5.6.3. (*a*) Wallace Tree Multipliers

(*i*) all partial products are calculated in parallel

(*ii*) then added according to carry save principle.

(*iii*) final resulting carry and sum vectors are merged in CPA.

In a Wallace tree multiplier we work forward from the multiplier inputs compressing the number of signals to be added at each stage. The full adder can be viewed as a 3:2 compressor. The Fig. 5.6.3 (*a*) (*i*), (*ii*) shows a 6 bit Wallace tree multiplier. To form P_5 we must add 6 summands (S_{05}, S_{14}, S_{23}, S_{41}, S_{32}, and S_{50}) and 4 carries from the P_4 column. We add these in stages 1-7 compressing from 6 : 3 : 2 2 : 3 : 1 : 1.

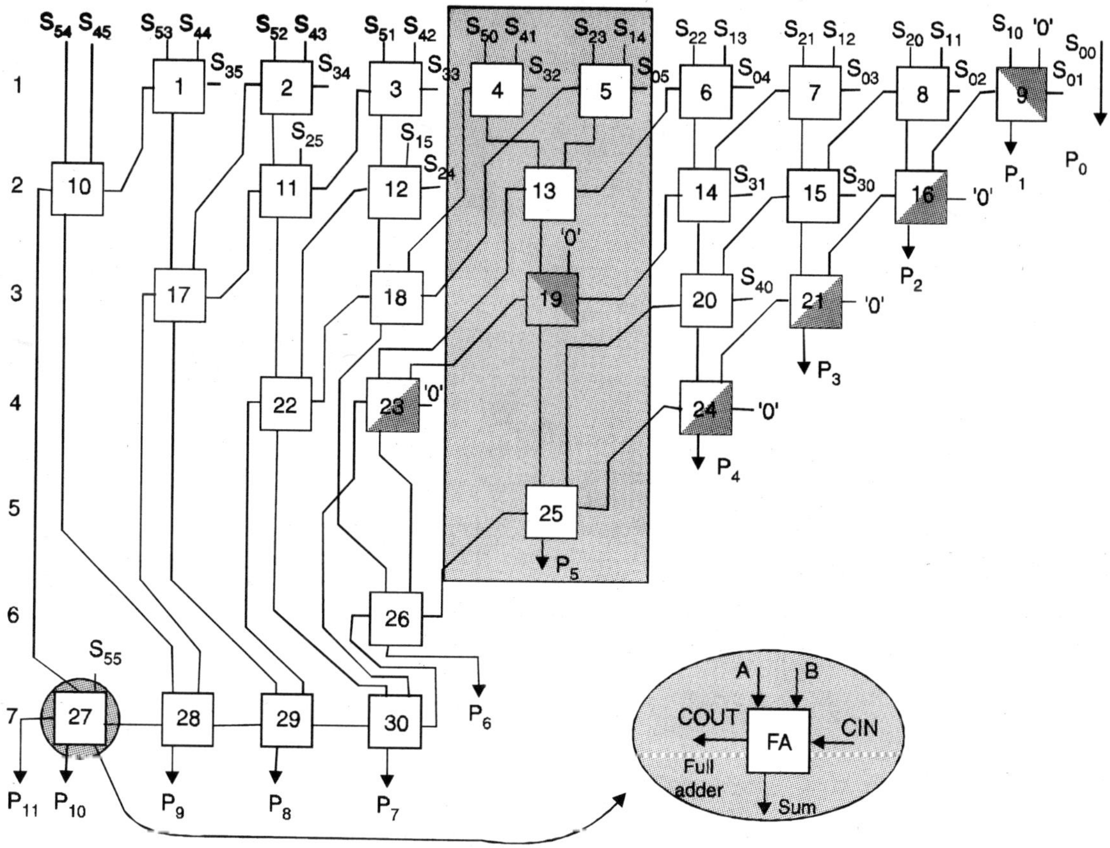

Fig. 5.6.3. (*a*) (*i*) A six bit Wallace tree multiplier.

In stage 5 we wait to get a carry from P_4. So we get an expansion (rather than compression) between stages 3 and 5. The maximum delay through the CSA array is 6 adder delays. (from 6 stages of CSA). Finally the delay of 7th stage which is the CPA is to be added. In total there are 22 full adders, 6 half adders all in CSA and 4 adders in CPA.

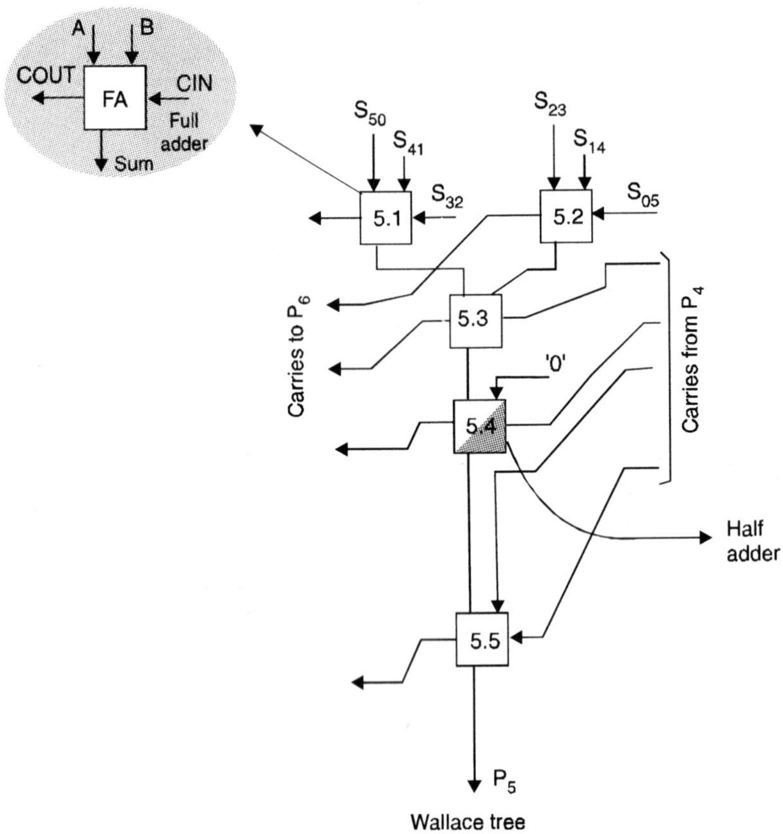

Fig. 5.6.3 (*a*) (*ii*) Sum bit chain P_5 of a Wallace tree.

5.6.3. (*b*) Dadda Multipliers

In this we work backward from the final product. Each successive stage is 3/2 times larger than the previous stage. [Fractions are rounded down to an integer]. The Fig. 5.6.3 (*b*) shows a 6 bit dadda multiplier. It requires only 3 stages of CSA. But there are 10 carry propagate adders. We have 15 FA and 4 HA in CSA, and 8 FA and 2 HA in CPA. This is somewhat faster and smaller than a Wallace tree multiplier.

The overall multiplier speed depends on the size and architecture of the final CPA. But, this may be optimized independently of the CSA array. So Dadda multipliers are always at least as fast as Wallace tree.

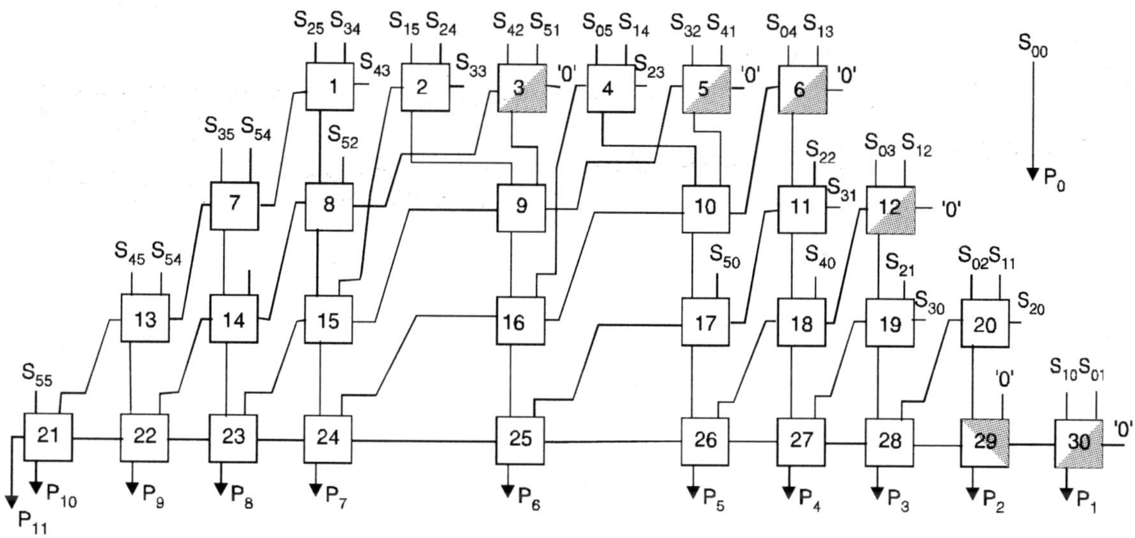

Fig. 5.6.3 (*b*). A six bit Dadda Multiplier.

5.7. ALU

This is the last important part of the data path. The ALU must provide arithmetic and logic operations on data furnished from the data path. The ALU accepts two operands performs a specified operation and outputs the result. A simple diagram of a 32 bit wide ALU is shown in Fig. 5.7.

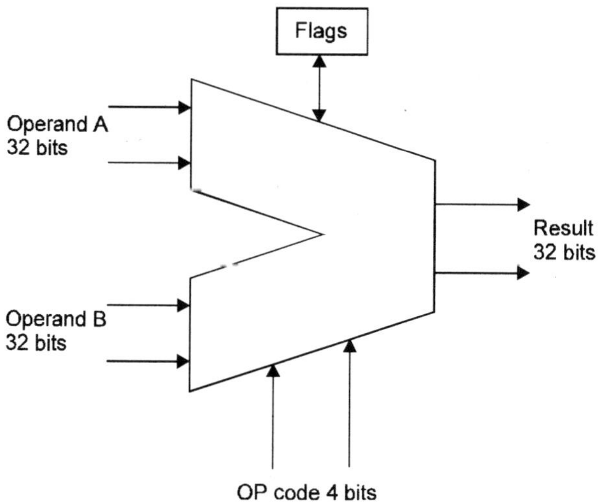

Fig. 5.7. Arithmetic Logic Unit.

For storing the 32 bit input operand, a dual port register array is required. Two parallel 32 bit buses should be provided in the data path between the register array and the ALU. So both inputs can be received simultaneously by the ALU. Again the same bus can be used to return the result evaluated by the ALU to the register array.

Normally an ALU in a microprocessor is expected to perform the following arithmetic and logical operations.

1. Addition with and without carry.
2. Subtractions with and without borrow.
3. Logical ANDing, ORing.
4. Comparing between two values.
5. Multiplication.
6. Shifting the datas.

For performing the above operations the basic units required are :

1. Adders
2. Subtractors
3. Gates
4. Multipliers
5. Comparators
6. Shift registers. The detailed study of all the above units have been already done.

6

PLDs and Memories

INTRODUCTION TO PROGRAMMABLE LOGIC DEVICES (PLDS)

A wide range of standard ICS are available in the market with a large number of logic functions and logic arrangements in a chip. Inspite of the ease in availability of the numerous type of ICS with large number of function on the chip, some design require hundreds and thousands of such ICS to be interconnected. This large number inter connections of ICS have inherent difficulties. For example a large circuit board space is required and cost of testing and soldering are high. The development of PLDs has come out with the solution to replace a large number of ICs with a single IC. A PLD is an IC having a large number of gates, FFs and registers internally interconnected on the chip. These connections are in the form of programmable fuses. The devices are programmable as few of the fuses can be broken selectively and others can be made intact. Even the users can blow the fuses as per the functions' requirement. This process is called programming.

Basically PLD consists of 2 types of arrays —

an array of AND gates for AND logic and

an array of OR gates for OR logic.

These arrays are programmably inter connected to produce the desired logic functions. The AND array produces product terms where as the OR array produces the sum of these product terms. We can program AND and OR array to get any desired function. Once all of the outputs have been programmed, the device will permanently generate the selected output functions.

6.1. PROGRAMMABLE LOGIC ARRAYS (PLA)

The implementation of logic function plays a key role in the design of digital system. One of the method in implementing the logic functions is to use a PLA. A PLA with 'n' inputs and 'm' outputs can realize m functions of n variables. In this we use two arrays. 1. AND array 2. OR array. Both are programmable.

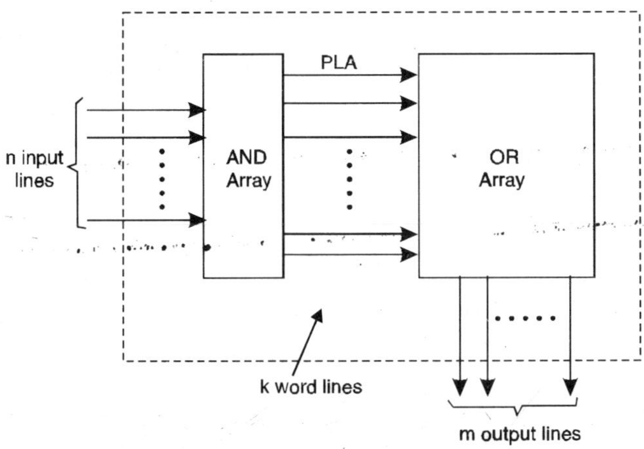

Fig. 6.0. Programmable logic array structure.

The AND array is used to realize the product terms of the input variables. The OR array ORs together the product terms needed to form the OUTPUT functions *i.e.* for generating SOP.

6.1.1. General Architecture of PLA

Consider a 3 input, 3 output PLA. With these three inputs and their complements, we can generate many number of product terms in AND array. Then, the different product terms are combined to get the SOP (sum-of-product) or logic function at the output. For getting 3 Outputs 3 OR gates are used. The input to the OR gates are product terms. Depending upon the number of product terms present in a logic function we can select a two input, or 3 input or more input OR gates.

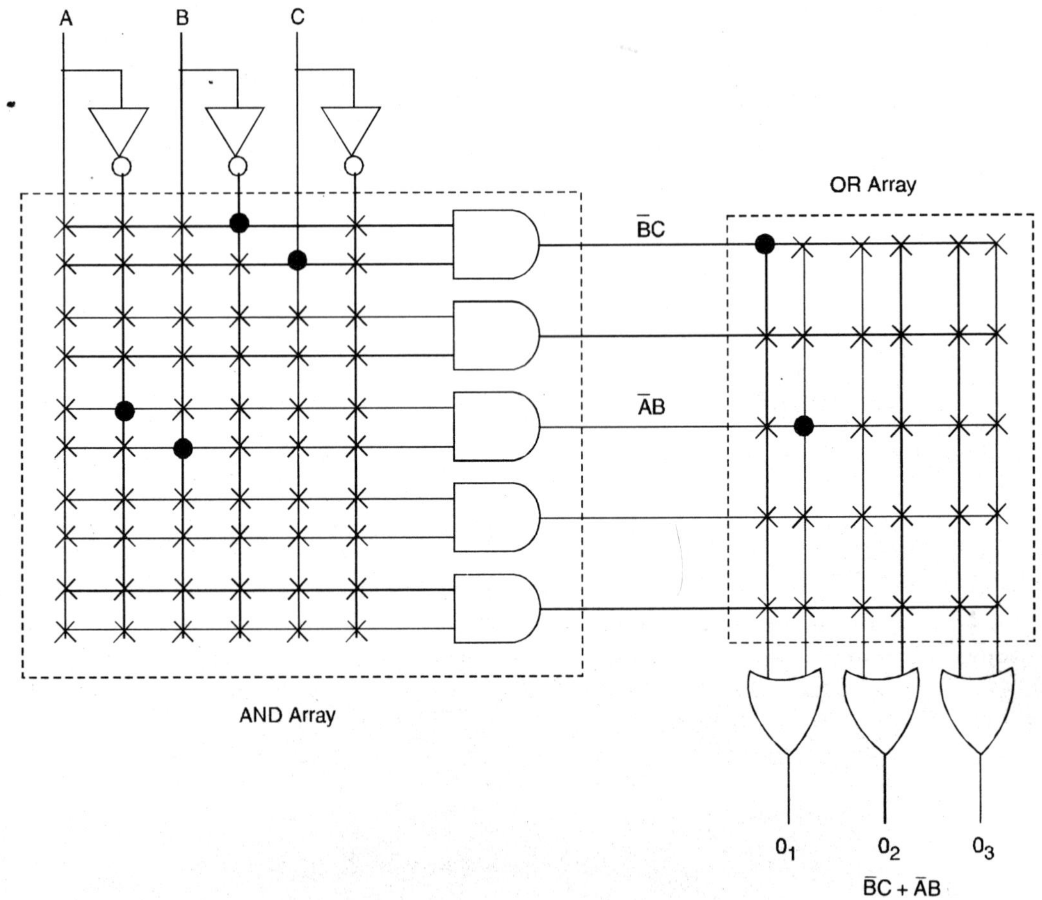

Fig. 6.1.1. Architecture of PLA.

Let the 3 Input variables be denoted as A,B, and C. And let the Output be denoted by O_1, O_2, and O_3. The architecture of PLA is shown in Fig. 6.1.1. In PLA both AND and OR array are programmable. The symbol 'X' shows there is no connection. The desired product terms can be obtained by making a connection at relevant position in AND array. Similarly the desired SOP or logic function can be obtained by making a connection at relevant position in OR array. For *e.g.* in order to realize the logic function $O_1 = \bar{B}C + \bar{A}B$ the connections are made as shown in Fig. 6.1.1. Note that the dark circles represent connections.

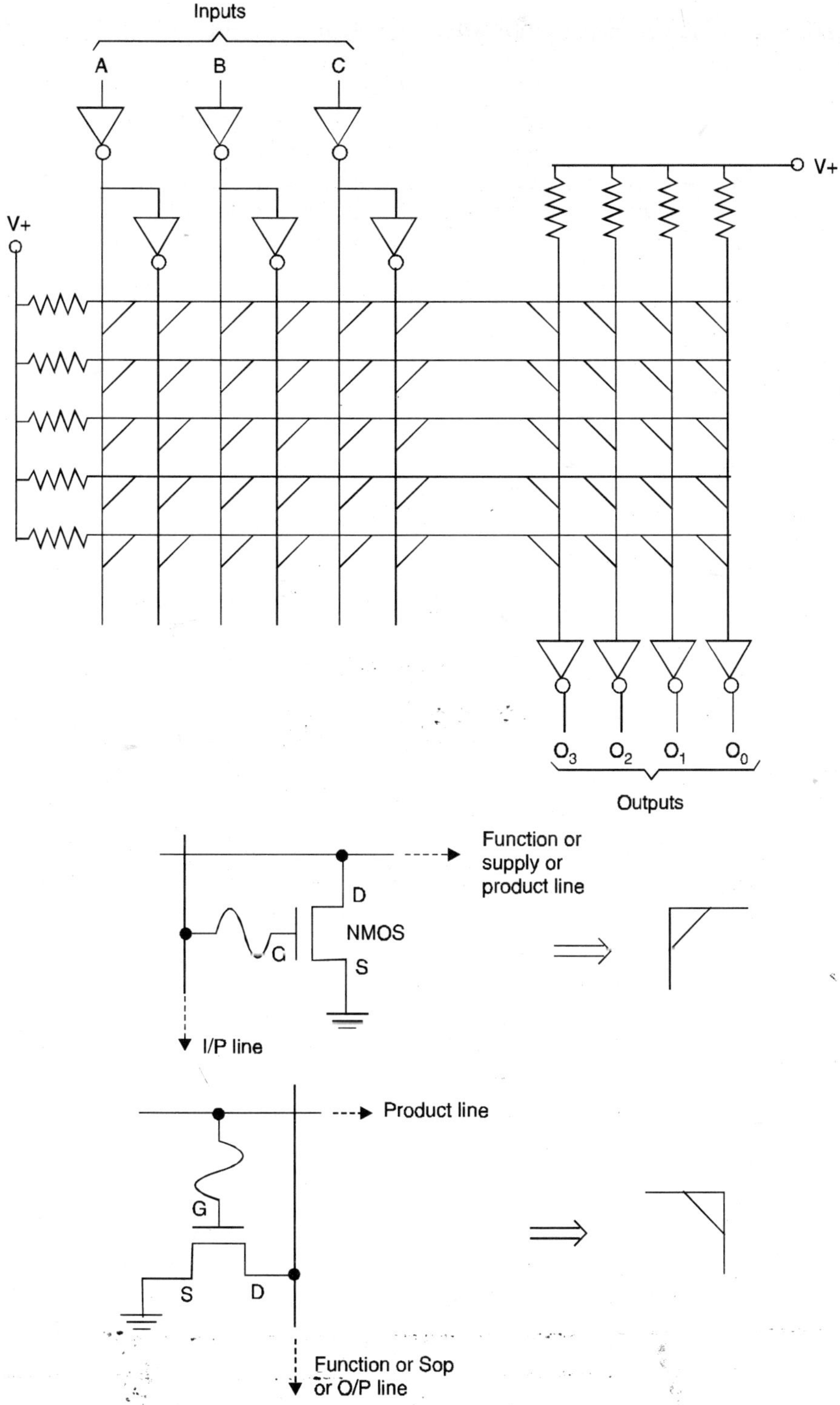

Fig. 6.1.2 (*a*). NMOS PLA architecture

The Fig. 6.1.2 (*a*). Shows NMOS PLA architecture. It has 3 input line and 4 output lines. From 3 inputs we get complemented and uncomplemented form. The input lines are connected to AND line through fuses and NMOS PLA. NMOS PLA is open only when its gate is given with '1', otherwise it is closed. Drain is connected to V^+ and source is connected to Gnd. If the fuse in series with gate is blown off then the corresponding input has no effect on AND line. When the fuse is intact the input has effect on AND line. For *e.g.* if input is '1', the NMOS is ON, it will pull the AND line to OFF. Next if input is '0', NMOS is OFF. So A '1' is sent through the AND line. This is similar to a NOR operation. To have AND function from NOR we insert an inverter at all inputs. This will function as an AND gate.

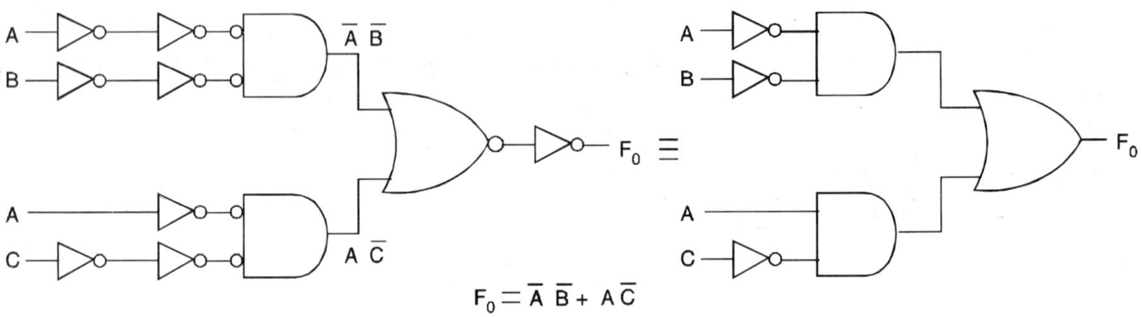

Fig. 6.1.2 (*b*) AND-OR from NOR-NOR.

Similarly for OR array also we have one NMOS in series with fuse at all the intersecting points of input line and product line. Here also its operation is similar as before and it also exhibits a NOR function. At the output we put one inverter to get OR function, from NOR which is the desired one. So the required AND – OR combination is obtained.

The Fig. 6.1.2 (*a*) shows unprogrammed NMOS PLA structure having 3 inputs, 5 product terms and 4 outputs. It can be programmed by blowing the corresponding fuses which is in series with NMOS. When fuses are blown link is cut and input has no effect on output. When link is intact output depends on corresponding input.

The following *e.g.* Explains how logic functions can be realised using NMOS PLA.

e.g. Realise using NMOS PLA the following :

$F_0 = \Sigma m(0, 1, 4, 6)$, $F_1 = \Sigma m(2, 3, 4, 6, 1)$, $F_2 = \Sigma m(0, 1, 2,6)$, $F_3 = \Sigma m(2, 3, 5, 6, 7)$

$F_0 = \Sigma m\ (0, 1, 4, 6)$

$\bar{A}\bar{B} + \bar{C}A$

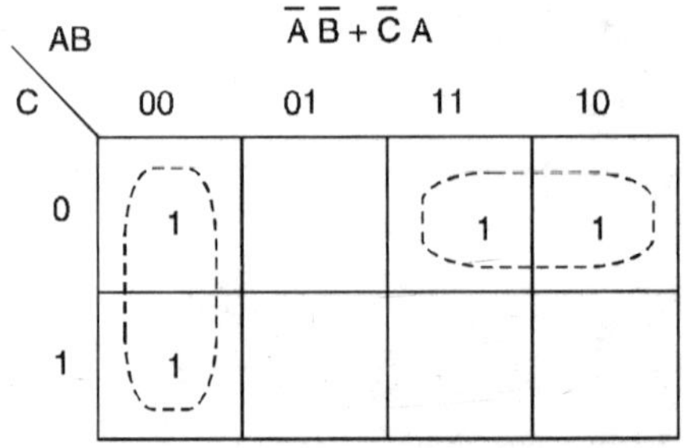

$F_1 = \Sigma m\ (2, 3, 4, 6, 7)$

$B + \bar{C}A$

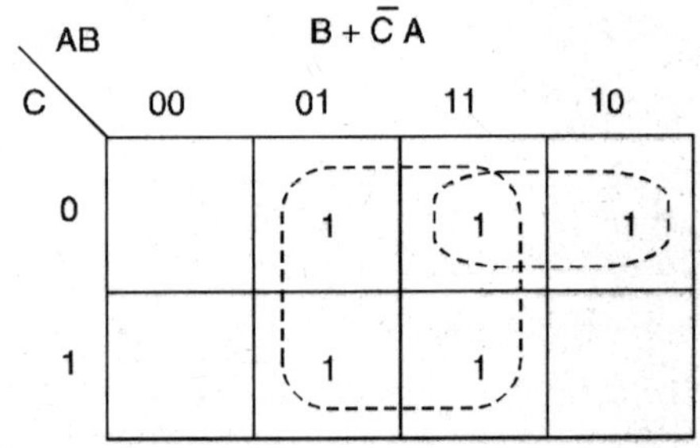

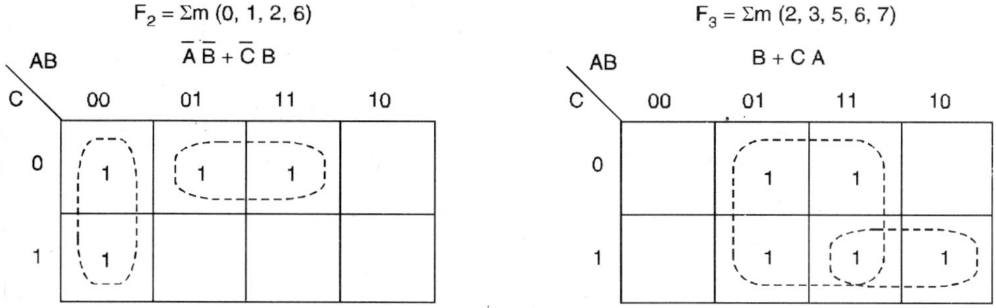

Fig. 6.1.3 (*a*) Karnaugh map.

6.1.3 (*b*) PLA Table

Product terms	*Outputs*
$\bar{A}\,\bar{B}$	F_0, F_2
$A\,\bar{C}$	F_0, F_1
B	F_1, F_3
$B\,\bar{C}$	F_2
$A\,C$	F_3

$F_0 = \bar{A}\,\bar{B} + \bar{C}\,A \qquad F_1 = B + \bar{C}\,A$

$F_2 = \bar{A}\,\bar{B} + \bar{C}\,B \qquad F_3 = B + CA$

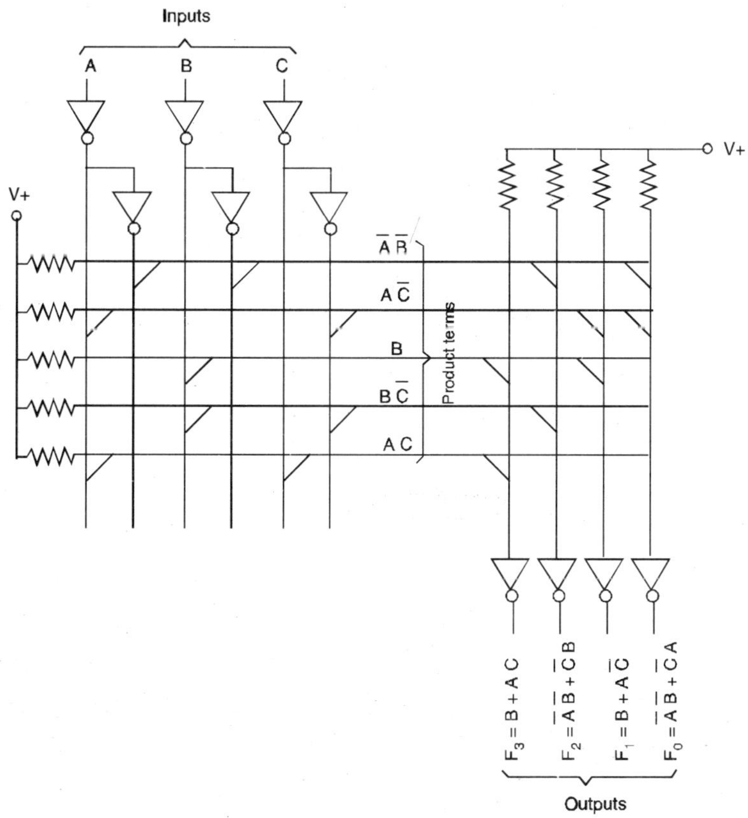

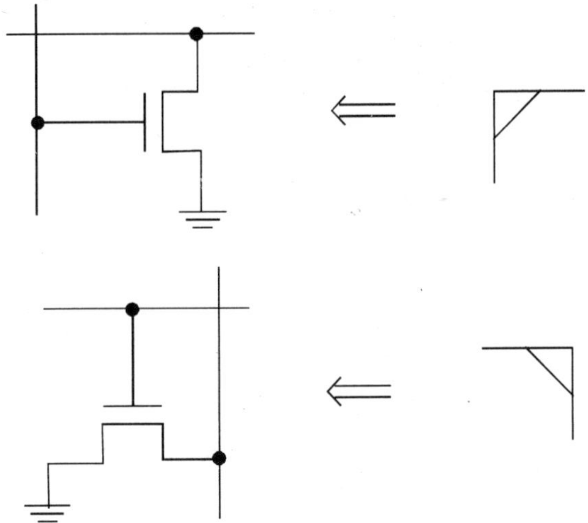

Fig. 6.1.4. Programmed PLA with 3 inputs, 5 product terms and 4 outputs.

In this both AND array and OR array are programmable. Wherever we want a connection, the particular NMOS is made intact using the fuses and others are disconnected from the circuit by blowing the fuses off. But, using NMOS we can realize only NOR function. In order to get AND function all inputs are inverted. Similarly to get OR function all outputs are inverted.

6.2. PAL AND PLDs

We have different kinds of PLDs. All have one AND array for generating product terms and one OR array for generating SOP terms.

In architecture

• – > Denotes hardwired

X – > Denotes fuse intact.

6.2.1. PLD PROM

In this AND array is hard wired and OR array is programmable. Let us consider a PROM with 3 inputs, 8 product terms and 4 outputs. Here as AND array is hard wired in such a manner that each of the 8 AND gates will generate each one of all the possible combination of inputs. This finds its application where all combinations of input variables are required for producing output functions. *e.g.* code converters and data storage table.

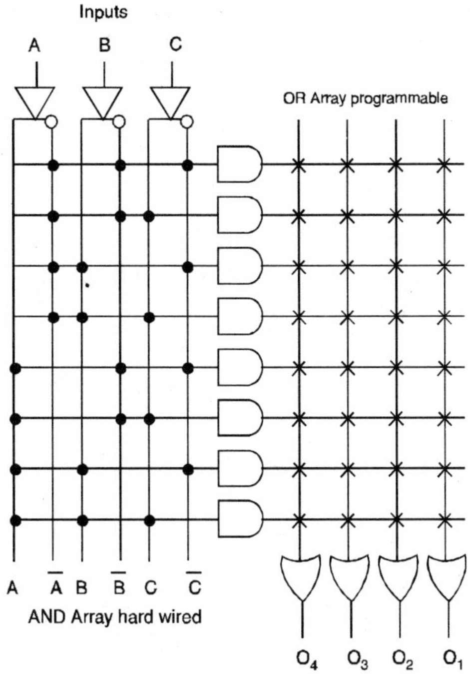

Fig. 6.2.1 (*a*) PROM architecture.

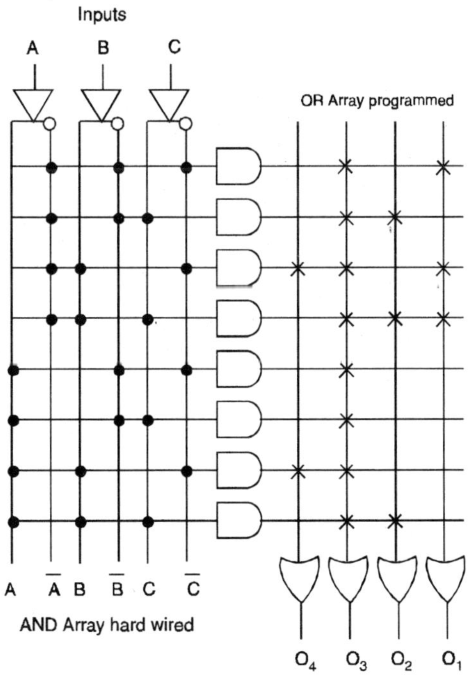

Fig. 6.2.1 (*b*) Programmed PROM.

The desired inputs or product terms for OR inputs can be got by blowing the unwanted links. When the link is intact output depends on input. When the link is open input has no effect on output.

For example, if we want to realize

$$0_1 = \bar{A}\,\bar{B}\,\bar{C} + \bar{A}\,B = \bar{A}\,\bar{B}\,\bar{C} + \bar{A}\,B\,C + \bar{A}\,B\,\bar{C}$$

$$0_2 = ABC + \bar{A}\,C = ABC + \bar{A}\,C\,\bar{B} + \bar{A}\,\bar{C}\,B$$

$0_3 = 1$ [comprises of all 8 input combinations]

$$0_4 = A\,B\,\bar{C} + \bar{A}\,B\,\bar{C}$$

Then the PROM can be programmed as shown in Fig. 6.2.1 (*b*).

6.2.2. PAL – Programmable Array Logic

In this AND array is programmable and OR array is Hard wired. It is used in applications where we don't require all the possible combination of inputs to generate output.

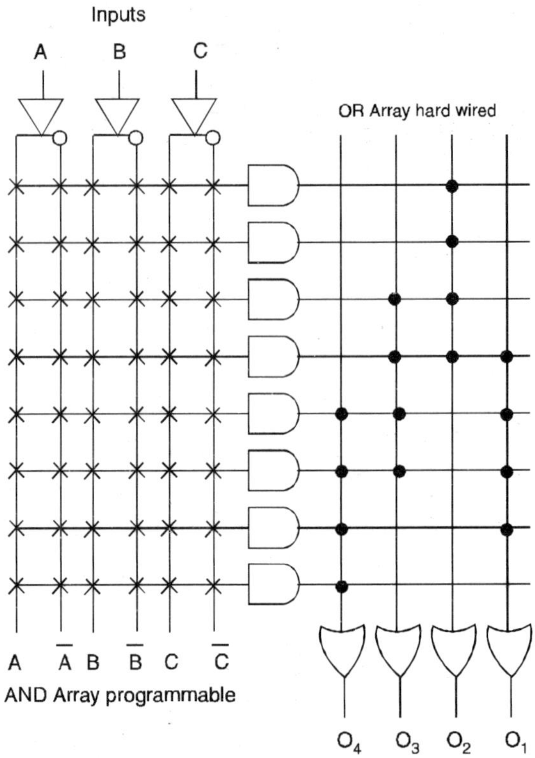

Fig. 6.2.2 (*a*) PAL architecture.

Here we can see that each OR gate has only 4 inputs. So a function requiring more than 4 product terms cannot be realized using this. But if we require less than 4 product terms for output, then the unwanted one can be made '0'.

For example, suppose we want to realize.

$$0_1 = \bar{A}\,\bar{B}\,\bar{C} + \bar{A}\,B\,C$$

$$0_2 = ABC + \bar{A}\,\bar{B}\,C + \bar{A}\,\bar{B}\,\bar{C} + \bar{A}\,B\,C$$

$$0_3 = \bar{A}\,B\,C + ABC$$

$$0_4 = \bar{A}\,\bar{B}\,C + \bar{A}\,\bar{B}\,\bar{C}$$

The above equations can be programmed as shown in figure below :

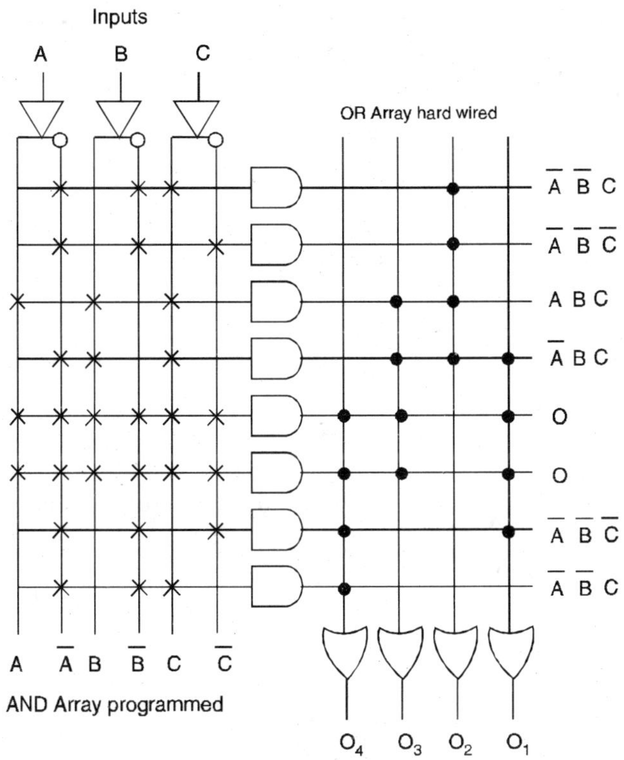

Fig. 6.2.2 (*b*) Programmed PAL.

Here if we want to have '0' product term, all the fuses at the input of AND gates should be made intact as $A\ \bar{A}.B\ \bar{B}.C\bar{C} = 0$

6.3. FOLDED PLAs

The improved form of standard PLA is called a folded PLA. The folded PLA is used under the following conditions.

1. The different product terms available in SOP equations should be functions of disjoint sets of input variable. In the following example. The product terms of 0_1, 0_2, 0_3 [A B $\bar{C}$, A $\bar{B}$ C...] are functions of input variables A, B and C, whereas the product terms of 0_4, 0_5 are functions of D, E and F only. Now the disjoint input sets can be spatially segregated. So, it is possible for two distinct input terms and their complements to share the same AND plane columns. Also inputs A, D share same AND plane column. Similarly EB and CF also. Now the width of PLA is reduced because of the reduction in number of AND plane columns. This is called AND plane folding.

2. The different output terms should be functions of disjoint sets of product terms.

In the Example outputs 0_1, 0_2 , 0_3 are functions of product terms A B $\bar{C}$, $\bar{A}$ B C, $\bar{A}$ B. These disjoint product terms can be spatially segregated. So, it is possible for two distinct output terms to share the same OR plane column. This reduces the width of PLA and is known as OR plane folding. **Also these operations reduces the area required by PLA. But the folding of different** groups of variables interact and it is difficult to get optimal PLA folding. Heuristics are normally used to find a good folded structure for a PLA.

The following *e.g.* illustrates how folded PLA structure can be implemented.

Let A,B,C,D,E,F be input variables and $0_1, 0_2, 0_3, 0_4, 0_5, 0_6$ be output terms.

Let

$$0_1 = AB\overline{C} + A\overline{B}C + \overline{A}BC + ABC$$
$$0_2 = AB + BC + AC$$
$$0_3 = \overline{A}B + A\overline{B}$$
$$0_4 = \overline{D}E + D\overline{E}$$
$$0_5 = DEF + \overline{D}\,\overline{E}\,\overline{F}$$
$$0_6 = ABC + DEF + D\overline{E}$$

For implementing this using standard PLA it requires 6 input columns (if complement terms are included it requires 12 input columns.) 13 product terms and 6 output columns. See Fig. 6.3.1 Whereas folded PLA requires only 3 input columns and four output columns.

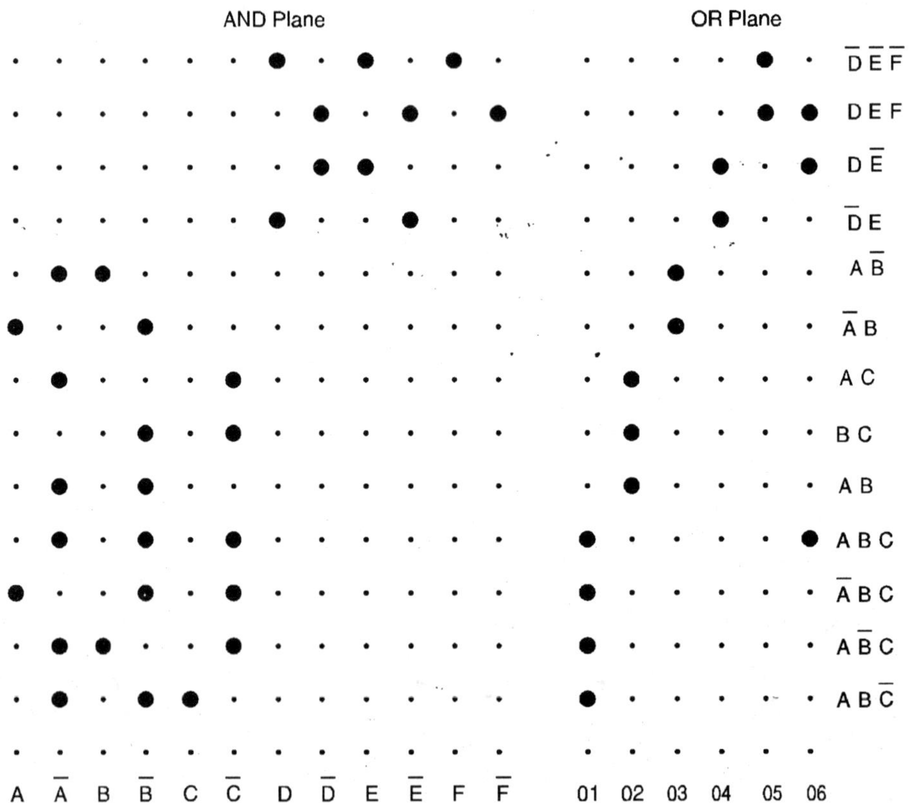

Fig. 6.3.1 Standard PLA implementation (6, 13, 6).

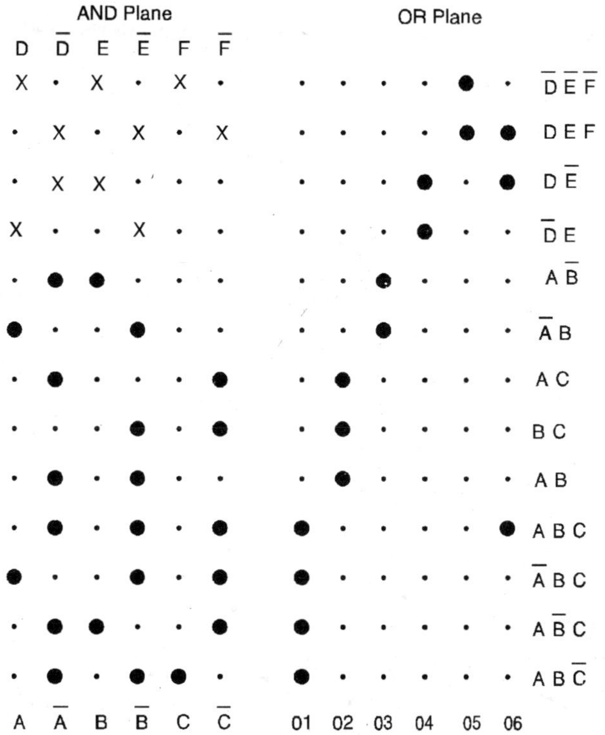

Fig. 6.3.2 Folded AND – plane PLA implementation (3, 13, 6).

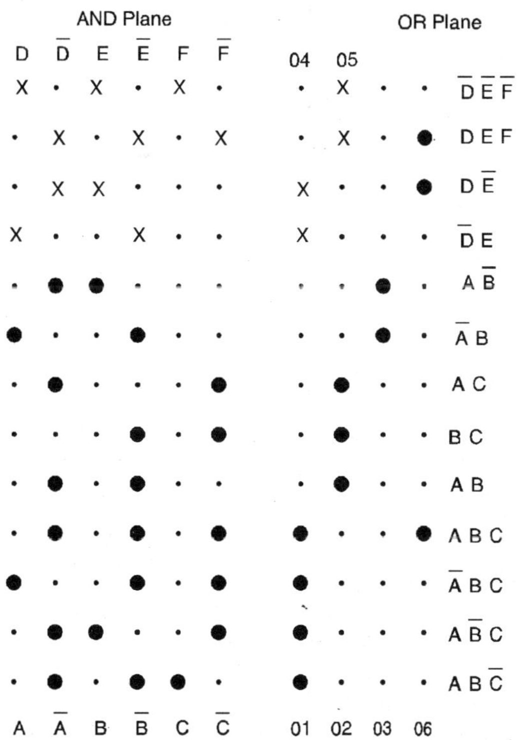

Fig. 6.3.3 Both AND and OR plane folded PLA implementation (3, 13, 4).

In Fig. '•' represent programming transistors associated with the lower input variables to the AND plane or the lower output terms of the OR plane. The 'X' represents programming transistors associated with the upper input variables to the AND plane or the upper output terms of the OR plane. See Figs. 6.3.2 and 6.3.3.

6.4. LARGE PLAs

PLAs offer an excellent mean to implement Logic equations. But the PLA structure becomes undesirable when it becomes very large. The reason is as follows:

1. As the size of a PLA grows, naturally number or interconnections in AND and OR planes will also increase. This in turn increases the interconnection capacitances which tends to slow down the operation of circuit. To compensate for the speed we can use large sized drivers, but this will lead to additional power consumption. Also higher area will be required.

2. In large PLAs most of the area go wasted as every output is likely to be a function of only few inputs.

So always it is desirable to group related logic signals into separate smaller PLAs rather than opting for one large PLA, through which we can enjoy a three fold advantage of reduction in area, power and delay.

6.5. FSM FINITE STATE MACHINE

The digital systems comprising combinational logic and memory in a combined form is called Finite State Machine. It can also be referred as a sequential machine. It is known that a sequential logic is nothing but a combinational logic with its output fed back to the input. To store the intermediate values a memory is required in feedback path. In digital systems, registers can be used as memory units. So a simple FSM can be realized by a PLA which serves as a combinational logic and a clocked shift register which serves as a memory in the feed back path.

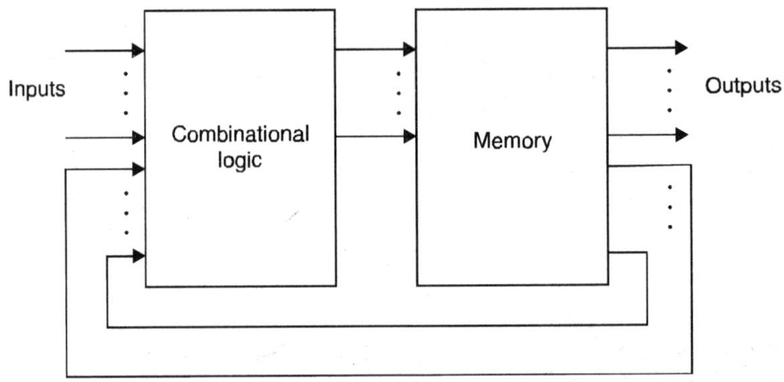

Fig. 6.5.1 Moore machine.

Fig. 6.5.1 shows the moore machine. This FSM is characterized by outputs that are isolated from momentary input changes by memory. The PLA based FSM is shown in Fig. 6.5.2. In this the PLA is augmented with pass transistors to gate its inputs and outputs. These pass transistors in combination with the output buffers and double-rail drivers form a clocked shift register. Now the next state in PLA OR plane is available as the present state at the inputs to the PLA double rail drivers after a ϕ_1, ϕ_2 clock sequence.

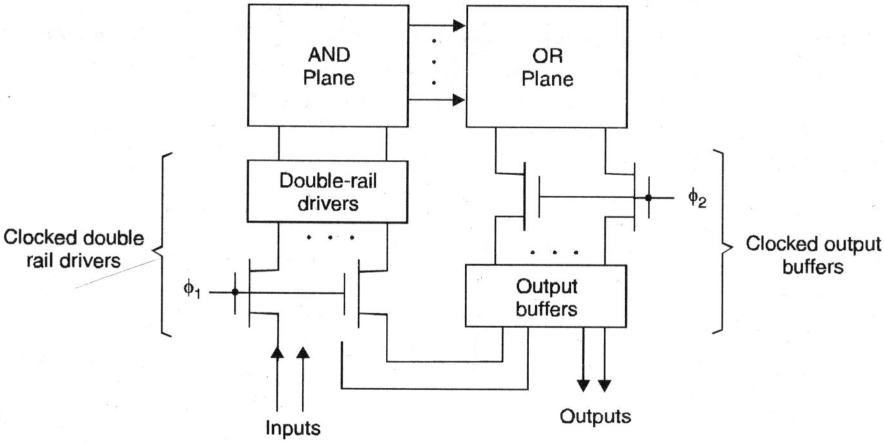

Fig. 6.5.2 PLA based FSM.

With automatic PLA generation programs a complete PLA layout can be created and programmed to realize correctly the specified logic functions. Then clocked input drivers and clocked output buffers can be included to form a clocked shift register. Such a PLA generator can be used in two modes

1. Simple PLA
2. A FSM (Finite State Machine)

State Machines (courtesy – old field)

State machines provide a formalism for sequential logic. Finite state machines (FSM), as the name implies, have a finite amount of memory. The next state of any FSM depends on its present state and its inputs. There are two types: the Moore, or state output machine, in which the output depends strictly on the present state, and the Mealy, or transition output machine, in which the output depends on both the current state and the inputs. A well-known example, the traffic light controller, is useful in bringing out the distinction.

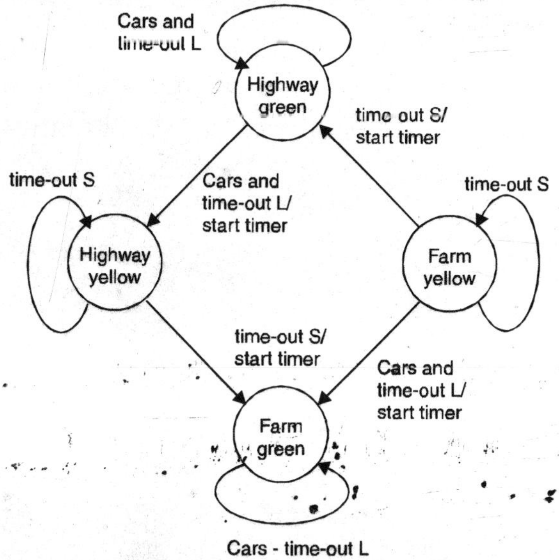

Fig. 6.5.3 State diagram for traffic light controller.

For example (*a*) A farm lane intersects a busy highway. There is a sensor to detect the presence of vehicles on the lane and traffic lights control both lane and highway. Fig. 6.5.3 shows the state diagram. Assuming there has been no farm traffic for a while, the detection of a vehicle on the lane will cause the highway traffic lights to switch from green to yellow, staying in yellow for a short time-say 8 seconds. Then they will change to red, while the farm lane lights display green. After a longer time-say 35 seconds-or if no cars are detected, the lights will show yellow for a short time to the farm lane while maintaining red to the highway traffic. Finally, the cycle completes by presenting green to the highway again. A longer time-out-35 seconds – prevents the cycle being repeated too frequently. The two time-outs, short and long, are produced by a separate timer. Note that it is started as a result of a transition, rather than presence in a state, and so the FSM is of the Mealy type. The outputs controlling the lights are directly associated with state, that is, they are of Moore-type, but overall the FSM is a Mealy machine.

PLA consists of combinational logic only. It can generate a FSM formed from a standard PLA plus clocked shift register feedback created by connecting the output of a clocked output buffer to the input of a clocked double-rail driver.

For example (*b*) To demonstrate the design of a FSM based on a PLA with clocked shift register feedback, consider the following example. (Courtesy – Geiger)

Assume that 16 magnetic switches are to be monitored remotely for a home security application. The 16 status bits corresponding to the state of the switches that are available from the remote location through an asynchronous serial data link as alternating bytes of data. The serial data is received in 8 bit groups by a UART (universal asynchronous receiver/transmitter) whose parallel output must be stored in two eight-bit registers that drive LED indicators. Each register is composed of 8 D-type flip-flops activated by a rising clock signal. The first byte of data received is displayed in one set of LED indicators, and the alternate byte is displayed in a second set of LED indicators. Thus, there are 16 LED indicators, one for each magnetic switch. This task can be accomplished with a sequencer (FSM) that alternately selects one of two display registers. A or B to store the received data bytes. To simplify the design, assume that the system is always synchronized with the first, third, and other odd bytes going to display register A and the even bytes to display register B. The FSM must also generate a data strobe signal (S) required by the UART to acknowledge that a byte is accepted. Of course, the UART generates a data ready signal (R) whenever a new status byte is available at its output. The logic components that compose the receiving system are shown in Fig. 6.5.4. Show the logical design for the PLA FSM block of this figure.

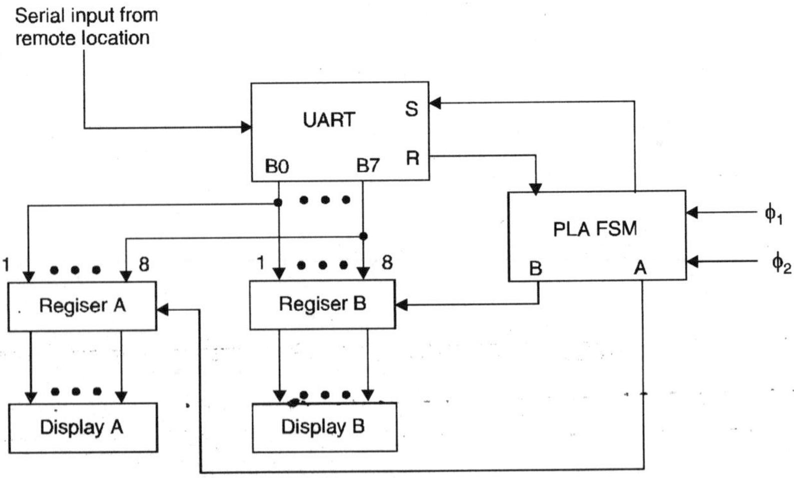

Fig. 6.5.4 Block diagram for system display.

Solution : A PLA FSM that satisfies these requirements is described by the state diagram of Fig. 6.5.5 and the state transition table of 6.5. The FSM waits in state '*a*' until the UART indicates that a data byte is ready by asserting the Data ready signal (R). Data ready causes a transfer to state '*b*' at the next clock, causing a load signal (A) to display register A and a data strobe signal (S) to the UART. It is important that the display register is loaded by the rising edge of load signal A. Then the FSM waits in state '*b*' until the UART removes the data ready signal (R) , causing a transfer to state '*c*'. At this point, one byte of data has been received and the value in register A updated. When a second byte from the UART is ready, the resulting data ready signal (R) causes a transfer to state '*d*' at the next clock, where the appropriate load signal (B) to display register B and data strobe signal (S) are generated. Later, after data ready is removed by the UART, the FSM returns to the first state and waits for new status data.

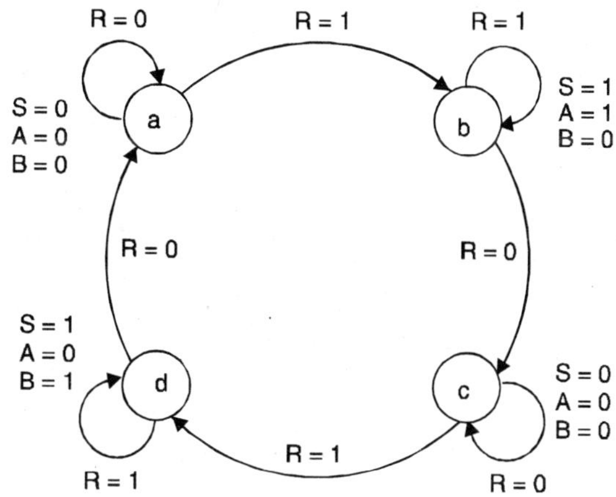

Fig. 6.5.5 State diagram for status sequencer.

Table 6.5. State Transition Table for Security Monitoring System

Input	*Present state*	*Next state*	*Outputs*		
R	*XY*	*xy*	*S*	*B*	*A*
0	00	00	0	0	0
1	00	01	0	0	0
0	01	11	1	0	1
1	01	01	1	0	1
0	10	00	1	1	0
1	10	10	1	1	0
0	11	11	0	0	0
1	11	10	0	0	0

The state transition table (Table 6.5) provides the information necessary to specify the logic operations to be performed by the PLA. Two state variables are required to specify four different states. Call the present state variables X and Y and the corresponding next state variables x and y. The states are encoded with a gray code (state a = 00, state b = 01, state c = 11, state d = 10) so that only one state variable changes for each state transition. The equations for the next state variables and the outputs are obtained from the state transition table and are given here.

$$x = \overline{R}Y + RX$$

$$y = \overline{R}Y + R\overline{X}$$

$$S = \overline{X}Y + X\overline{Y}$$

$$A = \overline{X}Y$$

$$B = X\overline{Y}$$

From these equations, it is easily determined that a (3, 5, 5) PLA is required-that is, three inputs by five product terms by the five outputs. Two outputs form the next state variables, while three other outputs generate the data strobe (S), load register A (A), and load register B (B) signals.

The ability to create a FSM automatically from a set of Boolean logic equations is an extremely powerful tool for digital system design. Small PLA based FSMs are frequently used as building blocks to construct larger digital systems such as microprocessors and communications processors. Large PLA based FSMs suffer from the two important limitations. A large PLA may be sparsely populated with programming sites, resulting in excessive area to realize a function. Also, large PLAs tend to be slower than alternative solutions when a large number of terms must be processed. One alternative is to use several small PLA FSMs rather than one large PLA FSM to implement required control logic;

For example (*c*) Draw the state diagram which can detect the sequence 1011.

The output should be equal to 1 when the sequence is detected. Otherwise it should be 0.

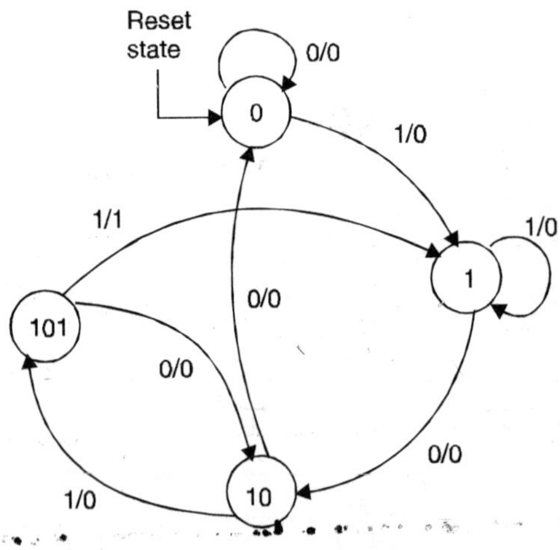

X/Y - Detected input/output produced

Fig. 6.5.6 State diagram of sequence detector.

For example (*d*) Obtain the state table from the given state diagram (Fig. 6.5.7) of a sequence detector.

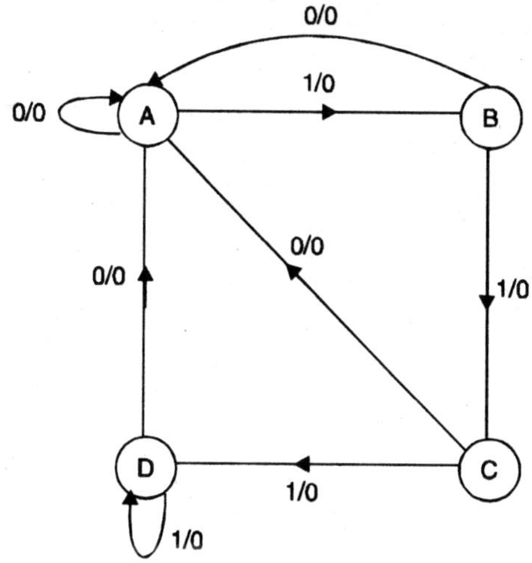

Fig. 6.5.7.

State Table

Present state	*Next state*		*O/P*	
	I/P '0'	*I/P '1'*		
A	A	B	0	0
B	A	C	0	0
C	A	D	1	0
D	A	D	0	0

6.6. MEMORIES

MOS Memory IC Technology

The MOS memory circuit has been the one IC application that accounts for most of the Ics produced. The design is simple. As the technology advances, the old processors may go obsolete. But the memory circuits can easily take up new technology advances.

Memory Organisation

Semiconductor memories are universally organized as arrays of single bit storage cells. The Fig.6.6.0 shows the block diagram of memory chip organization.

For having 64 memory locations, a memory array with 6 address lines are required. These 6 address lines provide selection of each of the memory locations. The memory array can be arranged in a linear manner, in which case a 6 to 64 line decoder would be required to allow individual selection of the bit addressed by the address lines. But, when the memory array is organized as a 8 row × 8 column square, two 3 to 8 decoders will only be required, one to select a row and a second to select a column. In the second type of organisation much area is saved and always it is preferable to have square organization of memory array.

The operation of the memory chip is as follows. Based on the output from row decoder, all the memory cells along that selected row will be enabled. The column decoder is used to select the particular column containing the desired data bit. This selected data bit will ultimately drive an output pin of the memory part. Some of the memory parts may be organized to provide access on n bits simultaneously. For this to happen, the data from n columns are to be selected and gated to n data output pins simultaneously.

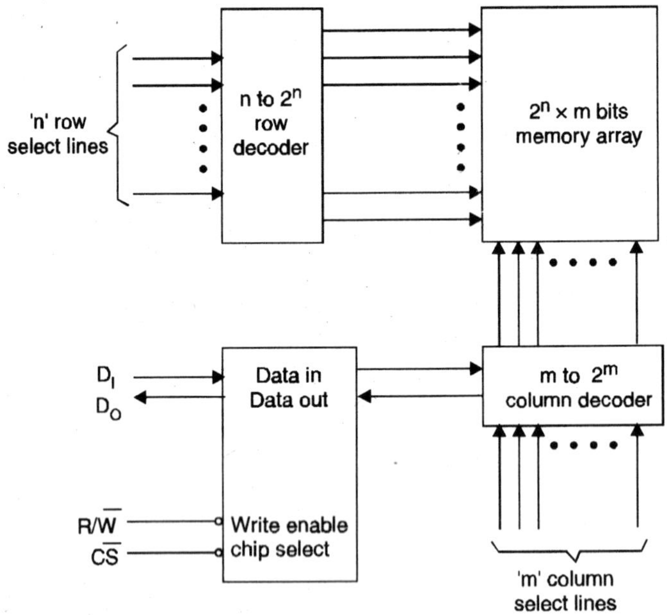

Fig. 6.6.0. Memory organisation.

Additional circuitry such as row address buffers, column address buffers and control logic unit together with row decoder, column decoder and memory array forms a complete memory chip. The size of memory storage cell and the resulting memory array determines the size of the complete memory chip. There are several different kinds of memory circuits. These include ROM, EPROM, EEPROM, SRAM, DRAM.

ROM. These are randomly addressable memories in which the data is stored permanently during the manufacturing process. ROMs usually have a write time much greater than their read time.

EPROM. The stored data can be erased through external means and these memories can be reprogrammed.

EEPROM. In this stored data is erased through electrical means. The stored data are non-volatile in ROM type memories.

RAM. Whenever required this allows data to be stored and accessed. They have similar read and write times. There are two kinds of RAM.

SRAM. This retains the data stored until the power is removed.

DRAM. This requires data in the cells to be periodically refreshed to recharge the stored bit of information.

6.6.1. Read only Memory – ROM

A ROM has n input lines or row select line and it has m output lines or column select lines. This ROM contains an array of 2^n words, and each word is m bits long. The row select line selects one of 2^n words. The column select line selects a particular bit from that word.

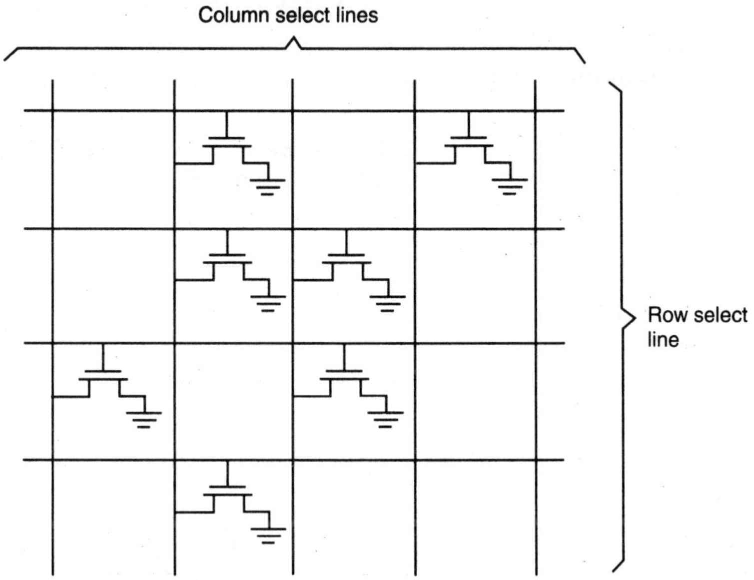

6.6.1. ROM memory array.

Storage cells are realized in the array by connecting *n*MOS switching transistor. This is shown in Fig 6.6.1. The transistors act as switches, so if the gate input is a logic 0, the transistor is OFF. If the gate input is a logic 1, the transistor in ON. Each storage cell transistor has its gate connected to a row line, its drain connected to a column line and its source grounded. Initially to all the column lines a positive V_{DD} is applied and all are pulled to a high level. Using a row decoder (n to 2^n decoder) an individual row line is selected and it is driven to a +ve Voltage. When a transistor is placed where the selected row crosses a column, that column will be pulled to a low level. Thus, each column will be high or low to reflect the stored data along the selected row line. The column decoder is then used to select the desired column or columns to provide the requested data.

The ROM memory contents are programmed by selectively placing transistor within the memory array. Placing transistor in row column intersection position is equivalent of storing a zero in that location.

6.6.2. Erasable Programmable ROM : (EPROM)

ROM is nonvolatile memory. It retains stored data even when power is off. This is a desirable characteristic in many applications. But modification of the data stored in ROM is often necessary during the development phases of a digital system,. EPROM fulfils this requirement. This is also a nonvolatile memory but it can be reprogrammed when it necessitates. Hence these memories are widely used in Microprocessor systems.

EPROM use a special storage mechanism to enable or disable the switching elements in memory array. It uses a unique storage cell consisting of transistors with two gates, one of which is isolated from the circuit and is called the floating gate. This floating gate some how acquires charge to represent a stored data value.

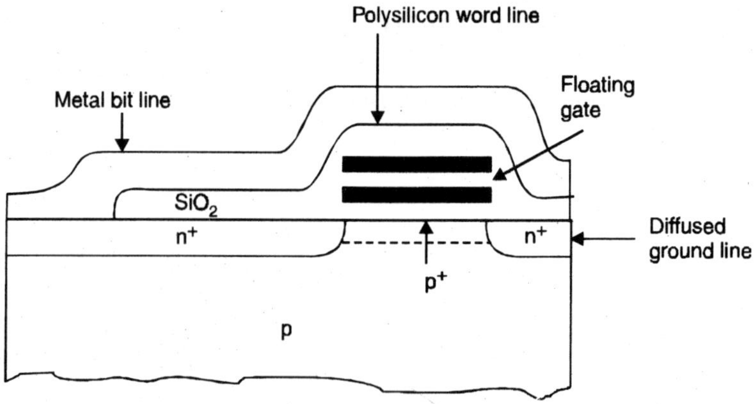

Fig. 6.6.2. EPROM storage cell.

The charge remains there for a long period of time upto 10 years. The presence or absence of the stored charge represents stored data. In order to store a new data a high voltage is applied to floating gate substrate region. Now avalanche injection of electrons take place on to the gate. So some amount of charge will be placed on the gate. Data is erased by removing the charge from the gate. This is achieved by exposing the floating gate to strong UV light of wave length 2536Å for a definite period of time and hence charge is removed from floating gate. Erase time is around 20 to 30 minutes. ROM is economically feasible if a large quantity is required. If only a small quantity of ROMs are required with a given data array EPROMs may be used.

6.6.3. Electrically Erasable Programmable ROM: (EEPROM)

It is similar to EPROM, except that erasure is a accomplished using electrical pulses instead of UV light. This can be erased and reprogrammed only a limited number of times. The erasure are done in two simple steps.

1. Instead of applying a high voltage externally, a charge pump circuit is used here. It generates the high voltage, which is necessary for programming from the standard 5V supply.

2. The erasure is achieved by allowing the charge to be removed from the floating gate which is inturn achieved by reversing the electron injection phenomenon.

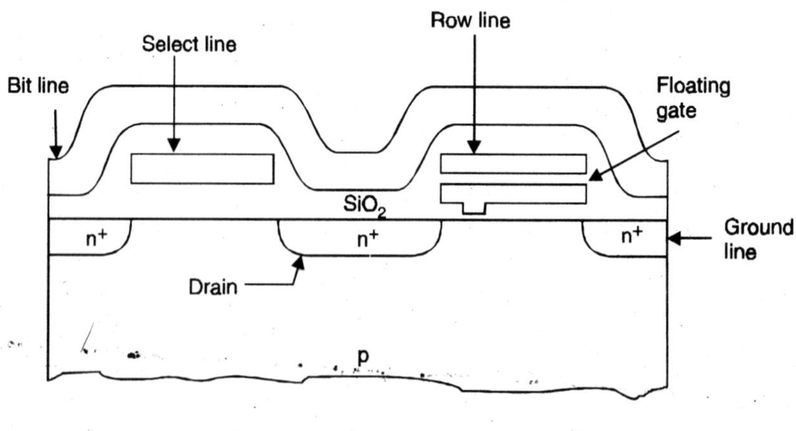

Fig. 6.6.3. EEPROM storage cell.

The basic memory cell of an EEPROM consists of a memory transistor and a select transistor. The memory transistor consists of two gates among which the bottom one is left floated. A small thin oxide isolates the floating gate and provides an injection area for electrons to and from the floating gate. Sending a short, high voltage pulse to the row line while grounding the drain causes tunnelling of electrons from the drain to the floating gates. Now erasing is achieved. Similarly grounding row line and applying pulse to drain causes tunnelling of electrons from the floating gate to the drain. Now storage is achieved. The additional select transistor per bit is used to realize a memory array. It avoids disturbing unwanted cells during erase or storage. As a result EEPROM memory is not as dense as EPROM memory.

The ROM is very dense, using only a single transistor as a storage cell; the EPROM is less dense, using a single transistor with select gate and a floating gate as a storage cell and the EEPROM is the least dense requiring two transistors for each storage cell.

6.6.4. RAM

Random access memory has an access time independent of the physical location of the data.

They can be further divided in to static and dynamic RAM

Static RAM	*Dynamic RAM*
1. Use flip flops or static storage cells for storage	1. Use capacitors for storage
2. Easier to design	2. Trouble some
3. Faster and larger	3. Slower and smaller
4. Need not be refreshed	4. Need to be refreshed periodically
5. Widely used in high-speed cache memories	5. Used in high density low cost/bit circuits

6.6.4 (A) SRAM

The common structure of SRAM also consists of (word) row select lines and column or (Bit) data select lines similar to ROM. At each intersection point of row and column lines a CMOS storage cell is found. The single unit of CMOS storage cell is shown in Fig. 6.6.4 (*a*).

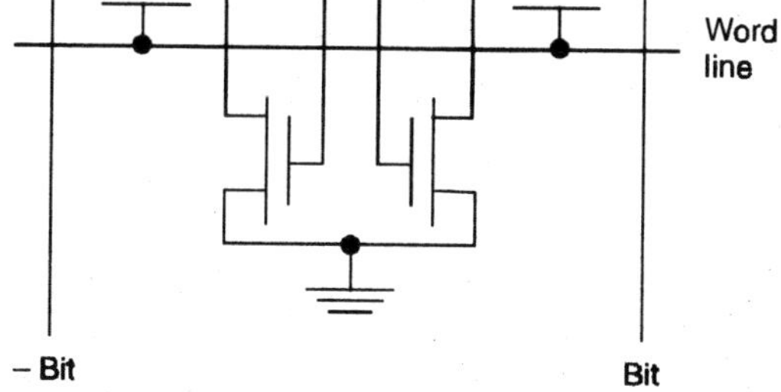

Fig. 6.6.4 (*a*). SRAM storage cell.

It shows a 6 transistor cross-coupled inverter circuit. The *p*-transistors may be replaced with high value polysilicon resistors. The row select line selects the desired row from the memory array. When it is high all memory cells in that corresponding row are selected. A pair of data lines are dedicated to each column to allow data from a selected cell on that column to be read or written.

SRAM Read

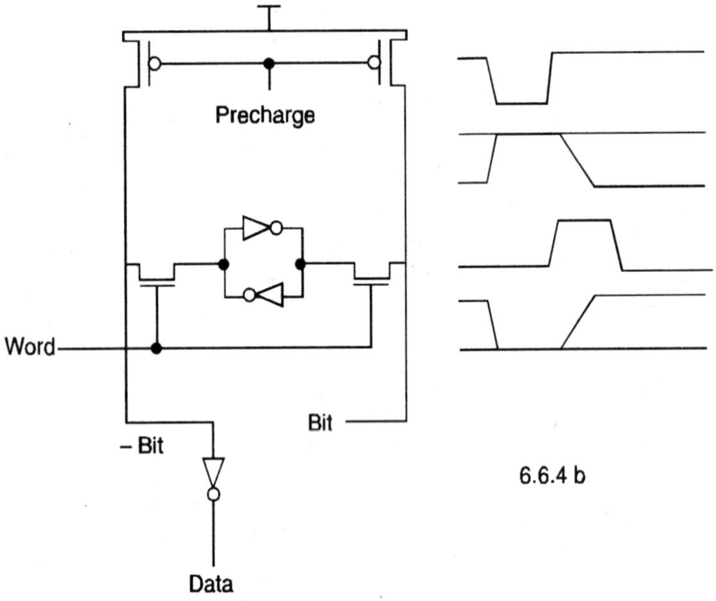

Fig. 6.6.4 (*b*) SRAM read operation with V_{DD} precharge.

After a word line is enabled, the one (1) node on the memory cell will attempt to pull the bit line up through access transistor and the *p*-load. The zero (0) node will attempt to pull the bit line down through the access transistor and the *n*-channel pull down. As *n*-channel transistor is poor at passing a one and the *p*-transistors in the RAM cell are generally small, design of RAM circuit concentrates on pulling the bit line from high to low. Thus one method of reading a RAM cell would be to precharge the bit lines high and then enable the word line decoder. So, for a given pair of bit lines, one RAM cell will attempt to pull down either the bit line (when data is '1') or -bit line (when stored data is '0') depending on the stored data. The bit line pull up circuit may use *p*-channel transistor to precharge each bit line. Fig 6.6.4 (*b*).

SRAM Write

In write operation we will apply voltages to RAM cell such that it will flip state. Fig. 6.6.4 (*c*) (*i*) shows the write circuit.

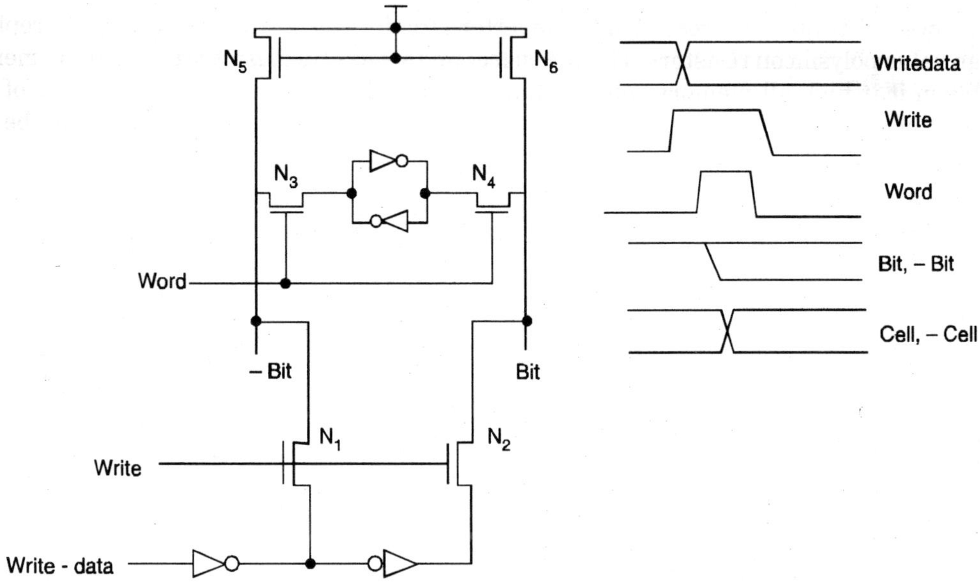

Fig. 6.6.4 (*c*) (*i*) SRAM write operation with pass transistors.

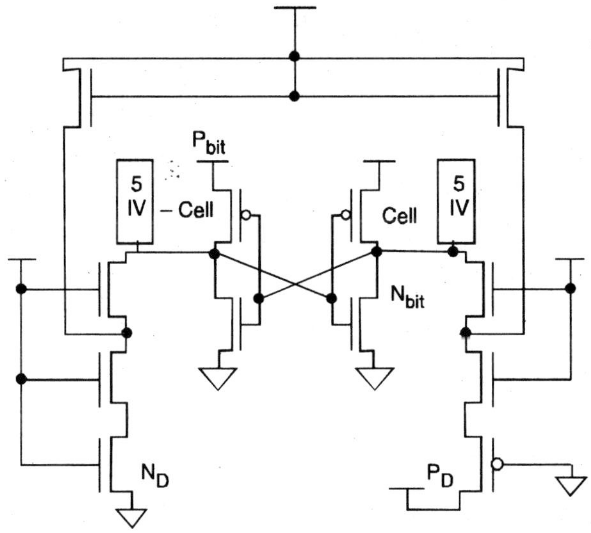

Fig. 6.6.4 (*c*) (*ii*) Circuit model during write.

In this circuit, the write enable transistors (N_1, N_2) are enabled to allow the data and complement to move to the bit lines. The word line is then enabled. Either the bit or-bit line is driven to V_{SS} while the other bit line is driven to a threshold down from V_{DD}.

The Fig. 6.6.4 (*c*) (*ii*) shows a detailed view. The Fig shows a zero stored in the cell. When a '1' is to be written, node-cell has to be pulled below the inverter threshold. Now the drive *n*-transistor N_D, write access transistor N_2, word access transistors N_3 have to pull P_{bit} below the inverter threshold. In addition bit-line pull up N_5 has to be pulled low by N_1 and N_D. And at the same time node cell has to be pulled above the inverter threshold.

∴ P_D, N_2 and N_4 have to pull N_{bit} as high as possible.

The (row) word select lines can be generated by the NOR decoder which is shown in Fig.6.6.4 (*d*) below :

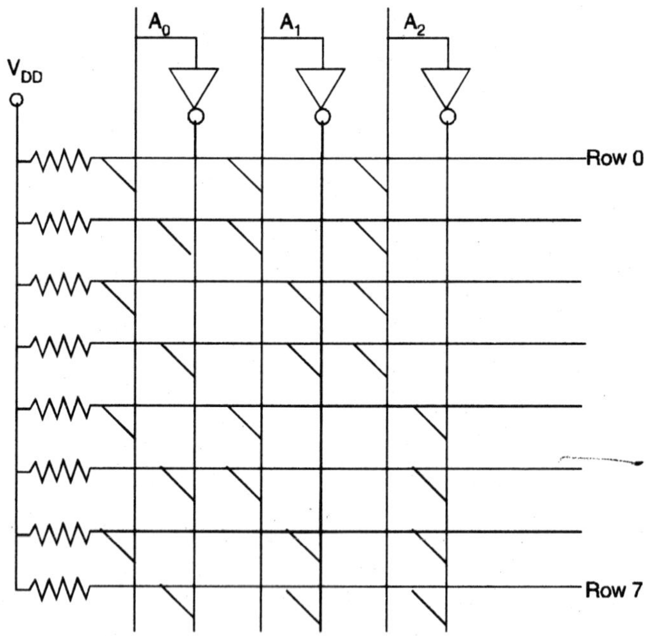

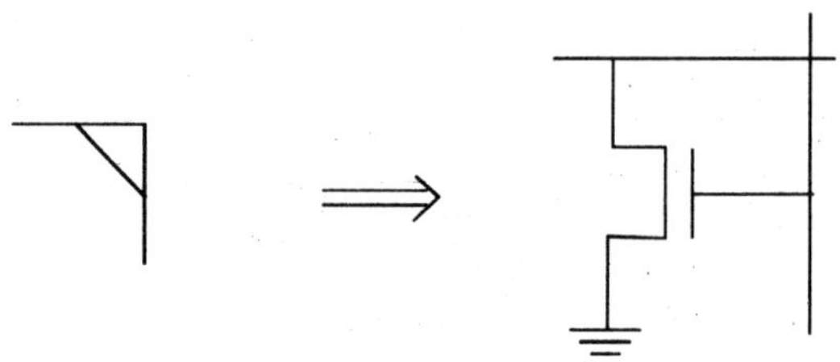

Fig. 6.6.4 (*d*) Address-to-row select decoder.

Here, the horizontal row lines are pulled high by the resistor to V_{DD}. From Fig 6.6.4 (*d*). The row 0 select line will be high only if A_0, A_1 and A_2 are all low. Similarly the remaining word or row select lines are selected by placing the NMOS transistors at proper intersection points. The row selection is very much similar to that of ***AND array programming in PLA*** which we have discussed earlier.

Once a given row is selected, data from all memory cell on that row are available on the column data lines. The Fig 6.6.4 (*e*) shows column selector circuit. It shows a 3 to 8 column decoder.

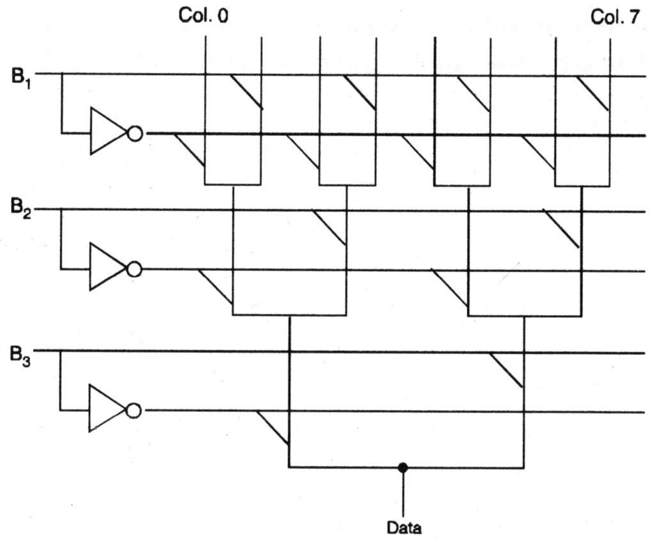

Fig. 6.6.4 (*e*) Address-to-column selector.

In this Fig. Lines B_1, B_2, B_3 are used to choose desired columns. These three are inputs to column decoder. B_1 selects odd columns, the complement of B_1 selects even columns. After B_1 stage, adjacent columns are paired. Now number of available column paths are reduced to half. Now B_2 selects odd pair and complement of B_2 selects even pair. Subsequent stages of selection each divide the number of column paths by two. In this example after 3 select stages only a single column line is left which will contain the desired data. To reduce the time between selection of a memory cell and availability of data at the chip output, sense amplifiers may be included in column select structure.

6.6.4 (B) DRAM

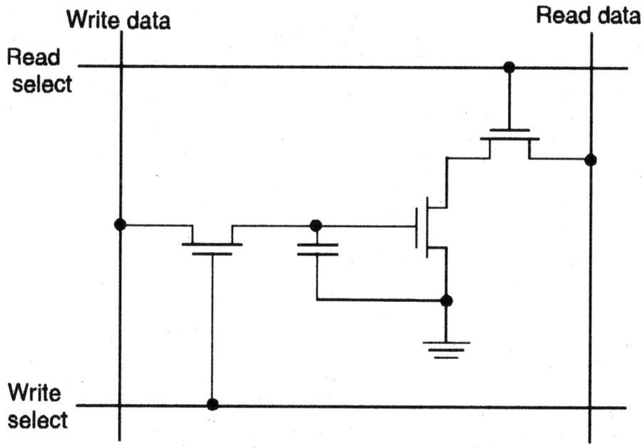

Fig. 6.6.4 (*f*). Three transistor DRAM cell.

It uses charge storage on a capacitor to represent binary data values. Only a few transistors (even one is enough) are required to select the cell and access the stored data. Because SRAM requires more transistors per memory cell, SRAM cannot be manufactured with the high memory density of DRAM. In this charge stored on the capacitors leaks off with time. To prevent loss of data, memory refreshing is to be done periodically.

Three transistor cell DRAM is shown in Fig. 6.6.4 (*f*). In this a capacitor is used to store charge. One transistor acts as a buffer to drive the read data line, second acts as read select switch and the third acts as a write select switch. All transistors are of minimum size and area required is less. The three transistor cell requires four bus lines for operation Viz. read select line, write select line, read bit line, write bit line. When a buffer transistor is made to drive the data line, stored charges will not deteriorate during read operation. However to prevent charge leaking off from capacitor a memory refresh is performed by executing a read operation followed by a write operation.

One Transistor DRAM. A single select transistor for pass transistor and capacitor for storage is shown in Fig. 6.6.4 (*g*). The transistor passes the stored value to a data bus when select line is enabled. The select line simultaneously selects all transistors along the same row, causing data to be placed on the column lines corresponding to each selected cell. One column is finally selected for connection to the OUTPUT.

To keep the size of a dynamic memory cell small, both the select transistor and storage capacitor must be small. As the length of the data is large, its capacitance is usually large when compared with storage cell capacitance. Undesirably, the charge on the larger data capacitance, will have more effect on the final data line voltage than the charge from the small memory cell capacitance. So, intelligent techniques and sensitive circuits are necessary for reliably sensing the stored value of DRAM cell. In this one transistor DRAM cell also, sense amplifiers can be used. These amplifiers sense the small change in voltage that results when a particular cell is switched on to the bit line. This type of cell forms the basis for most high density DRAMs.

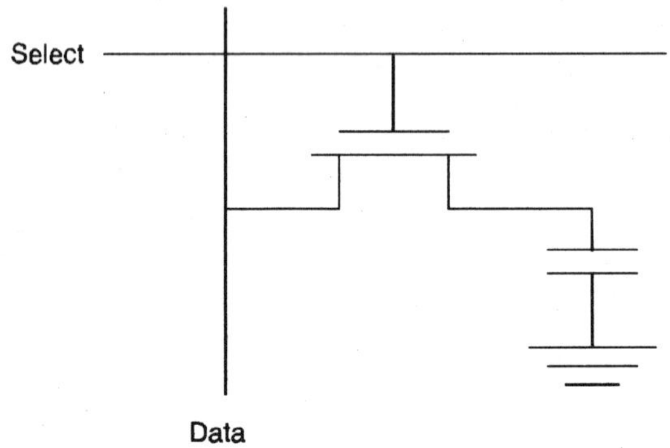

Fig. 6.6.4 (*g*) One transistor DRAM cell.

7

Digital Design Using Hardware Description Language

INTRODUCTION

A hardware description language(HDL) is similar to a computer programming language except that an HDL is used to describe hardware rather than a program to be executed on a computer. HDL is supported by all vendors that provide digital hardware technology and is officially endorsed as an Institute of Electrical and Electronics Engineers(IEEE) standard. The IEEE is a worldwide organization that promotes technical activities to the benefit of society. The following two HDLs are accepted as IEEE standards.

1. VHDL (Very high speed integrated circuit Hardware Description Language).

2. Verilog HDL. We use VHDL here because it is more popular than Verilog. In comparison to performing schematic capture, using VHDL offers a number of advantages. Because it is supported by most companies that offer digital hardware technology, VHDL provides design portability that a circuit specified in this can be implemented in different types of chips and with CAD tools provided by different companies, without changing the VHDL specification.

7.1. DIGITAL SYSTEM DESIGN PROCESS

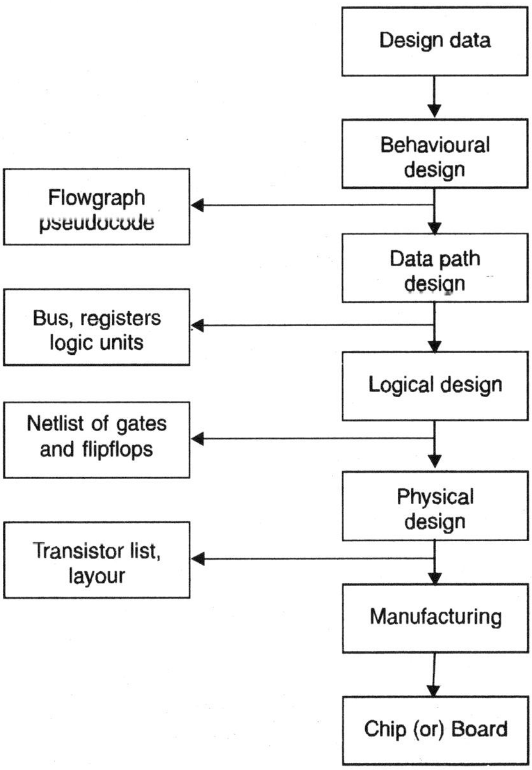

Fig. 7.1 (*a*) Digital system design process.

Design of any digital system goes through several transformations before its hardware implementation is obtained. Initially this process starts with design idea for a particular system. From this basic design Idea, the complete definition of the hardware implementation for the above system is developed.

Behavioural Design. It specifies the overall functionality of the system and the details about the required input and output ports for the required system. The product (or) output of this design stage may be a flow chart, a flow graph or pseudocode. The next stage of the design process is data path design.

Data path Design. In this phase, the designer specifies the registers and logic units necessary for the implementation of the system. These registers and logic units can be interconnected using either bi-directional or unidirectional buses. Based on the behaviour of the system, the procedure for controlling the movement of data (or) flow of data through buses between the registers and logic units is developed.

Thus, this design phase (or) data path design provides the architectural design of the system with specification of the control flow. (*i.e.* It gives the details of the bus, registers and logic units). But no information about the implementation of the controller, *e.g.*, hardwired, encoding technique, or microprogrammed — is given in this phase.

Result of the data path design phase

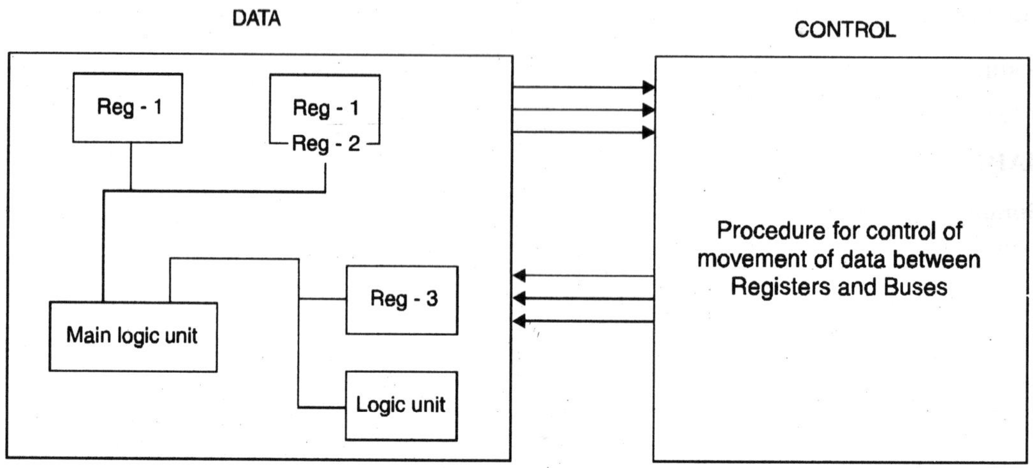

Fig. 7.1 (*b*)

Logic Design. After completion of data path design the next step is the logic design. It involves the use of the primitive (undeveloped) gates and flip-flops for the implementation of the data registers, buses, logic units and their controlling hardware. But gate technology and gate level details of the flip-flop are not included in the stage. Thus, the result of this logic design stage provides a net list of gates and flip-flops. *i.e.*, components used and their interconnections are specified in the netlist.

Physical Design. This stage involves the replacement of gates and flip-flops required for the system with their equivalent Transistors (or) library cells. This stage considers the loading and timing requirements in its cell (or) transistor selection process.

Manufacturing. It uses the transistor list to burn the fuses of a field programmable device (or) to generate the masks for Integrated circuit fabrication.

Note: At each step of transformation, the designer checks the result of the last transformation and if the results are not satisfactory the designer adds more information to it and passes it to the next step of transformation.

Summary

Behavioral design ⇒ It deals with function of the system and register input and outputs.

Data path design ⇒ (*i*) required registers and logic units for the above (behavioral description) are specified

(*ii*) Flow of the data through buses between registers and logic units are controlled.

Logic design ⇒ Data path units are replaced with required primitive gates and flipflops

Physical design ⇒ It replaces the gates and flipflops using transistors.

Manufacturing ⇒ Final implementation is done (or) Manufacturing of IC's for the above details specification is carried out.

Design Automation. In the digital system design process, most of the process involves transforming a design from one stage into another stage and the verification of the OUTPUT of the every design stage. Those functions (or) operations can be done by softwares by using computers. This process is called Design automation (DA) process. The tools used under DA are

– simulation tools (or) simulators

– synthesizers

7.2. HARDWARE SIMULATION

Simulation. It is the process of verifying the conduction of the model for the given circuit (or) given hardware without implementing practically.

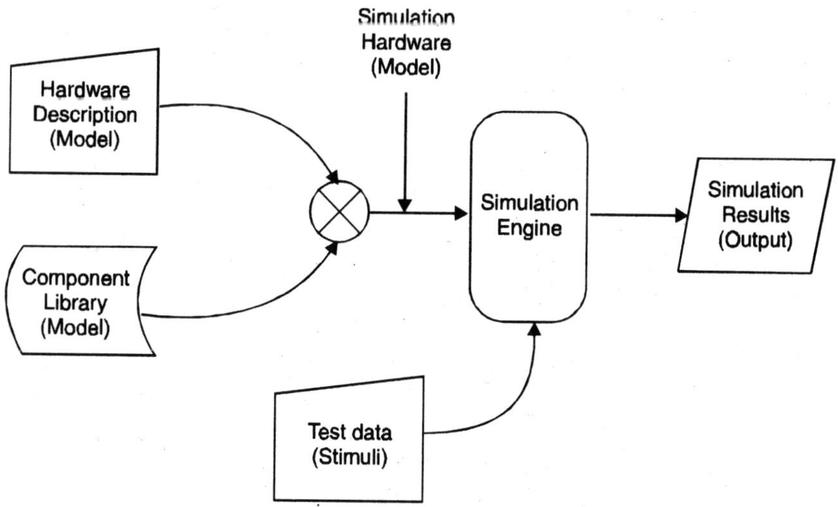

Fig. 7.2. Hardware simulation.

The hardware description for the model to be stimulated uses, (or) combined with component models and definitions from the simulation library provides a simulatable hardware model, which is given as input to the stimulation engine. In addition to the circuit description, simulator needs a set of simulation datas (or) stimuli. The simulation program applies this data to the simulation engine along with the stimulatable hardware at the specified times. The simulator, which produces simulation results, which indicates the conduction of the model for the given circuit for the given set of test data. The simulation engine is capable of producing detailed timing and signal voltages or it may be able to produce high-level function information. But the details of the results obtained from simulation engine depend on the level of details given in the hardware description (or) hardware model.

For example, if we want to know the detailed study of timing and signal voltages, that information should be included in the hardware description of the model itself at the input side of simulation engine. Similarly to verify the functionality (or) functional details, the information regarding the behaviour of the model should be provided along with the hardware model at the input and stimulation engine. Therefore, the output generated by the simulator depends on the level of the details of the hardware model presented to it. The output of the simulation program may be illustrated by waveforms, timing diagrams (or) time-value tabular listings. Simulators may be used to verify the results of any of the design stages of digital system design process. (*i.e.*, behavioural design, data path design, logical design and physical design stages) The results of the every stage will be verified by the designer and the designer will determine whether to repeat a design stage or not depending on the simulation results.

At the upper levels of the design, process simulation provides information regarding the functionality of the system under design. Simulators for this purpose normally (undergo a very quick run on a host computer) do very fastly. Simulation at a lower level of design process for example gate level (logic design) (or) device level (physical design) runs much more slowly but provides more detailed information about the timing and functionality of the circuit. So to avoid the high cost low level simulation runs, simulators should be used to detect design flaws (imperfections) as early in the design process as possible. Regardless of the level of the design digital system, simulators have been classified into oblivious and event driven simulators. In oblivious simulation, each circuit component is evaluated at fixed time points, while in event driven simulation, a component is evaluated only when one of its inputs changes.

7.3. HARDWARE SYNTHESIS

The process of transforming a design description from one form to another form is called synthesis. Various synthesis tools are available to synthesis various stages of the digital system design process. The commercially available synthesis tools use the output of the data path design stage as input and produce a net list for the circuit. The FPGA synthesis tools, synthesize hardware descriptions in one or more formats to FPGA program file formats that can be fed to a programmer. Such tools concentrate on resource sharing, logic optimisation and binding. Fig. 7.3 shows a block diagram of a register transfer level synthesis tool. In this logic, units are resources and they are shared between same operations at multiple instances. An operation used in several instances with different inputs requires the use of multiplexers for providing it with appropriate sets of inputs. Resource sharing may be directed by a synthesis tool user through synthesis directives. In logic optimisation pass, Boolean expressions are simplified for better utilization of chip area. The Binding phase maps operations of data structure to predefined library cells of target hardware.

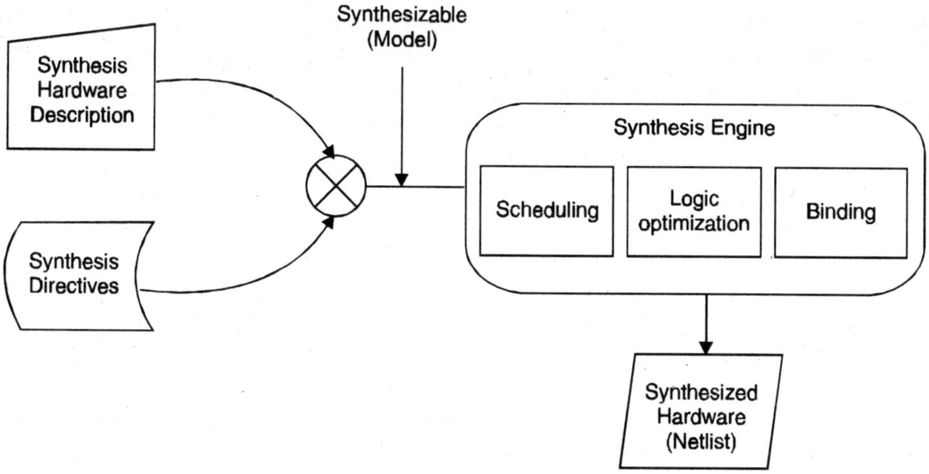

Fig. 7.3. Synthesis process.

Tools that generate layout from netlists have been available. These tools are commonly referred to as silicon compilers. The silicon compiler accommodates the given function in the least amount of chip area.

7.4. HISTORY OF VHDL

VHDL acronym for Very High Speed Integrated Circuits (VHSIC) Hardware Description Language. It is hardware description language that can be used to model a digital system (or) digital hardware device ranging from the algorithmic level to the gate level. The digital system can be described hierarchically. Timing can also be explicitly modelled in the same description. VHDL language can be regarded as an integrated amalgamation (integration) of the following languages.

Sequential language + Concurrent language + Net list language +

Timing specifications + Wave form generation language = VHDL

We can express the concurrent (or) sequential behaviour of a digital system with (or) without timing specification. It is also possible to model the system as an interconnection of components.

All the above constructs may be combined to provide the comprehensive description of the system. Initially, in 1981, many of the US companies developed VHSIC chips for the department of defence. (DOD). At that time, most of the companies were using different hardware description language to design and develop the integrated circuits.

1. First the department of defence (DOD) assigned the job to a team of three companies namely IBM, Texas, Instruments and Intermetrics to develop a version of the language in 1983.

2. Version 7.2 of VHDL was developed and released to the public in 1985.

3. After the release of version 7.2 there is an increasing need to make the language an industry wide standard so that it was given to IEEE for standardization of language in 1986.

4. After the refinement (or) enhancement is made to the language by the industry, university and DOD representatives, the language was standardized by the IEEE in 1987. Then, it was also upgraded with new features, that new version of the language is known as IEEE STD 1076-1993. This version is used now.

5. Different vendors supported different packages. Different packages uses different logic levels like 46 value logic, 7 value logic, 4 value logic and so on.

So that a committee was set up to standardize such a package. The output of this committee was the development of a 9-value logic package. This package, called STD_LOGIC_1164 was then approved to become an IEEE standard labelled as. IEEE STD 1164-1993.

This package is compiled in a design library named IEEE. All designs using this package must use the two statements shown below for making this package and its contents accessible.

The two statements are

(*i*) Library IEEE;

(*ii*) Use IEEE. std-logic_1164.all;

The nine Value Logic levels are as follows :

U → un initialised	X → unknown
0 → Forcing 0	1 → Forcing 1
L → weak low	H → weak High
– → Don't care	Z → High impedance
W → weak unknown	

In designs where simple '0' and '1' values of BIT types are not sufficient for expressing logic level behaviour of a circuit, the std_logic type can be used. Its usage was approved by IEEE and was named as std-logic_1164.

7.5. VHDL REQUIREMENTS

The VHDL was developed in order to overcome the flaws that were found in previous HDLs. The critical factors which were not been fulfilled by other HDLs are made as requirements of HDL by DOD. (Departments of Defence). The other HDLs did not support design Hierarchy, Type declaration etc. But those were fulfilled by VHDL.

The main requirements VHDL are as follows

1. Concurrency. The DOD requirement specifies that the VHSIC HDL should be a language for design and description of hardware. In any Actual system, all large and small units are active simultaneously and perform functions concurrently. So concurrency is heavily emphasized in VHDL. Concurrency means the transfer statements, description of components and component instantiation should be executed in such a way that they should appear to have been executed simultaneously.

2. Support for design hierarchy. A design consists of an interface description and a separate part for describing its operations. Several descriptions may exist for describing the operation of a design, all correspond to the same interface description. Instead of trying to design a whole system together, a divide and conquer technology is carried out here. First the system is divided into subsystems for which descriptions may be written either in behavioural or in structural. Then the subsystem can be divided further and the division can be done until we arrive at a basic unit.

3. Library support. User and system defined primitives and descriptions reside in a library system. The VHDL should provide a mechanism for accessing various libraries.

4. Sequential statement. After a system is partitioned into concurrent components or subsystems, the internal operation should be carried out sequentially. For fulfilling this requirements, the languages should provide constructs such as case, if then else, loop etc.

5. Generic design. In addition to inputs and outputs, other physical parameter may influence the operation of a system. These include ambient temperature, timing, loading of component etc. For *e.g.* the NAND gate, 7400 logic families has so many series (LS, F, ALS) They are functionally equivalent but differ only in timing and loading characterises. So VHDL should allow a generic description which should be configurable for size, physical characteristics, timing, loading and environmental condition.

6. Type declarations and usage. A language for the description of hardware at various levels of abstraction should not be limited to 'bit' or 'Boolean' types. VHDL requirements specified that the language should allow integer, floating point and enumerate types, as well as user-defined types. Types defined by the system or by the user should be placed in the library of the language environment, and their use should be transparent to the user.

7. Use of sub programs. The VHDL should allow the use of functions and procedures. Sub programs can be used in a similar way as used in common programming languages.

8. Timing control. The ability to specify timing at all levels is another requirement for the VHDL language. VHDL should allow the designer to schedule values to signals and delay the actual assignment of values until a later time.

7.6. LEVELS OF ABSTRACTION

7.6.1. Behavioural Description (Sequential)

It describes the function of the design in a software like procedural form and provides no detail as to how the design is to be implemented. The behavioural level of abstraction is the most appropriate for fast simulation of complex hardware units, verification and functional simulation of design ideas, modelling standard components and documentation. For simulation and functional analysis, a behavioural model is useful since the details of the hardware, which may not be known to the user, are not required in the behavioural model. (*i.e.*, hardware details are not required) This description can be accessible to non-engineers as well as to the end users of a hardware component.

1. Behavioural description : [Example – Full adder]

```
Library IEEE;
Use IEEE. Std_logic_1164.all;
Entity fadd is
   Port(a,b,c : IN STD_LOGIC; sum, carry : OUT STD_LOGIC);
end fadd;
architecture fadd-arch of fadd is
begin
   process (a,b,c)
   variable Temp1, Temp2 : STD_LOGIC;
begin
   Temp1 : = a xor b xor c;
   If (Temp1 = '1') then
            Sum ⇐ Temp1;
```

```
else
            Sum ⇐ '0 ';
end if;
   Temp2 : = (a and b) or (b and c) or (c and a);
if (Temp2 = '1') then
            carry ⇐ Temp 2;
else
            carry ⇐ '0 ' ;
end if;
end process;
end fadd-arch;
```

7.6.2. Data Flow Description (Concurrent Operation)

A data flow description is a concurrent representation of the flow of control and movement of data. Data flow description implies architecture and a unique hardware. (It describes the model of the circuit through registers and buses). Simulation of these descriptions involves the movement of data through registers, buses, and therefore is slower than the behavioural descriptions. This descriptions are abstract enough for technically oriented designer to describe the components to be synthesized and cannot serve as an end-user (or) non-technical documentation.

2. Data flow description

Example : - full adder:

```
Library IEEE;
Use IEEE.Std_logic_1164.all;
Entity FADD is
Port
   (a,b,c : IN STD_LOGIC;
   sum, carry : OUT STD_LOGIC);
end Fadd;
architecture fadd-arch of fadd is
begin
   sum ⇐ a xor  b xor c;
   carry ⇐ (a and b) or (b and c) or (c and a);
end fadd-arch;
```

7.6.3. Structural Descriptions (Concurrent)

It includes a list of concurrently active components and their interconnections. The hardware is clearly implied by these descriptions and a simple wire router (or) PCB layout generator can easily produce the described hardware. A structural description that describes wiring of logic gates is said to be the hardware description at gate level.

3. Structural description for full adder

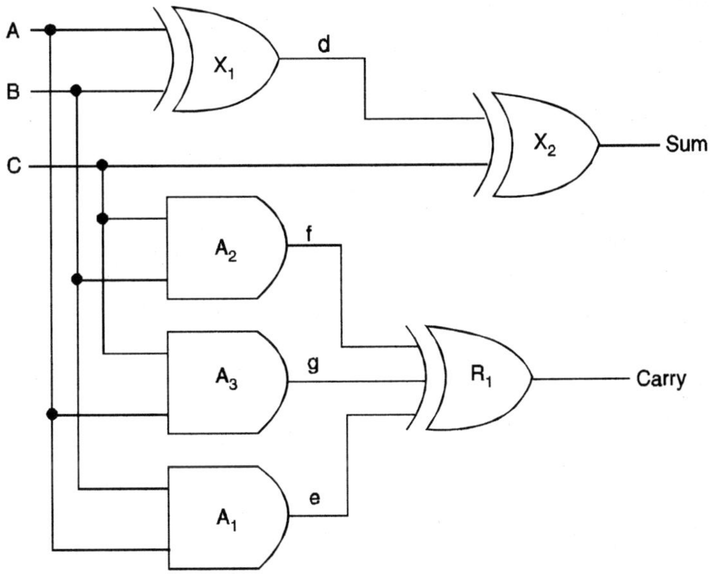

Fig. 7.6.3. Full adder.

$$S = A \oplus B \oplus C$$
$$C = AB + BC + CA$$

1. XOR gate

```
Library IEEE;
Use IEEE.STD_Logic_1164.all;
Entity EXOR IS
Port
   (p, q : IN STD_LOGIC;
   R : OUT STD_LOGIC);
End EXOR;
Architecture EXOR_ arch of EXOR is
Begin
   R ⇐ P XOR Q;
End EXOR _ arch;
```

2. AND gate

```
Library IEEE;
Use IEEE.STD_Logic_1164.all;
Entity AND–gate is
   Port
         (M,N : IN STD_LOGIC;
```

```
    T : OUT STD_LOGIC);
End AND-gate;
Architecture AND gate-arch of AND-gate IS
Begin
    T ⇐ M AND N;
End AND gate-arch;
```

3. OR gate

```
Library IEEE;
Use IEEE.STD_Logic_1164.all;
Entity ORG is
    Port
            (U,V,W : IN STD_LOGIC;
            Y : OUT STD_LOGIC);
End ORG;
Architecture ORG-arch of ORG is
Begin
    Y ⇐ U or V or W;
End ORG-arch;
```

Complete coding for full adder using structural description

```
Library IEEE;
Use IEEE.STD_Logic_1164.all;
Entity FA is
    Port
            (A,B,C : IN STD_LOGIC;
            Sum, Carry : OUT STD_LOGIC);
End FA;
Architecture FA-arch or FA is
Signal d,e,f,g : STD_LOGIC;
Component EXOR is
    Port
            (P,Q : IN STD_LOGIC;
            R : OUT STD_LOGIC);
End component;
Component AND-gate is
    Port
            (M,N : IN STD_LOGIC;
            T : OUT STD_LOGIC);
End component;
```

```
Component ORG is
   Port
          (U,V,W : IN STD_LOGIC;
          T : OUT STD_LOGIC);
End component;
Begin
   X1 : EXOR Port Map (A,B,d);
   X2: EXOR Port Map (d,c, Sum);
   A1 : AND-Gate Port Map (A,B,e);
   A2 : AND-Gate Port Map (B,C,f);
   A3 : AND-Gate Port Map (A,C,g);
   R1 : ORG Port Map (e,f,g, carry);
End FA-arch;
```

7.7. ELEMENTS OF VHDL

Constructs of the VHDL language are designed for describing hardware components, packaging parts and utilities and use of libraries and for specifying design libraries and parameters. This section presents language features related to each of these applications.

7.7.1. Components description

VHDL language is used to describe hardware components and systems. The description of VHDL consists of interface specification and architectural specifications.

Interface Specification

It specifies the – name of the entity, name of the interface ports, their mode (*i.e.*, direction in, out…), the type of ports (bit, Std_logic)

i.e., The interface description begins with the ENTITY keyword and contains the input-output ports of the component. Other characteristics of a component, such as time and temperature can also be included in the interface description of the component. The name of the component comes after the ENTITY keyword and is followed by IS, which is also a VHDL keyword.

```
ENTITY component name is
        input output ports physical and other parameters
END component name
```

Each interface part can have one of the following modes:

IN. The value of an input port can only be read with in the entity model.

OUT. The value of an output port can only be updated with in the entity model. It cannot be read.

INOUT. The value of a bi-directional port can be read and updated within the entity model.

BUFFER. The value of a buffer port can be read and updated with in the entity model.

However, it differs from INOUT mode in that it cannot have more than one source and the only kind of signal that can be connected to it can be another buffer port or a signal with at most one source.

LINKAGE. The value of linkage port can be read and updated. This can be done only by another port of mode linkage. The usage is not very clear and is thus not recommended.

Example : Interface Specification for AND GATE :

```
ENTITY And Gate IS
Port
   (A,B : IN Std _logic;
   C : Out Std_logic)
END And gate;
```

A, B ⇒ input port, through which inputs can be given.

C ⇒ output port, through which output can be read.

7.7.2. Architecture Specification

An architecture specification begins with the ARCHITECTURE Keyword, which describes the functionality of a component. This functionality depends on input-output signals and other parameters specified in the interface design. A functional description of the component starts after the BEGIN keyword.

Several architectural specification with different identifiers can exist for one component with a given interface description. For example, a component's architectural specification can be explained in three ways namely, behavioural level, structural level and data flow level.

For simulations where the functionality of system is to be verified, the behaviour architectural specification is used but where the exact timing response of the component is to be verified, the architectural specification at the structural level is used.

```
ARCHITECTURE identifier OF component name IS
        declarations.
BEGIN
        Specification of the functionality of the component
        interms of its input lines and influenced by physical
        and other parameters.
END identifier;
```

Example : Architectural specification for AND GATE :

```
   ARCHITECTURE AND gate-arch of and gate IS
BEGIN
          C ⇐ A AND B;
END Andgate – arch;
```

7.8. PACKAGES

For a hardware design, it is necessary to group components or utilities used for description of components. Examples of VHDL constructs for describing utilities are type definitions and subprograms. Format for the package declaration is shown in Fig. 7.8 (*a*), which contains components and utilities to become visible, by Entities and Architectures. Format for the package body, also shown in Fig. 7.8 (*b*), which contains subprogram definitions and utilities used by subprograms.

```
PACKAGE Package name IS
                component declarations
                sub-program declaration
END package name;
```

Fig. 7.8 (*a*) Format for package declaration.

```
PACKAGE BODY package name IS
                type definitions
                sub programs
END package name;
```

Fig. 7.8 (*b*) Format for package body.

7.9. LIBRARIES AND BINDING

Binding of subcomponents of a design to elements of various libraries and the use of libraries are allowed in VHDL. Constructs for such applications include a library statement and configurations Figure 7.9 (*a*) shows library specification and a configuration for configuring a design.

```
LIBRARY library-name;
CONFIGURATION Configuration name of component name IS
   binding of entities and Architectures
           specifying parameters of a design
           binding component of a library to subcomponent
END CONFIGURATION;
```

Fig. 7.9 (*a*) Design binding.

7.10. OBJECTS AND CLASSES

An entity that has a value of a given type is a VHDL object. Objects may be explicitly declared or they may be implicitly created in the language. Port signals, loop index variables, signals for interconnecting components, temporary variables and files are some of the objects in VHDL. An object may be of one of the four classes of signal's variables, constants and files.

Signals. This represents hardware wires and has timing associated with them. Values assigned to a signal are placed on signal driver and will appear on the signal after specified time delay. Signal can be declared only in concurrent bodies. However, can be assigned or used both in concurrent and sequential bodies. The '$\Leftarrow$' symbol is used for signal assignment *e.g.*, $a \Leftarrow 3$;

Variable. Variables are used for temporary storage. They are used in similar way as is used in software programming languages. Variables can only be declared, or have values assigned to them in sequential bodies of VHDL. The ': =' symbol is used for assigning values to variables. *e.g.*, $a := 3$; The shared variables are subclass of variables and can be declared in concurrent or sequential bodies.

Constant. They represent constant values of a given type. They can be declared, and they are available for use in concurrent and sequential bodies. The value given to the constant can not be altered.

File. As with the constants, files can be declared and used in concurrent and sequential bodies.

The Table 7.11 summarises declaration and used of object classes in concurrent and sequential bodies.

Table 7.10. Objects in VHDL

Objects in VHDL	*Body*					
	Concurent			*Sequential*		
	Declare	Assign to	Use	Declare	Assign to	Use
Signal	Yes	Yes	Yes	No	Yes	Yes
Variable	No	No	Yes	Yes	Yes	Yes
Constant	Yes	—	Yes	Yes	—	Yes
File	Yes	—	Yes	Yes	—	Yes

7.10.1. Signal Assignments and Delay Mobility

Assignment of values to signals in sequential and concurrent bodies is an important issue in VHDL.

In its simplest form, a signal assignment consists of a target signal on the left-hand side of a left arrow and an expression for defining a waveform on the right-hand side (with no AFTER clause). Such an assignment specifies that the right-hand side be assigned to the left-hand side a delta time later. Physically this time is 0 s, but it has nonzero scheduling significance. For example, an assignment that is scheduled to occur two delta times later will be done after an assignment that is scheduled to occur after one delta, and the result of the later assignment will not be available for the earlier assignment. Both assignments, however, occur before the smallest physical time unit.

Optionally, a signal assignment can include an AFTER clause specifying that a physical time delay is to occur before the assignment to the left-hand side takes place. If this time delay is zero, the simple form described above will apply.

Signal assignments can have *inertial* or *transport* delays. Inertial delays can have an additional *reject* specification. The way delays are handled in signal assignments is referred to as **delay modeling.** The delay mechanism for a signal assignment may be specified on the right-hand side of the assignment. The default mechanism is inertial. Transport delay mechanism must be explicitly specified.

The *delay* architecture of the *example* entity in Fig. 7.10 (*a*) includes assignments with inertial, inertial with reject, and transport mechanism. This architecture also includes assignments for comparing the three delay mechanisms and an assignment for creating a reference signal. In the first assignment, the INERTIAL keyword is optional. The second assignment specifies a reject value for an inertial delay mechanism. In this assignment, both REJECT and INERTIAL keywords are mandatory. The third assignment has a transport delay mechanism.

```
ENTITY example IS END ENTITY
--
ARCHITECTURE delay OF example IS
        SIGNAL waveform : BIT;
        SIGNAL target1, target2, target3: BIT;
        SIGNAL diff12, diff13, diff23 : BIT;
        -- This is a comment
BEGIN
        -- Illustrating inertial delay
        target1 <= waveform AFTER 5 NS; (First)
        target2 <= REJECT 3 NS INERTIAL waveform AFTER 5 NS; (Second)

        -- Illustrating transport delay
        target3 <= TRANSPORT waveform AFTER 5 NS; (Third)

        -- Comparing targets
        diff12 <= target1 XOR target2;
        diff13 <= target1 XOR target3;
        diff23 <= target2 XOR target3;

Creating waveform
waveform <=
        '1' AFTER 03 NS, '0' AFTER 08 NS, '1' AFTER 14 NS, '0' AFTER 18 NS,
        '1' AFTER 24 NS, '0' AFTER 27 NS, '1' AFTER 33 NS, '0' AFTER 35 NS,
        '1' AFTER 41 NS, '0' AFTER 47 NS, '1' AFTER 52 NS, '0' AFTER 58 NS,
        '1' AFTER 62 NS, '0' AFTER 68 NS, '1' AFTER 71 NS, '0' AFTER 77 NS,
        '1' AFTER 79 NS, '0' AFTER 85 NS,
END delay;
```

Fig. 7.10 (*a*) VHDL description for the demonstration of delay mechanism.

7.10.2 (*a*) Inertial Delay Mechanism

Inertial delays can be used to model capacitive networks, such as the one shown in Fig. 7.10 (*b*), which corresponds to the first assignment of Fig. 7.10 (*a*). If a pulse whose width is less than 5 ns occurs on *waveform,* it will be rejected and does not appear on *target1.* A pulse of exactly 5 ns will not be rejected.

Delays through capacitive networks and through gates with threshold values can be more accurately modeled with an inertial delay mechanism and a pulse rejection value that is less than the value of the inertial delay. The second assignment is an inertially delayed signal assignment with a reject value of 3 ns. As long as pulses on the right-hand side *waveform* signal are larger than 3 ns, they appear on *target2* with a 5-ns delay. A pulse that is less than, or equal to, 3 ns will not appear on *target2*.

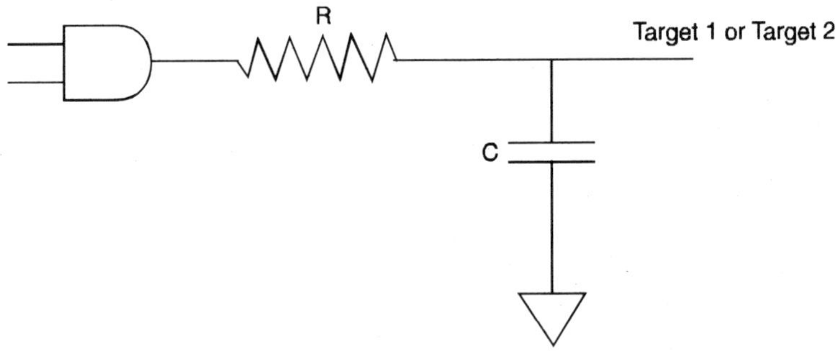

Fig. 7.10 (*b*) The RC delay is best represented by an inertial delay mechanism.

7.10.2 (*b*) Transport Delay Mechanism

Delays through transmission lines and networks with virtually infinite frequency response can be modeled by the transport delay mechanism. Regardless of the width of a pulse, a signal assignment with a transport delay mechanism delays the right-hand-side signal by the specified transport delay value. The third assignment causes *target3* to be a 5-ns-delayed duplicate of *waveform*.

7.10.2 (*c*) Comparing Inertial and Transport

In addition to three assignments demonstrating the format of various delay mechanisms, Fig. 7.10 (*a*) includes three assignments for comparing the waveforms on the three target signals and a fourth assignment that produces a waveform on the *waveform* signal.

On the right-hand side of the *waveform* signal, a waveform consisting of four positive and four negative pulses of 5, 4, 3 and 2 ns are produced. Positive pulses are separated by 6-ns negative pulses, and negative pulses are separated by 6-ns positive pulses. The first line of Fig 7.10. (*c*) shows the waveform produced on *waveform*.

The same *waveform* signal appears on the right-hand side of three signal assignments to *target1, target2* and *target3* signals. Delay values for these three assignments are 5 ns. Assignments to *taraget1* and *target2* are inertial, while assignment to *target3* is of the transport mechanism. Assignment to *target2* specifies a reject value of 3 ns.

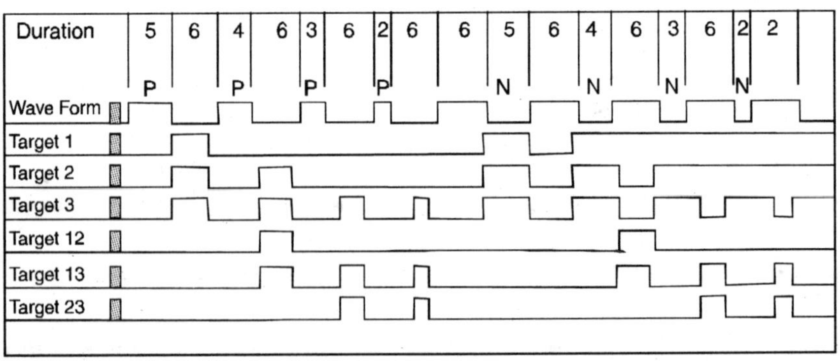

Fig. 7.10 (c) Illustrating differences between delay mechanisms in VHDL.

As shown in Fig. 7.10 (*c*), positive or negative pulses of 5 ns or more appear on all three targets. A 4-ns pulse (positive or negative) appears on *target2* because it is larger than the reject value of this assignment, and it appears on *taraget3* because of the transport mechanism. The two 4-ns pulses on *waveform* are rejected by the inertially delayed assignment to *target1.* Fig. 7.10 (*c*) also shows that pulses that are less than 3 ns only appear on *target3,* which has a transport delay mechanism on its right-hand side.

Differences between *taraget1, target2* and *taraget3* signals are generated on *diff12, diff13 and diff23* signals. Assignments to these *difference* signals use XOR of the signals they are comparing on their right-hand sides. Because no delay value is specified with these assignments, a 0-ns delay is assumed, and with this delay value there will not be any difference in waveforms produced by inertial or transport delays.

The *diff12* signal shows that there is a difference between *target1* and *target2* only when a 4-ns pulse (larger than reject and smaller than inertial delay) appears on *waveform*. Differences between left-hand side *taraget1* and *target3* signals occur when pulses less than the inertial delay value of 5 ns appear on *waveform.*

An expression followed by the AFTER clause, as delay value, and an optional delay mechanism specification is referred to as a *waveform element.* Assignments to *target* and *diff* signals in Fig. 7.10 (*a*) all have a single waveform element with the specified delay mechanism. On the other hand, the assignment to *waveform* signal in this figure has 18 waveform elements. When multiple waveform elements appear on the right-hand side of a signal assignment, the specified delay mechanism only applies to the first waveform element, and all others are assumed to have transport delay mechanism.

The effect of a delay mechanism can be produced by using other mechanisms. For example, an inertial delay with a delay value *d* and a reject value of 0 ns is equivalent to a transport delay with delay d. An inertial mechanism with equal reject and delay values is equivalent to an inertial mechanism with no reject specification. Other forms of pulse delay and/or pulse rejection can be generated by Boolean combination of signals with different delay formats.

7.11. VARIABLE ASSIGNMENT STATEMENT

A variable is assigned a value using the variable assignment statement, which has the general form as,

Variable-object : = expression ;

The given expression is evaluated when the statement is executed and the computed value is assigned to the variable object instantaneously at the current time of simulation time itself. Variables can be declared inside a process statement. Consider the following process statement.

```
Process (Q)
Variable Q : INTEGER;
Begin
        Q : = Q + 1;
End process;
```

At the start of simulation, the process is executed once. The variable Q is initialised to the value 1 and then incremented by one. At every time, an event occurs on signal Q, the process is activated and the variable assignment statement is executed, which makes the variable to be incremented. Finally, at the end of simulation, the variable Q contains the total numbers of events that occurred on signal-Q.

A variable can be declared outside of a process or a subprogram. Such a variable can be read and updated by more than one process. The variables are called shared variables.

7.11.1. Signal Assignment Statement

Signal is assigned a value using a signal assignment statement. The general form of a signal assignment statement is

Signal – object ⇐ expression (after delay value)

Such an assignment specifies that the right hand side be assigned to the left-hand side a delta time later.

The expression is computed when a signal assignment statement is executed and the computed value is assigned to the signal after a specified time delay. The expression is evaluated at the time the statement is executed and not after the specified delay.

A signal assignment can have an AFTER clause specifying that a physical time delay is to occur before the assignment to the left-hand side takes place. If no AFTER clause (AFTER delay-value) is specified the delay is assumed a default delta delay.

Example :

A ⇐ A + "0010"; (Assignment after a delta delay [delta delay is a very small delay])

X ⇐ A XOR B after 12 ns; (assignment after a delay of 12 ns.)

7.12. SEQUENTIAL STATEMENT

7.12.1. Case Statement

The format of a case statement is,

```
CASE expression IS
        WHEN choices ⇒ sequential statements
        WHEN Choices ⇒ sequential statements
        — Can have any number of choices
        WHEN OTHERS ⇒ sequential statements
END CASE;
```

The expression that follows the CASE keyword is checked against the choices of all the case statement alternatives. When a match is found, the sequence of statement that follows the arrow is executed. If the value of the expression is not matched with any of the choices, the statement that follows the keyword WHEN OTHERS is executed.

For example, a 4-by-1 multiplexer using case statement is given below :

```
Library IEEE;
Use IEEE.STD_LOGIC_1164.all;
Entity MUX is
Port
   (I0, I1, I2, I3, : IN STD_LOGIC;
   Sel : IN STD_LOGIC –VECTOR (1 down to 0);
   Y : OUT STD_LOGIC);
End MUX;
```

```
Architecture mux-arch of MUX is
    Process (I0, I1,I2,I3)
Begin
Case sel is
    When "00" ⇒  Y ⇐ I0;
    When "01" ⇒  Y ⇐ I1;
    When "10" ⇒  Y ⇐ I2;
    When "11" ⇒  Y ⇐ I3;
    When others ⇒ null;
End Case;
    End process;
    End mux-arch;
```

When the expression sel = 00, the value of I_0 is assigned to the output Y, for 01, the value of I_1 is assigned to the output Y, for 10, the value of I_2 is assigned to the output Y, for 11, I_3 is assigned to the output and if the expression sel is not matched with any of the above value of the expression, the null statement is executed.

7.12.2. Null Statement

The statement is,

Null. When this statement is executed, no action will take place and the execution continues with the next statement. This statement may be useful in an if statement (or) case statement where, for certain conditions, it may be useful to explicitly specify that no action needs to be performed.

7.12.3. If Statement

Based on the value of a condition, an if statement selects a sequence of statements for the execution. The condition can be any expression that evaluates to a Boolean value. The general form of an if statement is,

```
IF Boolean-expression THEN
    (sequential statements)
Elsif Boolean-expression then
    (sequential statements)
Else
    (sequential statements)
End if;
```

If the condition is true, the following sequential statement is executed. The if statement can have zero (or) more elsif clauses and an optional else clause. Here are some examples.

Example 1.

```
If A< = 50 then
            A : = A + 10;
end if;
```

Another examples for usage of 'if' statement can be referred from behavioural descriptons (7.6.1) where descriptons of FULL adder is given using IF statement.

7.12.4. Loop Statement

A loop statement is used to iterate through a set of sequential statements.

The syntax of a loop statement is,

[loop-lable:] iteration-scheme **loop**

(Sequential – statements)

end loop [loop-label];

There are 3-types of iteration schemes.

1. For iteration scheme :

It has the form as,

For identifier **in** range

An example of this iteration scheme is,

SUM : = 1;

For NUMBER **in** 2 to N **loop**

SUM : = SUM * NUMBER;

end loop;

In this example, the body of the for loop is executed (N–1) times, with the loop identifier, NUMBER, being incremented by one at the end of each iteration. The object NUMBER is implicitly declared with in the for loop to belong to the integer type whose values are in the range 2 to N.

The range in a for loop can also be a range of an enumeration type such as the one shown below :

Type HEXA is ('0', '1', '2', '3', '4', '5', '6', '7', '8', '9' 'A', 'B', 'C', 'D', 'E', 'F');

For NUM in HEXA ('9') down to HEXA ('0') loop

— *Num will take values in type HEXA from '9' through '0'.*

end loop;

for CHAR in HEXA loop

— *CHAR will take all values in type HEXA from '0' through 'F'.*

end loop;

2. While iteration scheme :

It has the general form as, **while** Boolean expression

An example of the while iteration scheme is,

SUM = 5; i:= 0;

While i < 10 loop

SUM : = SUM * 2;

i : = i+3;

end loop;

The statements with in the body of the loop are executed sequentially and repeatedly as long as the loop condition, i<10, is true. If the loop condition becomes false, it come out from the loop and it continues the execution of the statements outside of the loop.

3. Third iteration scheme using exit statement example :

```
        SUM : = 1; i = 0;
        L2 : loop – this loop has a label L2;
                i: = i + 20;
                SUM : = SUM * 10;
                exit when SUM>100;
end loop L2;
```

In this, all the statements in the loop body are executed repeatedly until some other action causes the loop to exit. It is caused by an exit statement, a next statement or a return statement. In the above example, exit statement is used.

An complete example of a full adder using for loop statement.

Loop (for loop) Full adder

```
Library IEEE;
Use IEEE.Std_logic_1164.all;
Entity FADD is
Port
   ( a,b : in Std_logic_vector (3 down to 0);
   C in : in STD_LOGIC;
   S : out STD_LOGIC –VECTOR ( 3 down to 0);
   Carry : out STD_LOGIC);
End FADD;
Architecture FADD-arch of FADD is
   Singal Ca : Std_logic_vector (0 to 4);
Begin
   Process (a,b)
   Begin
         Ca (0) ⇐ C in;
         For i in 0 to 3 loop
   S (i) ⇐ (a (i) xor b (i)) xor Ca (i);
   Ca (i+1) ⇐  (a(i) and b (i)) or (a (i) and Ca (i)) or (b(i) and Ca (i))
   end loop;
   carry  ⇐ Ca (4);
   end process;
end FADD –arch;
```

7.12.5. Exit Statement

It is a sequential statement that can be used only inside a loop. It causes execution to jump out of the innermost loop (or) the loop whose label is specified.

The syntax is,

Exit [loop-label] [when condition];

If no loop-label is specified, the innermost loop is exited. If the when clause "when condition" is used, the specified loop is exited only if the given condition is true; otherwise execution continues with the next statement.

Example 1.

```
SUM : = 1; i = 0;
    L2 : loop
    i = i + 20;
    SUM : = SUM * 10;
    if SUM > 100 then
    exit L2;
    end if;
    end loop L2;
```

Example 2.

```
loop
wait on A, B;
exit when A = B;
end loop;
```

In this, when the condition A = B becomes true, the exit statement causes the execution process to come out from the loop.

7.12.6. Wait Statement

When a process statement with sensitivity list is used, the execution is suspended after executing the last sequential statement in the process. An alternate way to suspend the execution of a process is carried out using wait statement. There are three basic forms of the wait statement.

Wait on sensitivity-list;

Wait until Boolean-expression;

Wait for time-expression;

The above statements can be combined as shown below. For example,

Wait on sensitivity-list **until** Boolean-expression **for** time-expression;

Some examples of wait statements are,

Wait on A, B, C; — Statement 1

Wait until A = B; — Statement 2

Wait for 5 ns; — Statement 3

Wait on clock for 30 ns — Statement 4

Wait until SUM > 100 **for** 30 ns; — Statement 5

Wait on clock **until** sum>100; — Statement 6

The execution of the wait statement in statement-1 causes the execution to wait until an event occurs on signal A, B or C. once that happens, the process resumes execution from the next statement onwards. If the wait statement was the last statement in the process, the process resumes execution from the first statement.

The execution of the wait statement in statement-2 causes the execution to wait until the condition becomes true. When an event occurs on signal A (or) B, the condition is evaluated and if it is true, the process resumes execution from the next statement. Otherwise, it suspends again. When statement 3 is executed, the enclosing process suspends for five ns after then, it resumes execution from the next statement. The execution of statement 4 causes the enclosing process to suspend and then wait for an event to occur on signal clock for a time-out interval of 30 ns. If no event occurs with in 30ns, the process resumes execution with the statement following wait.

In the statement 5, the process suspends for a maximum of 30 ns until the value of signal sum is greater than 100. The condition is evaluated every time there is an event on signal sum. If the condition is not satisfied for three ns, the process resumes from the statement following the wait statement.

In the last statement, the wait statement causes the execution to wait until an event occurs on clock, then the check if SUM > 100. If SUM < = 100, continue to wait. If SUM > 100, resume from the statement following the wait statement.

7.12.7. Assertion Statement

The syntax of an assertion statement is,

Assert Boolean-expression

[**report** string-expression]

[**severity** expression];

An assertion statement is useful in modelling constraints of an entity.

From the syntax format, if the value of the Boolean expression is false, the report message is printed along with the severity level (a pre-defined enumeration type with values NOTE, ERROR, WARNING and FAILURE). The severity level is used by a simulator to initiate appropriate actions depending on its value. For example, if the severity level is ERROR, the simulator may abort the simulation process and provide relevant information.

7.12.8. Report Statement

The syntax of a report statement is,

report string-expression

[severity expression];

A report statement is used to display a message. It is similar to an assertion statement, but with out the assertion check.

When a report statement is executed, it causes the specified string to be printed and the severity level to be reported to the simulator for further action.

Example :

1. **if** CLK = 'Z' **then**
   report "Signal CLR has a high impedance value.";
   —default severity level is NOTE.
   end if;
2. **if** CLK / = '0' and CLK / = '1' **then**
   report "CLK is neither a '0' nor a '1'"
   severity ERROR;
   end if;

7.12.9. Process Statement

A process statement contains sequential statements that describe the functionality of a portion of an entity in sequential terms:

Process [(sensitivity-list)]

A set of signals to which the process is sensitive is defined by the sensitivity list. In other words, each time an event occurs on any of the signals in the sensitivity list, the sequential statements with in the process are executed in a sequential order, that is, in the order in which they appear. The process then suspends after executing the last sequential statement and waits for another event to occur on a signal in the sensitivity list.

The syntax of a process statements is;

```
Process [(sensitivity-list)] [IS]
        [Process-item-declarations]
        begin
        sequential statements; these are →
        variable –assignment – statement
        signal-assignment-statement
        wait-statement
        if-statement
        case-statement
        loop-statement
        null-statement
        exit-statement
        assertion-statement
        report-statement
        procedure-call-statement
        return-statement
        end process;
```

7.13. USAGE OF SUBPROGRAMS

Subprograms are used to simplify coding, modularity and readability of description. VHDL uses the subprograms for these applications as well as for those that are more specific to hardware descriptions.

VHDL allows two forms of subprograms, Functions and Procedures. Functions return values and cannot alter the values of their parameters. On the other hand, a Procedure is used as a statement that can alter the values of its parameters.

7.13.1. VHDL Functions

A function executes a sequential algorithm and returns a single value to the calling program.

The general form of a function declaration is

Function function-name (formal parameter-list)

Return return-type is (declarations)

Begin sequential statements

– *must include* ***return*** *return-value;*

End function name;

The general form of a function call is

Function-name (actual-parameter-list)

The number and type of parameters on the actual-parameters-list must match the formal parameter-list in the function declaration. The parameters are considered as input values and cannot be changed during execution of the function.

Example :

Decimal to binary and binary to decimal converters using subprogram-FUNCTION

```
        Library IEEE;
        Use IEEE. Std_logic_1164.all;
        Use IEEE.Std_logic_unsigned.all;
        Use IEEE.Std_logic_arith.all;
        Entity MUL is
                Port (B : IN bit-vector (3 down to 0);
                C : OUT integer range 0 to 15);
        end MUL;
        Architecture MUL-arch of MUL is
Function bi-dec (BI : bit-vector (3 down to 0)) return integer is
variable Dec:integer;
begin
        Dec : = 0;
  For J in 0 to 3 loop
    If BI (J) = '1' then
        Dec : = dec +2 ** J;
    End if;
  End loop;
        Return dec;
        End bi-dec;
```

```
Begin
        C ⇐ bi_dec (B);
End MUL-arch;
```

Here the function name is bi_dec (binary to decimal conversion program), the formal parameter is BI and the return-type is integer. Variable Dec is used to hold the intermediate value during the calculation. When the function (subprogram) is called the value of the input is applied to the subprogram and it returns the output value to the outport of the main coding.

7.13.2. VHDL Procedures

Procedures facilitate decomposition of VHDL code into modules. Unlike functions, which return only a single values using output parameters

The general form of a procedure declaration is,

```
Procedure procedure-name (formal-parameter-list) is
[declarations]
begin
sequential statements
end procedure-name;
```

The formal-parameters-list specifies the inputs and outputs to the procedure and their types. A procedure call is sequential (or) concurrent statement of the form.

```
Procedure-name (actual-parameter-list);
```

With in the procedure declaration, the class, mode, and type of each parameters must be specified in the formal-parameter-list. The class of each parameters can be signal, variable, or constant. If the class is omitted, constant is used as the default. If the class is a signed, then the actual parameters in the procedure call must be a signal of the same type. For a formal parameters of class variable, the actual parameters must be a variable of the same type. However, for a constant formal parameter, the actual parameter can be any expression that evaluates to a constant of the proper type. This constant value is used inside the procedure and cannot be changed. Thus a constant formal parameter is always of mode in, signals and variables can be of mode in, out (or) inout. Parameters of mode out and inout can be changed in the procedure, so, they are used to return values to the caller.

The formal parameter-list in a function declaration is similar to that of a procedure, except parameters of class variable are not allowed. Further, all parameters must be of mode in, which is the default mode. Parameters of mode out or inout are not allowed, since a function returns only a single value and this value cannot be returned through a parameter.

Example :

Decimal to binary and binary to decimal converters using subprogram-procedure:

```
Library IEEE;
Use IEEE.Std_logic_1164.all;
Entity DTB is
        Port
        (Byte : IN Std_logic_vector (7 down to 0);
        int : OUT integer range 0 to 255);
```

```
End DTB;
Architecture DTB-arch of DTB is
        procedure by-int (By: In Std_logic_vector (7 down to 0); inte:OUT integer) is
              variable result : integer : = 0;
        begin
        for I in 0 to 7 loop
        If By (I) = '1' then
        Result : = Result + 2**I;
        end if;
        end loop;
        Inte : = Result;
        End By-int;
        Begin
        Process (Byte)
        Variable OT : INTEGER;
        BEGIN
        By-int (Byte, OT);
        Int ⇐ OT;
        End process;
              End DTB-arch;
```

7.14. COMPARISON BETWEEN VHDL AND VERILOG

Introduction. There are now two industry standard hardware description languages, VHDL and Verilog. The complexity if ASIC and FPGA designs has meant an increase in the number of specialist design consultants with specific tools and with their own libraries of macro and mega cells written in either VHDL or Verilog. As a result, it is important that designers know both VHDL and Verilog and the EDA tools vendors provide tools that provide an environment allowing both languages to be used in unison. For example, a designer might have a model of a PCI bus interface written in VHDL, but wants to use it in a design with macros writer in Verilog.

Verilog. Verilog HDL is a hardware description language that can describe hardware not only at the gate level and the register-transfer level (RTL), but also at the algorithmic level. This makes translating a design described in Verilog HDL to logic gates a non-trivial process.

Background. VHDL (Very high speed integrated circuit Hardware Description Language) became IEEE standard 1076 in 1987. It was updated in 1993 and is known today as "IEEE standard 1076 1993". The Verilog hardware description language has been used far longer than VHDL and has been used extensively since it was launched by Gateway in 1983. Cadence bought Gateway in 1989 and pened Verilog to the public domain in 1990. It became IEEE standard 1364 in December 1995.

There are two aspects to modelling hardware that any hardware description language facilities; true abstract behaviour and hardware structure. This means modelled hardware behaviour is not prejudiced by structural or design aspects of hardware intent and that hardware structure are capable of being modelled irrespective of the design's behaviour.

1. Capability. Hardware structure can be modelled equally effectively in both VHDL and Verilog. When modelling abstract hardware, the capability of VHDL can sometimes only be achieved in Verilog when using the PLI. The choice of which to use is not therefore based society on technical capability but on: Personal preferences, EDA tool availability and Commercial business and marketing issues.

2. Compilation

VHDL. Multiple design-units (entity/architecture pairs) that reside in the same system file, may be separately compiled if so desired. However, it is good design practice to keep each design unit in it's own system file in which case separate compilation should not be an issue.

Verilog. The Verilog language is still rooted its native interpretative mode. Compilation is a means of speeding up simulation, but has not changed the original nature of the language. As a result, care must be taken with both the compilation order of code written in a single file and the compilation order of multiple files. Simulation results can change by simply changing the order of compilation.

3. Data types

VHDL. A multitude of language or user defined data types can be used. This may mean dedicated conversation functions are needed to convert objects from one type to another. The choice of which data types to use should be considered wisely, especially enumerated (abstract) data types. This will make models easier to write, clearer to read and avoid unnecessary conversion function that can clutter the code. VHDL may be preferred because it allows a multitude of language or user defined data type to be used.

Verilog. Compared to VHDL, Verilog data types are very simple, easy to use and very much geared towards modelling hardware structure as opposed to abstract hardware modelling. Unlike VHDL, all data types used in a Verilog model are defined by the Verilog language and not by the user. There are net data types, for example wire, and a register data type called reg. A model with a signal whose type is one of the net data types has a corresponding electrical wire in the implied modelled circuit. Objects, that are signals, of type reg hold their value over simulation delta cycles and should not be confused with the modelling of a hardware register. Verilog may be preferred because of its simplicity.

4. Design reusability

VHDL. Procedures and functions may be placed in a package so that they are available to any design unit that wishes to use them.

Verilog. There is no concept of packages in Verilog. Functions and procedures used within a model must be defined in the module. To make functions and procedures generally accessible from different module statements the functions and procedures must be placed in a separate system file and included using the include compiler directive.

5. Easiest to Learn. Starting with zero knowledge of either language. Verilog is probably the easiest to grasp and understand. This assumes the Verilog compiler directive language for simulation and the PLI language is not included. If these languages are included, they can be looked upon as two additional languages that need to be learned. VHDL may seem less intuitive at first for two primary reasons. First, it is very strongly typed; a feature that makes it robust and powerful for the advanced user after a longer learning phase. Second, there are many ways to model the same circuit. Specially those with large hierarchical structures.

6. Forward and back annotation. A spin-off from Verilog is the Standard Delay Format (SDF). This is a general-purpose format used to define the timing delays in a circuit. The format provides a bi-directional link between, chip layout tools, and either synthesis or simulation tools, in order to provide more accurate timing representations. The SDF format is now an industry standard in it is own right.

7. High level constructs.

VHDL. There are more constructs and features for high-level modelling in VHDL than there is in Verilog. Abstract data types can be used along with the following statements:

* Package statements for model reuse.
* Configuration statements for configuring design structure.
* Generate statements for replicating structure
* Generic statements for generic models that can be individually characterized, for example, bit width.

All these language statements are useful in synthesizable models.

Verilog. Except for being able to parameterise models by overloading parameter constants, there is no equivalent to the high-level VHDL modelling statements in Verilog.

8. Language Extension. The use of language extensions will make a model non standard and most likely not portable across other design tools. However, sometimes they are necessary in order to achieve the desired results.

VHDL. Has an attribute called 'foreign that allows architectures and subprograms to be modelled in another language.

Verilog. The programming Language Interface (PLI) is an interface mechanism between Verilog models and Verilog software tools. For example, a designer, or more likely, a Verilog tool vendor, can specify user defined tasks or functions in the C programming language, and then call them from the Verilog source description. Use of such tasks or functions make a Verilog model non-standard and so may not be usable by other Verilog tools. Their use is not recommended.

9. Libraries

VHDL. A library is a store for compiled entities, architectures, packages and configurations. Useful for managing multiple design projects.

10. Structural replication

VHDL. The generate statement replicates a number of instances of the same design-unit or some sub part of a design, and connects it appropriately.

Verilog. There is no equivalent to the generate statement in Verilog.

8

Analog VLSI and High Speed VLSI

INTRODUCTION

Although digital counter parts replaced many of the analog circuits, the need for good analog circuit design is still there. As most of naturally encountered signals are analog in nature, analog-to-digital and digital-to-analog converters are always needed. So to interface with the real world analog circuitry is included in modern integrated circuit. Although analog circuits occupy only small area of the chip, the design is difficult. So enough care should be taken while designing. This chapter discusses about various integrated circuit building blocks, required in designing analog circuits.

8.1. SWITCHES

The switch finds many applications in IC design. Under this topic the characteristics of switches that are compatible with ICs are studied. Most of the basic electronic devices can be made to operate as a switch. [*e.g.* Resistor, Diode, BJT, MOS.] In this chapter after knowing the ideal characteristic of the switch, we will see how the Bipolar and MOS transistor functions as a switch.

8.1.1. Ideal Voltage-controlled Switch

The Fig. 8.1.1 (*a*) shows the model of a switch. Let terminals 1 and 2 represent the switch terminals and Let 3 is the controlling terminal.

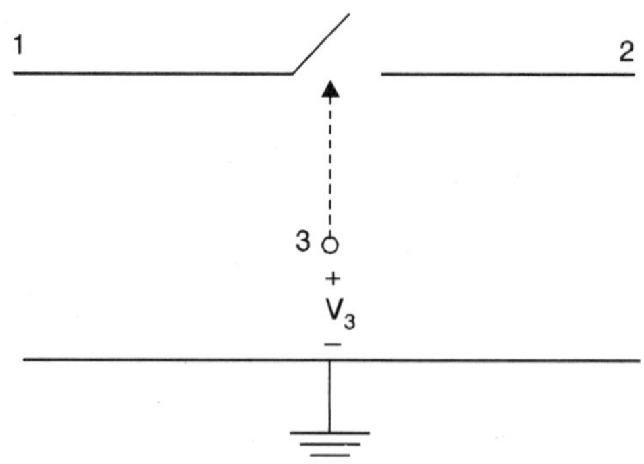

Fig. 8.1.1 (*a*) Switch model.

Ideal switch has two states. ON state and Off state. In the ON state, the ideal switch presents a short circuit between terminals 1 and 2. In the off state, it is open between terminals 1 and 2. This ideal switch cannot be realised in practice. The practical switch implementations only approximate the ideal. Fig. 8.1.1 (*b*) shows the practical switch where R_{on} is a small finite resistance between terminals 1 and 2 when the switch is ON, and R_{off} represents a large resistance between terminals 1 and 2 when the switch is off. To approach towards ideal characteristics R_{on} should be as small as possible and R_{off} should be as large as possible.

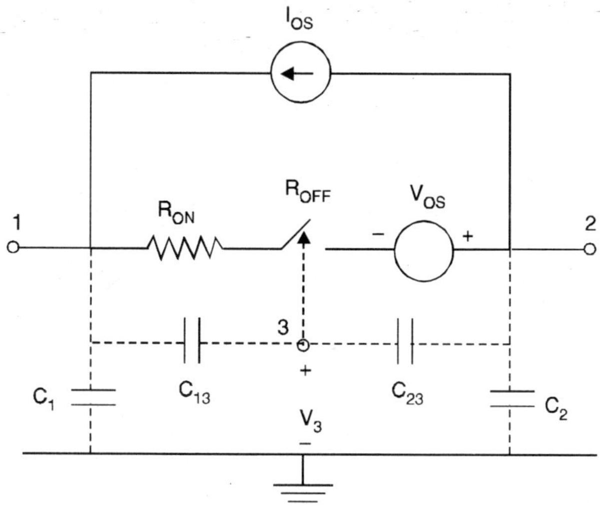

Fig. 8.1.1 (*b*) Practical switch model.

In ON state there should not be any voltage drop between terminals 1 and 2. But in practice an off-set voltage exists and is modelled by an independent voltage source Vos. Similarly when the switch is off the current through the terminal should be zero. But an offset current exists and is modelled by an independent current source $I_{OS.}$ Apart from these, various parasitic capacitances also exist between various switch terminals and ground, which are also shown in Fig. 8.1.1 (*b*).

The graphical characterisation of a switch is shown in Fig. 8.1.1 (*c*). This helps in determining how well the switch can be implemented by a bipolar or MOS transistor.

$V_1 - V_2 = V_{12}$ – > Voltage across the switch

I_1 – > Current in to the terminal 1.

I_2 – > Current out of terminal 2.

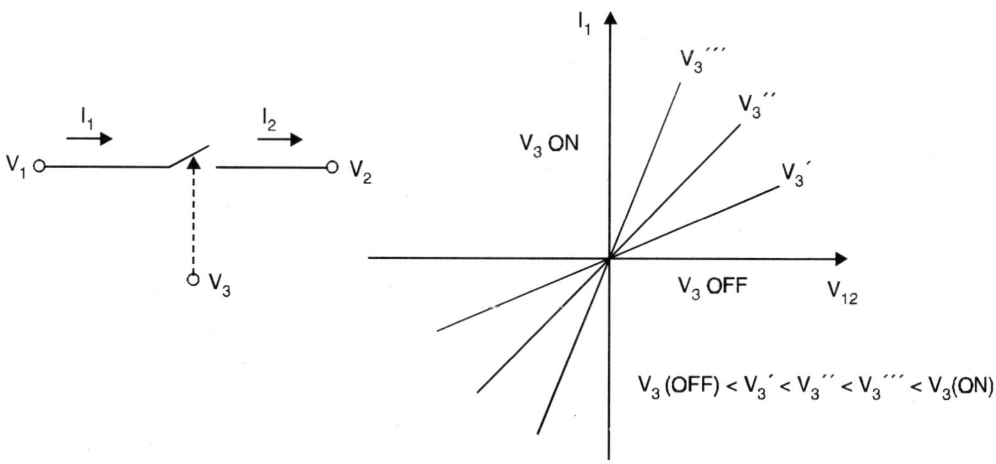

Fig. 8.1.1 (*c*) Graphic Characteristics of a switch.

When V_3 is ON, for an ideal switch $V_{12} = 0$ and I_1 or I_2 can be of any value. Ideal switch is a short circuit in ON state. When V_3 is off, $V_{12} \neq 0$ and $I_1 = I_2 = 0$. The ideal switch is now identical to an open circuit. The graph shows the values of V_3 which cause the switch to be between its on and off states.

8.1.2. Bipolar Transistor as a Switch

The Fig. 8.1.2 (*a*) shows the Bipolar transistor implemented as a switch.

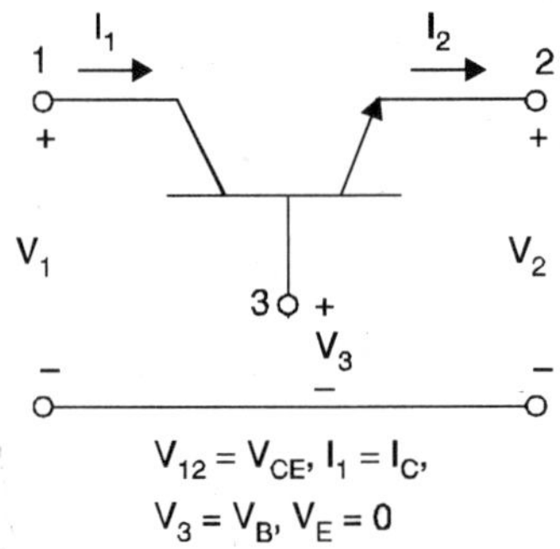

Fig. 8.1.2 (*a*) Bipolar transistor switch.

In Fig. 8.1.2 (*a*) $V_{12} = V_{CE}$

$I_1 = I_C$

$V_3 = V_B$

Keeping emitter voltage $V_2 = 0$ the characteristics are shown in Fig. 8.1.2 (*b*).

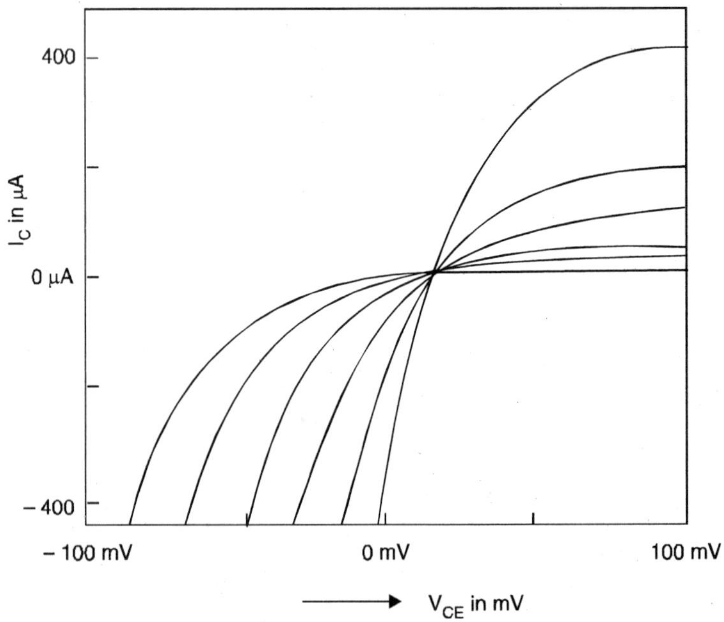

Fig. 8.1.2 (*b*) Characteristics of BJT Switch.

The short comings of BJI switch are listed below :

(*i*) Origin of the characteristics for $I_1 = 0$ is not zero and offset voltage is ≅ 16 mv.

(*ii*) Even with a constant Base voltage, the curves seem to be non-linear. By this dynamic range of the switch is affected.

(*iii*) The switch characteristics are not symmetrical in the first and third quadrants.

Also the BJT requires a small current flowing in the controlling terminal, causing the current flowing in terminal 1 to be different from that flowing out of terminal 2.

8.1.3. MOS Transistor as a Switch

The Fig. 8.1.3 (*a*) shows a MOS transistor implemented as a switch.

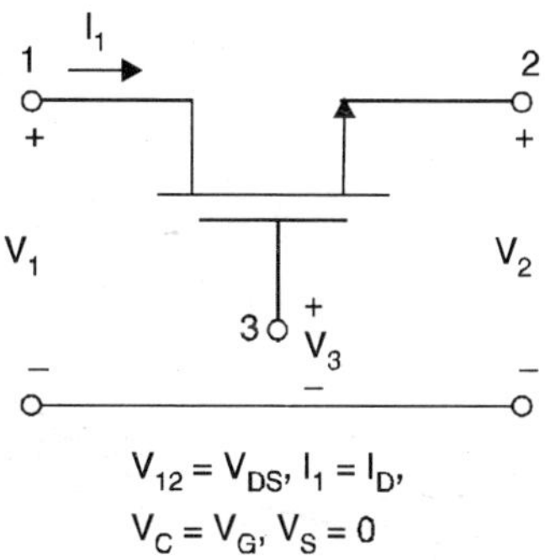

Fig. 8.1.3 (*a*) MOS transistor as a switch.

In Fig. 8.1.3 (*a*) $V_{12} = V_{DS}$

$I_1 = I_D$

$V_3 = V_G$

The graphical characteristics of MOS as a switch is shown in Fig. 8.1.3 (*b*).

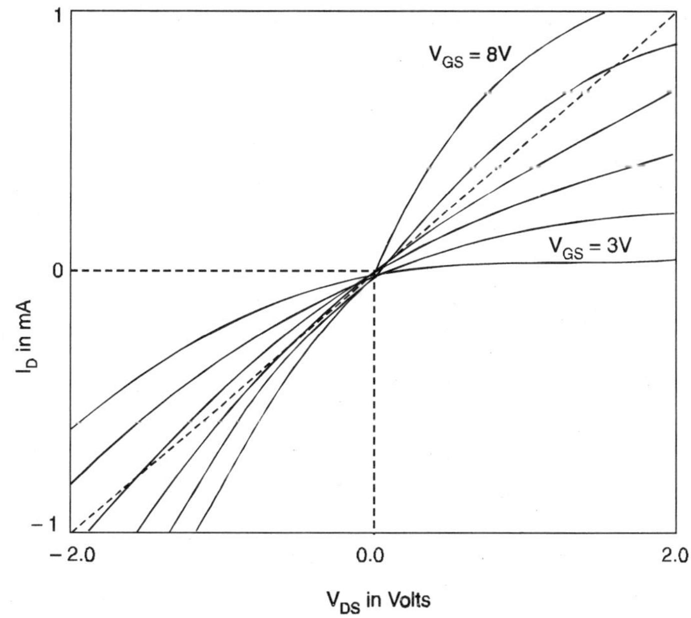

Fig. 8.1.3 (*b*) Graphical characteristics of MOS.

Keeping source voltage equal to zero the characteristics is drawn between V_{DS} and I_D for various values of V_{GS}. From the characteristics, it is obvious that it better approximates to an ideal switch. So it is recommended that MOS can be well be used as a switch. The Fig. 8.1.3 (*c*) shows a MOS transistor, which can be used as switch.

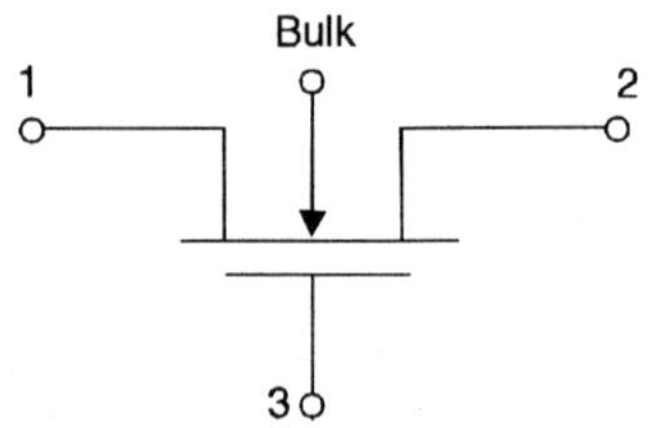

Fig. 8.1.3 (*c*) MOS as a switch.

The Bulk (substrate) terminal is of importance here as it influences the switch operation. For realising the performance of this switch compare large signal model of the MOS transistor with Fig. 8.1.1.(*b*).

The terminals 1 and 2 can be either the drain or source. The R_{ON} consist of the series combination of drain resistance R_D, source resistance R_S and the channel resistance.

To find On Channel Resistance. In ON state, Voltage across switch is very small and V_{GS} should be large. Now the device may be said to be in ohmic region. Ignoring channel length Modulation effects drain current in given by

$$I_D = \frac{K'W}{2L}\,[2(V_{GS} - V_T)\,V_{DS} - V^2_{DS}]\;;\qquad 0 < V_{DS} < V_{GS} - V_T$$

Assuming no offset voltage, large signal channel resistance when $V_{GS} > V_T$ is

$$R_{ON} = \frac{1}{\partial I_D / \partial V_{DS}} = \frac{1}{(K'W/L)\,(-V_{GS} - V_T - V_{DS})}$$

Fig. 8.1.3 (*d*) below illustrates the above equation for very small values of V_{DS}

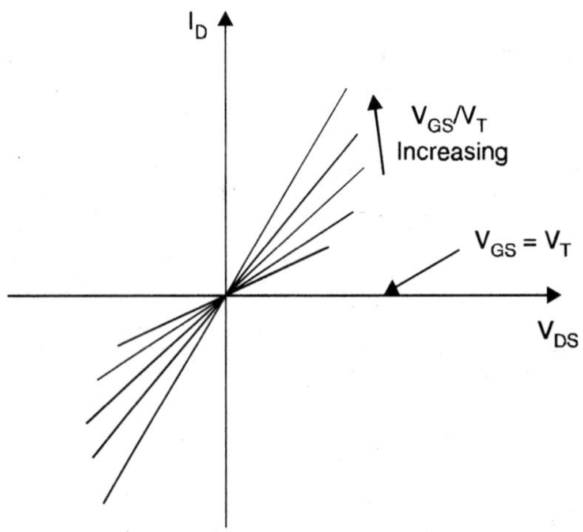

Fig. 8.1.3 (*d*) Illustration of the on state of a switch.

When $V_{GS} < V_T$ and $V_{DS} = 0$, R_{ON} is infinite. For large V_{DS}, the curves will start to decrease in slope. For small V_{DS}, a plot of R_{ON} as a function of V_{GS}/V_T is shown in Fig. 8.1.3 (*e*) for various values of W/L. A lower value of R_{ON} is achieved for large values of W/L.

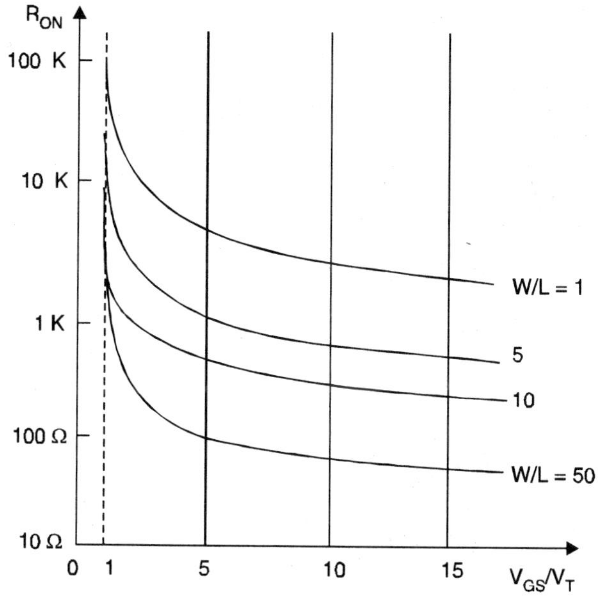

Fig. 8.1.3 (*e*) Illustration of the R_{ON} Characteristics of a MOS transistor with various W/L.

Switch in Off State. In this state $V_{GS} \leq V_T$ and transistor is in cutoff region. R_{off} is ideally infinite. ($\simeq$ in the range of 10^{12} Ω) ∴ a leakage current will flow from drain and source to substrate. This current influences much the operation of the switch. Hence this is an important parameter than R_{off} and it is considered seriously. This leakage current is a combination of sub threshold current, surface leakage current and the package leakage current. Also this is temperature dependent. The resistances representing this leakage current R_1 and R_2 can be connected from terminal 1 and 2 ground in Fig. 8.1.1. The offset voltage of the MOS device is zero. But the parasitic capacitors between the different terminals of MOS determines the commutation rate of MOS switch.

Typical Voltages used for an *n*-channel MOS Switch. In the n-channel MOS, to bring the switch in ON state gate voltage must be larger than either the drain or source voltage. Also bulk is kept at most negative potential for the *n*-channel MOS switch. Suppose that the on voltage of the gate is the positive power supply V_{DD}, with the bulk connected to ground this should keep the MOS switch on until the signal on the switch terminals approach $V_{DD} - V_T$. This often introduces an undesired nonlinear distortion in the transmission of analog signals. Typical voltages used for an *n*-channel MOS switch are shown in Fig. 8.1.3 (*f*) where the switch is connected between two circuits and the power supplies are ± 5 V.

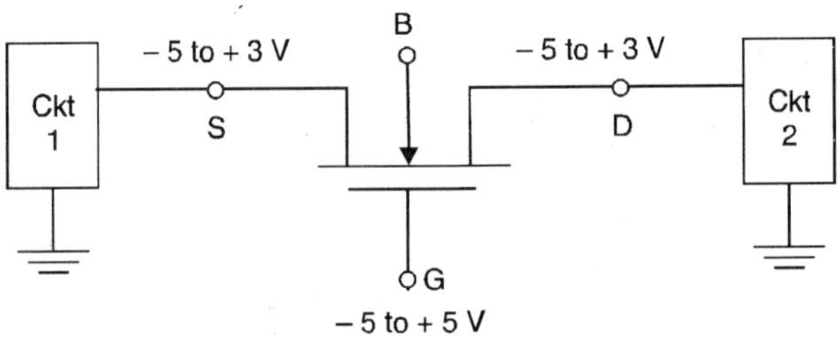

Fig. 8.1.3 (*f*) N channel MOS as a switch.

8.1.4. An Application of a MOS Switch

Let us consider an application of a switch to charge an Integrated capacitor. As in Fig. 8.1.4 (*a*).

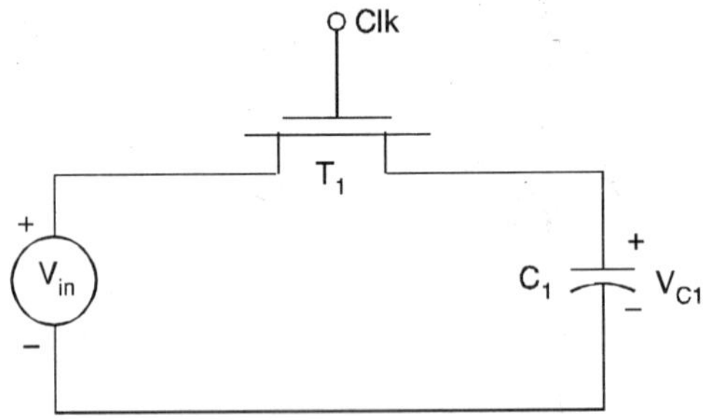

Fig. 8.1.4 (*a*) An application of MOS switch where clk – clock T_1 – MOS transistor.

1. Switch in ON state. When Clk goes high T_1 is ON and C_1 charges towards V_{in} with time constant R_{on} C_1. The equivalent circuit is shown in Fig below.

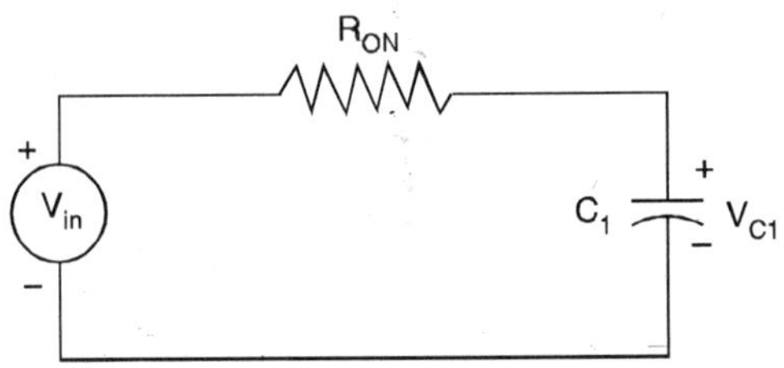

Fig. 8.1.4 (*b*) Switch in ON State.

For successful operation, The time constant R_{ON} C_1 should be very much less than T (T is the time when clk is high). Also the value R_{ON} is not constant during charging of C_1 because the value of V_{GS} decreases as the capacitor charges. The maximum acceptable value of R_{ON} will determine the required values, width and length of MOS transistor T_1. Generally small capacitors are only used. ∴ switches with a large R_{ON} can also perform satisfactorily. One of the most serious limitations of switches is clock feed through that can occur between the switch control signal and the switch terminals. This coupling occurs through the parasitic capacitors of the switch. The parasitics associated with the switch are illustrated in Fig. 8.1.4 (c).

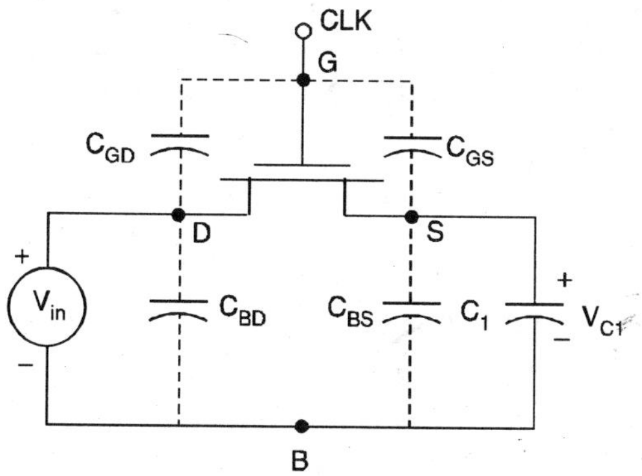

Fig. 8.1.4 (*c*) Parasitics associated with the switch.

The effect of clock feed through can be partially cancelled by using a dummy Mos transistor T_D, to which an inverse clock $\overline{clk}$ is applied. This produces an opposite clock feed through against T_1. So partially, the clock feed through effect is cancelled. Fig. Below shows the use of a dummy transistor to cancel clock feed through.

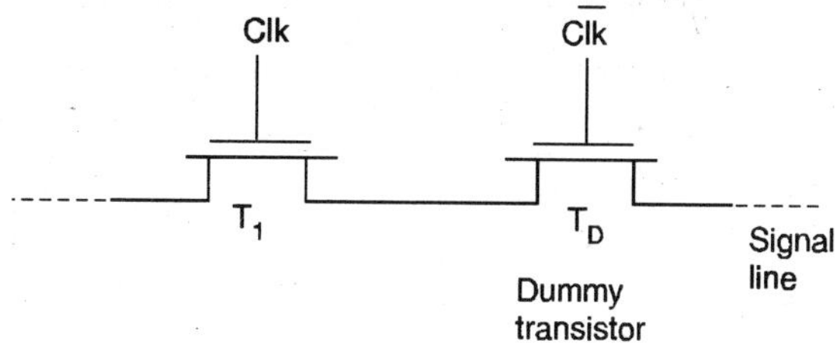

Fig. 8.1.4 (*d*) Use of a dummy transistor to cancel clock feed through.

In conclusion, MOS transistor makes a good switch realization for Ics. It requires small area, dissipates very little power and provide reasonable values of R_{ON} and R_{off} for most applications.

8.2. ACTIVE RESISTORS

In analog signal, processing the resistors and Capacitors plays a vital role. The gain of an amplifier is determined by ratios of resistors or capacitors while the time constant of a filter is determined by product of resistors and capacitors. Hence, the performance of analog systems largely depend upon these components.

The capacitors can be implemented as a paralled plate configuration. The MOS capacitance uses thinox as the insulator. The conducting plates can be of metal or polysilicon. The only problem with capacitors is the influence of parasitics which can be ignored in many applications.

But the resistors implemented by IC technologies were found to be lacking in several areas of performance.

1. Large values of resistors requires large areas
2. They become Non linear with high value of resistance
3. Tolerance, temperature co-efficient and voltage co-efficient are poor.

Hence, we shall discuss two ways of implementing resistors without having these inherent disadvantages.

1. **To use active devices such a BJT, MOS in implementing resistors: This minimizes the area required for resistors.**

2. To use switched capacitors: Switched capacitors circuits are extremely popular for their accurate frequency response good linearity and dynamic range. If resistor is implemented using switched capacitors, accuracy can be improved.

The resistors can be classified as DC and AC resistors. A dc resistors produces a dc voltage drop of V_{dc} for a dc current of I_{dc}. An ac resistor produces an ac voltage. ΔV_Q for an ac current of ΔI_Q.

8.2.1. DC resistor Implemented through Active Elements (BJT and MOS)

We know both BJT and MOS are three terminal devices. By properly connecting these three terminals, the active device can be made as a two terminal active resistor. This can provide a dc voltage drop and act as a small signal resistance that is linear over a small range. Also this requires much less area than is required with an equivalent polysilicon or diffused resistor.

The active resister can be implemented by simply connecting the gate of an *n* or *p*-channel enhancement MOS device to the drain, as shown in Fig. 8.2.1 (*a*),(*b*). To eliminate bulk effect the source of *n*-channel should be kept at most negative supply voltage and source of *p*-channel should be kept at most positive voltage. Now V_{DS} is equal to V_{DS}. Fig. 8.2.1 (*c*) shows the transconductance curve (I_D V_S V_{GS}) of the MOS transistor, which characterises the large signal behaviour of the active resistor. The curve is not linear. When the signal giving is very small, the active resistor works very well. As the gate is connected to drain, the transistor is said to be operated in saturation region for $V > V_T$.

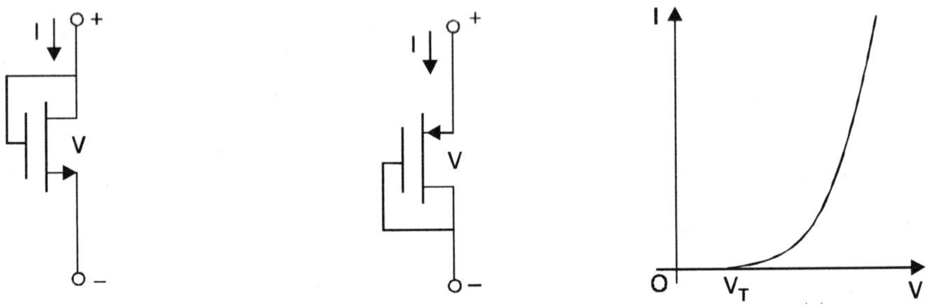

(*a*) *n*-channel enhancement active resistor. (*b*) *p*-channel enhancement active resistor. (*c*) VI characteristics of the MOS active resistor.

Fig. 8.2.1.

Now $$I_D = \frac{K'W}{2L}[(V_{GS} - V_T)^2] \qquad \text{... 8.2.1 } (a)$$

or $$V_{GS} = V_{DS} = V_T + \left(\frac{2I_D L}{K'W}\right)^{1/2}$$

Where $K' = \mu_0 C_{OX}$

When I_D is known V_{GS} can be found and Viceversa. As gate is connected to the drain V_{DS} controls I_D, and therefore the channel transconductance becomes a channel conductance. The small signal conductance can be found by differentiating equation 8.2.1(*a*) with respect to V.

$$g = \frac{\partial I}{\partial V} \cong \left(\frac{2IK'W}{L}\right)^{1/2} = \frac{K'W}{L}(V - V_T)$$

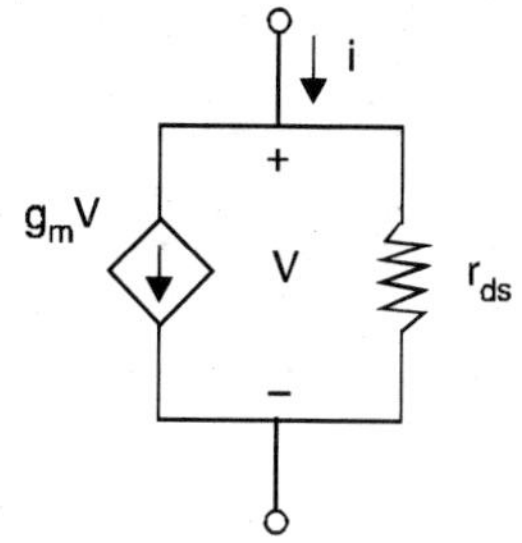

Fig. 8.2.1 (*d*) Small signal model of MOS transistor.

The Fig. 8.2.1 (*d*) shows small signal model of MOS transistor. From the Fig. the conductance *g* can found as

$$g = \frac{1 + g_m rds}{rds} \cong g_m \quad \text{as} \quad g_m rds >> 1$$

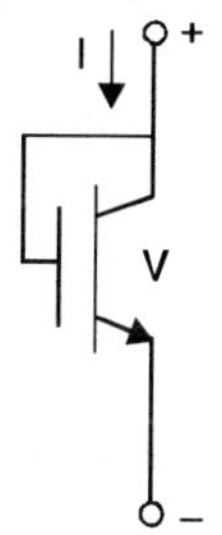

(*e*) *npn* BJT active resistor.

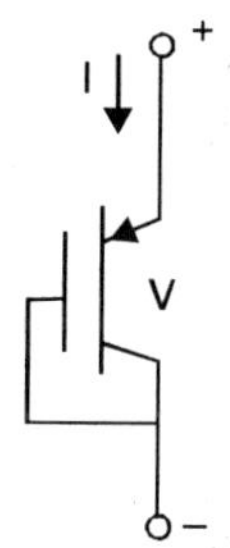

(*f*) *pnp* BJT active resistor.

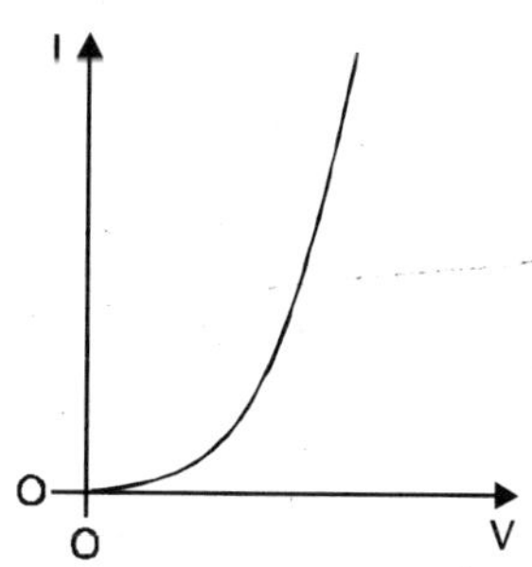

(*g*) VI characteristics of BJT active resistor.

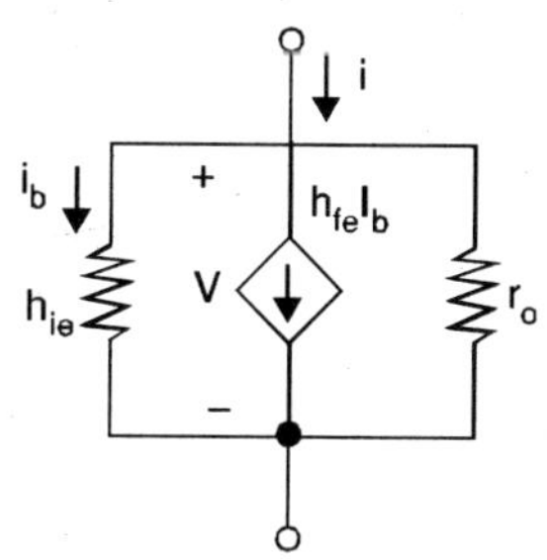

(*h*) Small signal model.

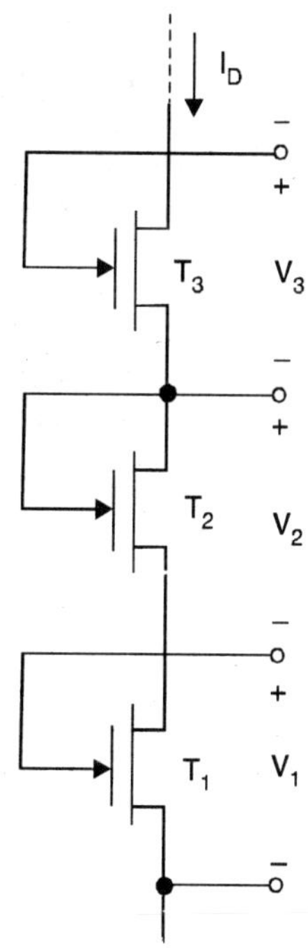

(*i*) Devices cascaded to create large voltage drops.

Fig. 8.2.1.

In a similar manner, an active resistor can be formed using BJT also. In BJT base and collector are tied together. Fig. 8.2.1 (*e*), (*f*) and (*g*) shows the BJT active resistor and its IV characteristics.

$$I \cong I_S \exp (V/V_t) \text{ and} \qquad \text{... 8.2.1 } (b)$$

$$V = V_t \ln (I/Is)$$

I_S = Reverse saturation current of emitter base junction.

$$V_t = \frac{kT}{q} = 0.026 \text{ V @ room temperature.}$$

Small signal conductance can be found by differentiating equation 8.2.1 (*b*) w.r.t to emitter base voltage.

$$g = \frac{\partial I}{\partial V} = \frac{|I|}{V_t}$$

Also it can be found from small signal model of BJT in Fig. 8.2.1 (*h*) as

$$g = h_{oe} + \frac{1 + h_{fe}}{h_{ie}}$$

h_{oe} → output conductance of CE configuration

h_{fe} → forward voltage ratio

h_{ie} → input impedance

For developing large dc voltage, W/L ratios much less than unity is required. This situation is circumvented by cascading more MOS devices as shown in Fig. 8.2.1 (*i*).

8.2.2. AC Resistor from Active Devices

For realising ac resistors Figs. 8.1.2 (*b*) and 8.1.3 (*b*) can be used. Comparing these two the Mos switch offers a linear output performance with zero dc offset. Linear ac resistance can be got by controlling the voltage between the gate and source. The linearity of ac resistor is limited by following two effects.

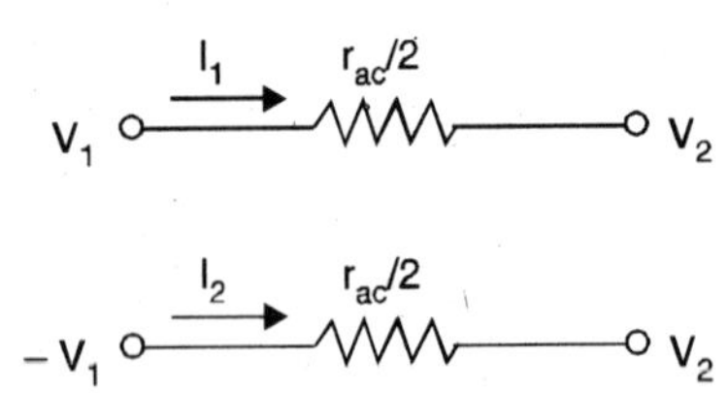

(*a*) Differential AC resistor configuration.

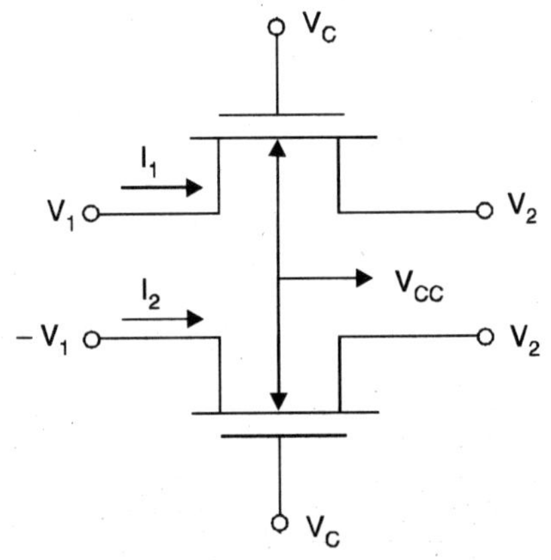

(*b*) Single MOS implementation of Fig (*a*).

Fig. 8.2.2.

(*i*) Negative values of V_{DS} is affected by V_{BS} and positive values V_{DS} is affected by V_{DS} itself.

(*ii*) The bulk effect which causes V_T to change whereby linearity is affected.

The influence of V_{DS} on ac resistor can be eliminated by using two identical devices biased in such a way that the effect of one will cancel the other. By using resistors in differential pairs the bulk effect can be cancelled which leads to increase in linearity. Fig. 8.2.2 (*a*) and (*b*).

Different configurations by which an ac resistor can be realised and the extent of linearity provided by them are tabulated below.

S. No.	*Configuration*	*Linearity*	*Controlling Factor*
1.	Single MOS	Poor	Gate Voltage, W/L
2.	Parallel MOS	Good	Vc, W/L
3.	Single MOS differential resistor	Good	Vc, W/L
4.	Double-Mos differential resistor	Good	$(V_{C1} - V_{C2})$, W/L

8.2.3. Switched Capacitor realisation of resistors

In this approach resistors are realised with switches and capacitors keeping the switching frequency very high.

There are several different approaches by which a resistor can be realised using capacitors and switches. They are

1. Parallel switched capacitor resistor realisation
2. Series Switched capacitor realisation
3. Series parallel Switched capacitor realisation
4. Bilinear Switched capacitor realisation

Fig. 8.2.3 (*a*) shows the parallel switched capacitor realisation where clk_1, and clk_2 are clocks which control the opening and closing of switches. V_1 and V_2 are independent dc voltage sources. The clocks used here are non-overlaping clocks. The term non-overlaping refers to two logic signals running at the same frequency and arranged in such a way that at no time both signals are high.

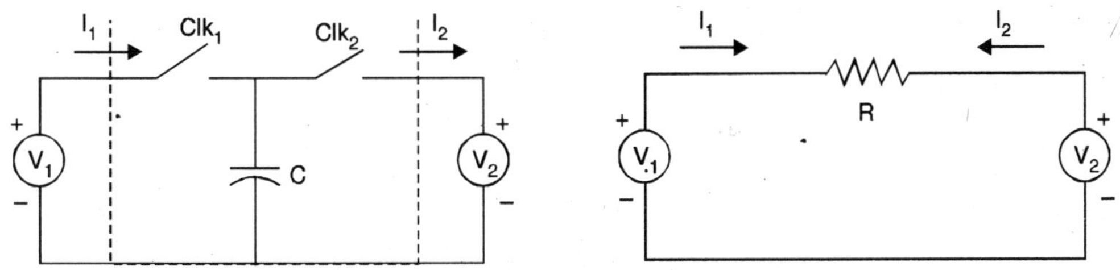

(*a*) Parallel switched capacitor resistor realisation. (*b*) A continuous resistor.

Fig. 8.2.3.

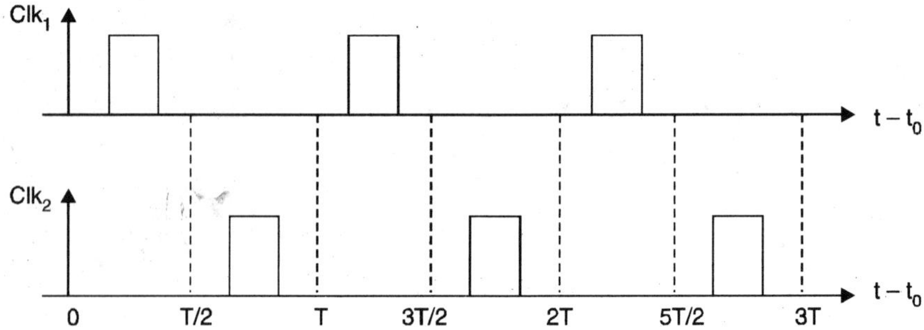

Fig. 8.2.3 (*c*). Clock wave forms for the switches of Fig (*a*).

We know charge on a capacitor q is given by

$$q = CV$$

C – Capacitance of capacitor

V – Voltage in volts

As we use non-overlapping clocks C is charged to V_1 and then to V_2 during each clock period. ∴ change in charge over one clock period, Δq, that is transferred from node V_1 into node V_2 is given by $\Delta q_1 = C (V_1 - V_2)$.

This charge transfer is repeated every clock period.

$$\therefore \text{ Average current due to charge transfer} = \frac{\Delta q_1}{\text{Clock period}}$$

$$I_{avg} = C\,\frac{(V_1 - V_2)}{T} = \frac{1}{T}\int_{t_0}^{t_{0+T}} I_1(t) \qquad \text{... 8.2.3 (a)}$$

Relating the above equation to an equivalent resistor circuit shown in Fig. 8.2.3 (*b*).

$$\text{We get,} \quad I_{avg} = \frac{V_1 - V_2}{R_{eq}} \qquad \text{... 8.2.3 } (b)$$

Equating 8.2.3 (*a*) and 8.2.3 (*b*) we get

$$C = \frac{V_1 - V_2}{T} = \frac{V_1 - V_2}{R_{eq}}$$

$$\text{Or} \quad R_{eq} = \frac{T}{C}$$

$$\text{But} \quad T = \frac{1}{f_c} \quad (f_c - \text{clock frequency})$$

$$\therefore \quad R_{eq} = \frac{1}{Cf_c}$$

This equation shows that switched capacitor resistor realisation is dependent upon the capacitor C and the clock frequency f_c. The Fig 8.2.3 (*c*) shows the clock waveforms.

Advantages of this implementation is that very large resistors can be realised with small area using switched capacitors. The other typos of switched capacitor realisations of resistors and the resulting equivalent resistance are given below.

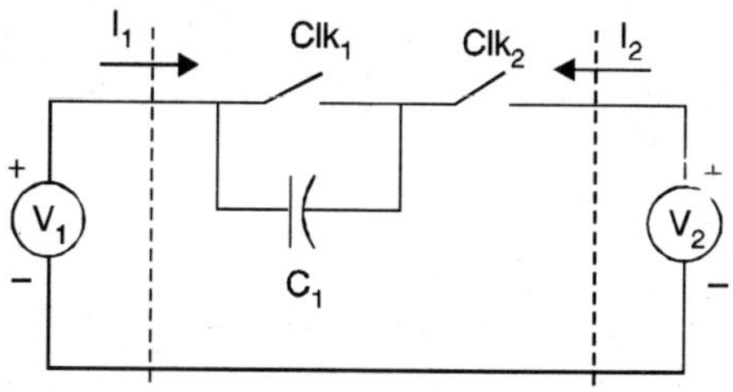

$$R_{eq} = \frac{T}{C} = \frac{1}{C f_c}$$

Fig. 8.2.3 (*d*) Series switched capacitor realisation of a continuous resistor.

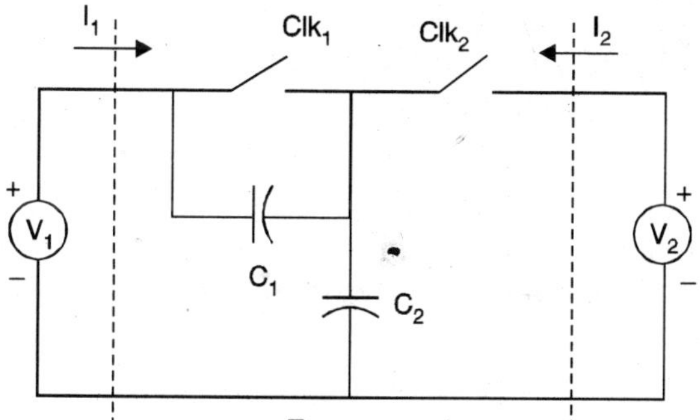

$$R_{eq} = \frac{T}{C_1 + C_2} = \frac{1}{(C_1 + C_2) f_c}$$

Fig. 8.2.3 (*e*) Series parellel switched capacitor realisation of a continuous resistor.

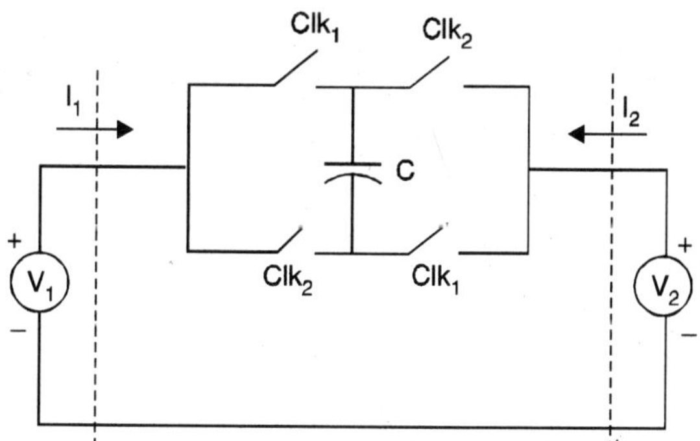

$$R_{eq} = \frac{T}{4C} = \frac{1}{4C f_c}$$

Fig. 8.2.3 (*f*) Bilinear switched capacitor realisation of a continuous resistor.

8.3. CURRENT SOURCES AND SINKS

This section describes how BJT and MOS are used to realise current sources and sinks. An ideal current source is a two terminal element whose current is constant for any voltage across the source. In most of the applications one terminal of current source is either connected to a most +ve or most –ve of the supply in which case it is referred to as current source or current sink respectively. See. Fig. 8.3 (*c*), (*d*), ($V_N = V_{negative}$, $V_P = V_{positive}$) referred as floating current source.

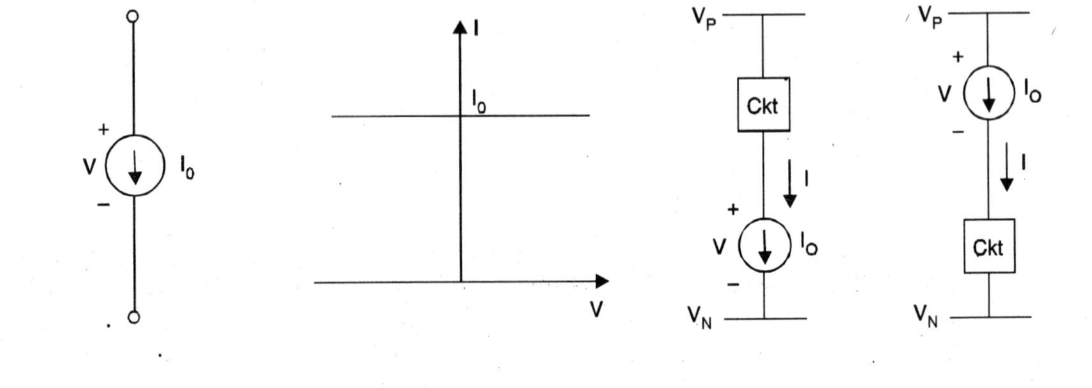

(*a*) Schematic symbol of a current source. (*b*) IV characteristics. (*c*) A current sink. (*d*) A current source.

Fig . 8.3.

Fig. 8.3 (*a*) and (*b*) shows an ideal current source and its V-I characteristic. But this ideal characteristic cannot be realised in practice. The Fig. 8.3 (*e*), (*f*) shows practical I-V characteristics of a current source and sink which only approximates the ideal.

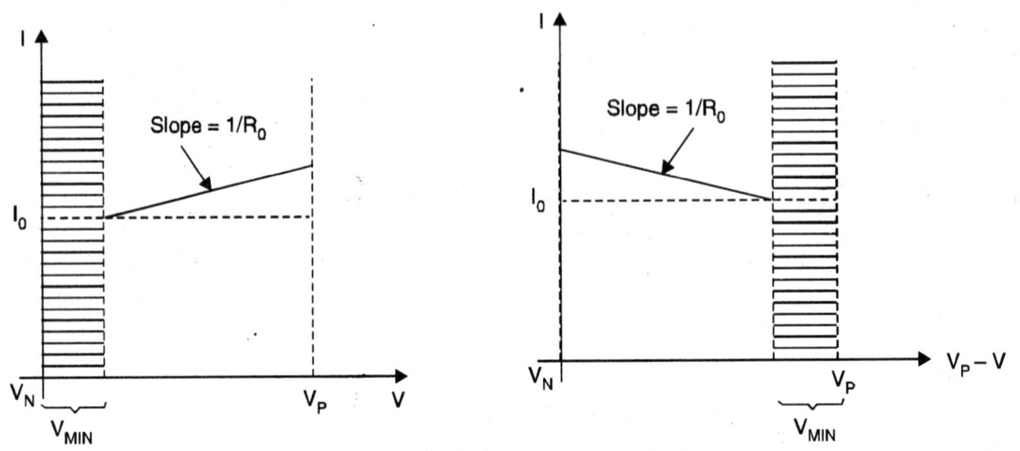

(*e*) Practical IV characteristics of a current Sink. (*f*) Practical IV characteristics of a current source.

Fig. 8.3.

The current source/sink are characterised by two important parameters V_{min} and R. where V_{min} is minimum voltage below which the current source/sink will not be a good approximation of ideal. Even in the region of good approximation, a slope of I/R exists. Here R. represents the parallel resistance of the current source/sink and is ideally infinite.

BJT realisation of a current sink

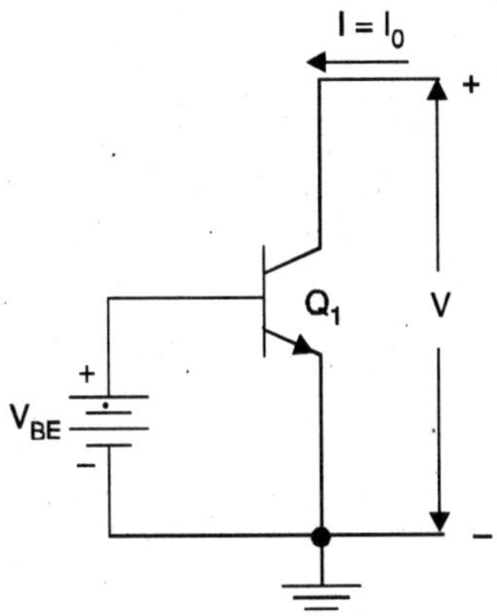

(*g*) BJT current sink.

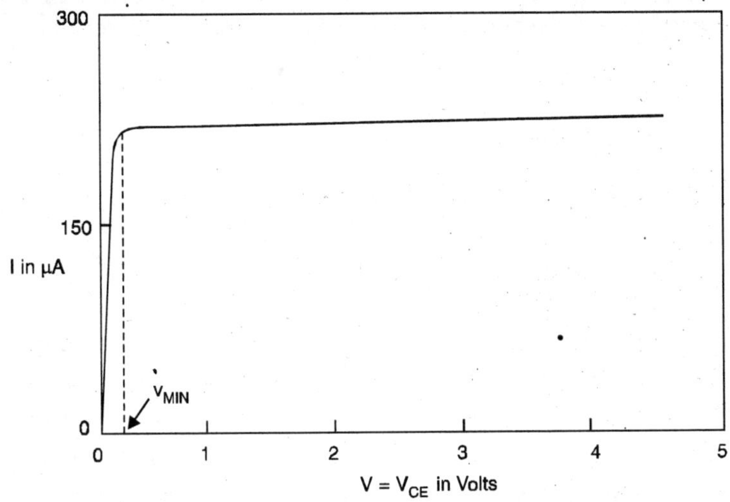

(*h*) IV characteristics.

Fig . 8.3.

Fig. 8.3 (*g*) and 8.3 (*h*) shows the BJT current sink and its I-V characteristics respectively. The transistor Q_1 is biased by battery V_{BE}. For $V > V_{CE}$ (Sat) the current is given by

$$I = I_S \exp\left(\frac{V_{BE}}{V_t}\right)$$

Where V_{CE} (Sat) → Collector-emitter voltage when Q_1 is in saturation region.

Also V_{CE} (Sat) $\simeq V_{min}$.

For $V > V_{min}$, the slope of the current is proportional to g_o, Thus, the output resistance of the current sink,

$$R_O \simeq 1/g_o$$

MOS realisation of current sink

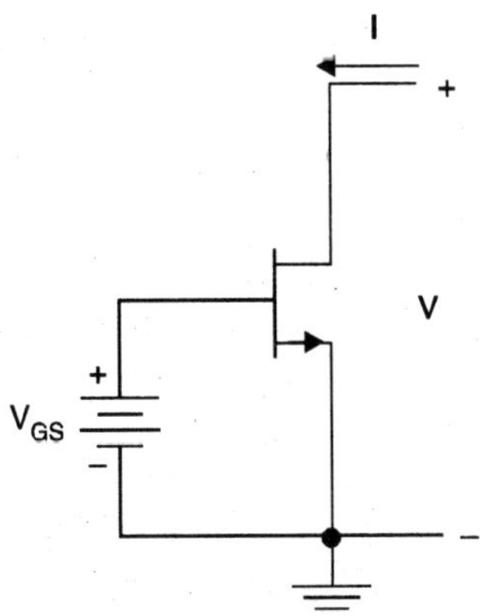

(*i*) MOS current sink.

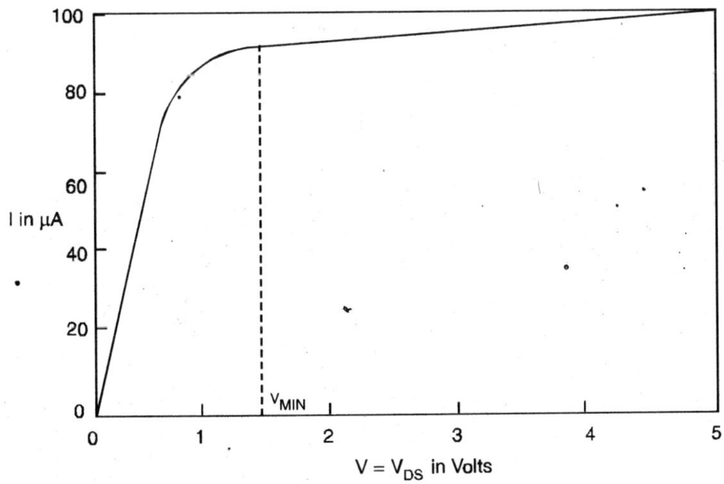

(*j*) IV characteristics.

Fig. 8.3.

Fig. 8.3 (*i*) and 8.3 (*j*) shows the MOS current sink and its I-V characteristics. The transistor is biased by V_{GS}. Keeping $\lambda = 0$, the expression for current is given by

$$I_D = \frac{K'W}{2L}\,[(V_{GS} - V_T)^2]$$

Here also $$V_{Min} \simeq V_{DE}(sat) = V_{GS} - V_T$$

For $V > V_{min}$, the slope of current is proportional to g_{ds} and the output resistance of the current sink is $R_0 = \frac{1}{g_{ds}}\,1/g_{ds} = \frac{1}{\lambda|I|}$

The BJT and MOS current source / sinks are good approximations of the ideal only for limited values of voltage. Comparing the two both has same values of output resistance, but BJT has much smaller value of V_{Min}. As the output resistance of simple BJT or MOS current source/sinks are not sufficiently large, feed back technique (either positive or negative) is used to improve the output resistance without affecting I. The technique of applying positive feedback is also called Bootstrapping and this works well with depletion devices. However, all these techniques increase V_{min}, which is undesirable. There comes a sink referred as regulated cascade current sink. This out performs all the above types by offering high output resistance and a low V_{min}.

8.4. CURRENT MIRRORS

In these circuits matching principle of similar devices are used for analysis. As the matching principle can be considered equivalent to looking into a mirror the circuits using this principle are called current mirror circuits.

Simple CMOS Current Mirror. As shown in Fig. 8.4.1 (*a*) I_{in} is the independent current source. Two transistors Q_1 and Q_2 act as current mirrors. Both are assumed to be in active region, and also it is assumed that both are of same size. Under this assumption, same current will flow in both the transistors with same V_{gs} applied. However, when finite output impedance is considered, the transistor with larger drain-source voltage will have a larger current. In addition, the finite output impedance of the transistors will cause the small signal output impedance of the current Mirror (Small signal impedance looking into the drain of Q_2) to be less than infinite.

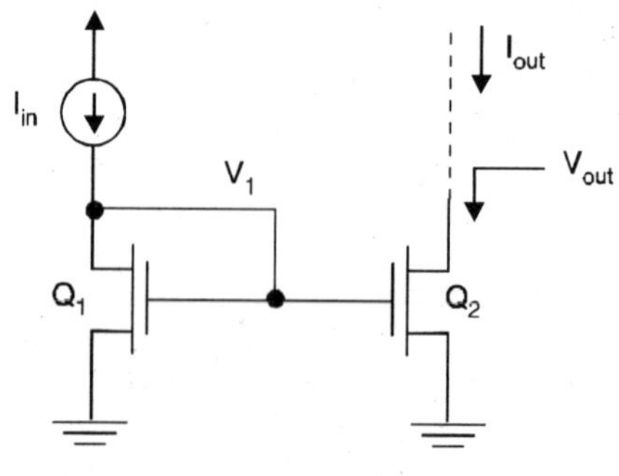

Fig. 8.4.1 (*a*) Simple CMOS current mirror.

To find the output impedance v_{out} of the current mirror

1. Get the small signal model of Q_1
2. At the output node of current mirror, insert a signal source V_x.

On analysing the above we can understand

$$V_{out} = \frac{V_x}{I_x} = \frac{\text{Signal voltage at output mode}}{\text{Curent flowing out of the source into the drain of } Q_2}$$

Let us first consider the small signal model of Q_1 alone which is shown in 8.4.1 (*b*) and (*c*). We see that, Q_1 is diode connected. (its gate and drain are connected) and I_{in} doesn't exist in the small signal model. ∴ I_{in} was replaced with an open circuit. Also the small signal model used for Q_1 is a low frequency model and hence all the capacitors are ignored in the model. The small signal model can be further reduced by finding thevenin equivalent circuit. The thevenin output voltage V_0 is zero as there is no input signal. The thevenin impedance R_{th} can be found by applying a test signal voltage. V_y, at node V_1 and measuring the signal current i_y.

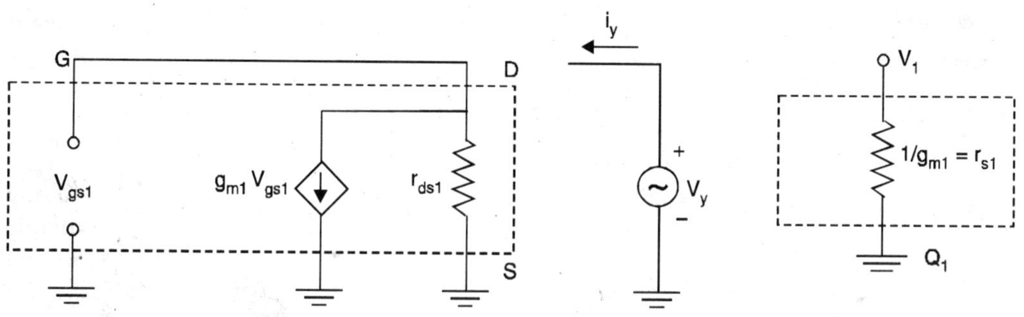

(*b*) Small signal model for Q_1.

(*c*) Equivalent model for Q_1.

Fig. 8.4.1.

From small signal model i_y is given by

$$i_y = \frac{V_y}{r_{ds1}} + g_{m1}, V_{gs1} = \frac{V_y}{r_{ds1}} + g_{m1} V_y$$

Rearranging $$i_y = V_y \left(\frac{1}{r_{ds1}} + gm_1 \right)$$

or $$\frac{v_y}{i_y} = \frac{1}{gm_1} \; || \; r_{ds1}$$

But $$r_{ds1} >> 1/g_{m1}$$

∴ output impedance $V_y/I_y \simeq 1/g_{m1} \simeq r_{s1}$

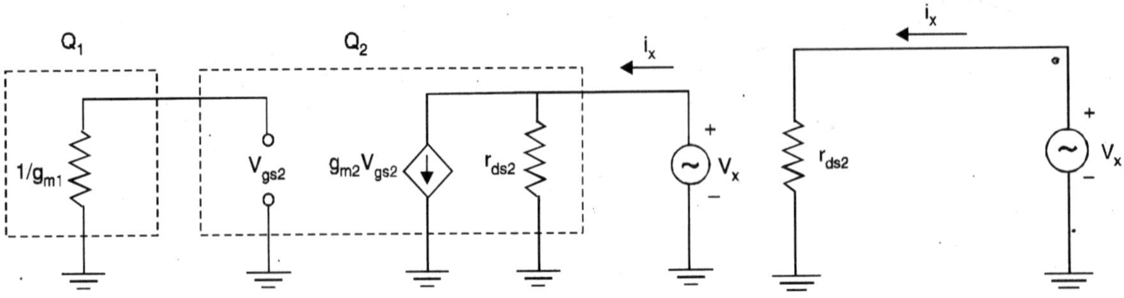

(*a*) Small signal model for current mirror. (*b*) Simplified model.

Fig. 8.4.2.

Now consider the small signal model of overall current mirror which is shown in Fig. 8.4.2 (*a*). In this V_{gs2} is connected to ground via a resistance of $1/g_{m1}$. Since no current flows through the $1/g_{m1}$ resistor, V_{gs2} always equals 0 irrespective of voltage applied at V_x. $\therefore g_{m2} V_{gs2} = 0$. Now the circuit reduces to one that is shown in Fig. 8.4.2 (*b*). From this it is clear that the small signal output impedance, r_{out}, is simply equal to r_{ds2}.

There are also current mirror circuits which offer very high output impedance. Examples are Cascode and Wilson current mirrors.

8.5. COMMON SOURCE AMPLIFIER

Fig. 8.5.1 (*a*) shows a *n*-channel common source amplifier with a *p*-channel current mirror used as active load. This finds its application where high input impedance is desired. By using the current mirror as an active load, high-impedance output load can be realised. Otherwise it requires excessively large resistors or a large power supply voltage for realising high impedance load. Also for a given power supply voltage the gain achieved through this active load is very high.

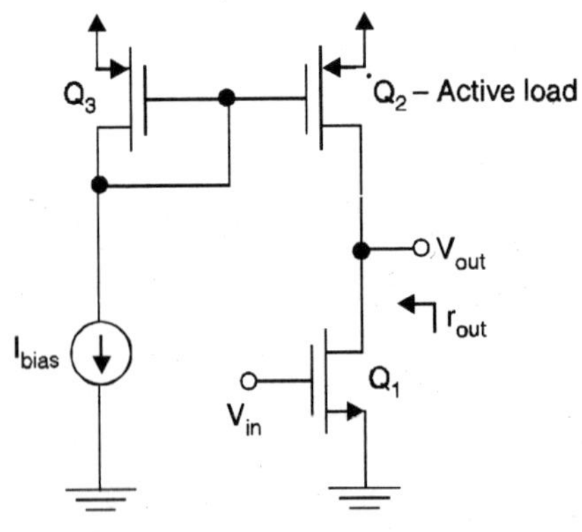

Fig. 8.5.1 (*a*) Common source amplifier with a current mirror active load .

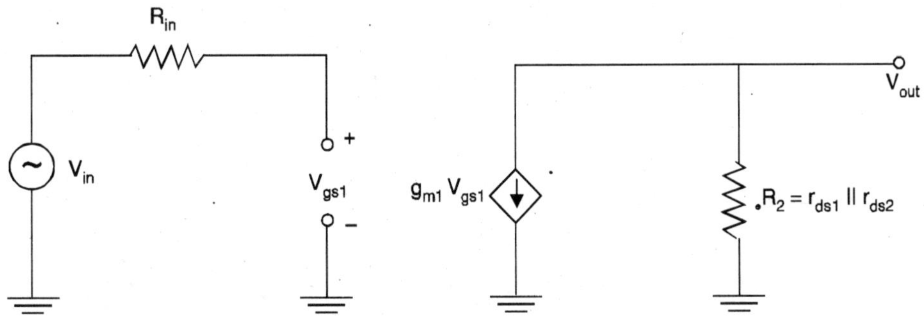

Fig. 8.5.1 (*b*) Small signal equivalent.

Fig. 8.5.1 (*b*) shows the small signal equivalent circuit for low frequency analysis of the common source amplifier. V_{th} and R_{th} are the Thevenin equivalent of the input source. Both the transistors are biased to be in active region. The output resistance R_2 is the parallel combination of r_{ds1} and r_{ds2}.

Where r_{ds1} = drain –to-source resistance of Q_1

• r_{ds2} = drain to source resistance of Q_2

Neglecting the body effect, the current through source of Q_1 is always zero. Using small signal analysis, we have $V_{gs1} = V_{in}$ and

$$\therefore \quad A_V = V_{out}/V_{in} = -g_{m1} R_2 = -g_{m1} (r_{ds1} \mid\mid r_{ds2})$$

The gain achieved from this circuit is in the range of –10 to –100. To achieve similar gain using resistive loads, much larger power-supply voltages are required and the power dissipation is also very high. But resistive load can be used for low gain, high frequency stages as it has less parasitic associated with them.

8.6. DIFFERENTIAL AMPLIFIER

A differential amplifier is one which amplifies differential signals. It consists of two input nodes to which signals V_1 and V_2 are fed. These two signals can be decomposed into difference mode signal, $V_D = V_1 - V_2$, and a common mode signal, $V_c = (V_1 + V_2)/2$. But the differential amplifier amplifies only the difference between two input signals.

CMRR (Common Mode Rejection Ratio). The ratio of the differential mode gain to the common mode gain is called the CMRR. The differential amplifier has become a very useful circuit because of its compatibility with IC technology.

8.6.1. MOS Differential Amplifier

A differential pair together with a biasing current source is shown in Fig. 8.6.1 (*a*). Its low frequency small signal model is shown in Fig. 8.6.1.(*b*).

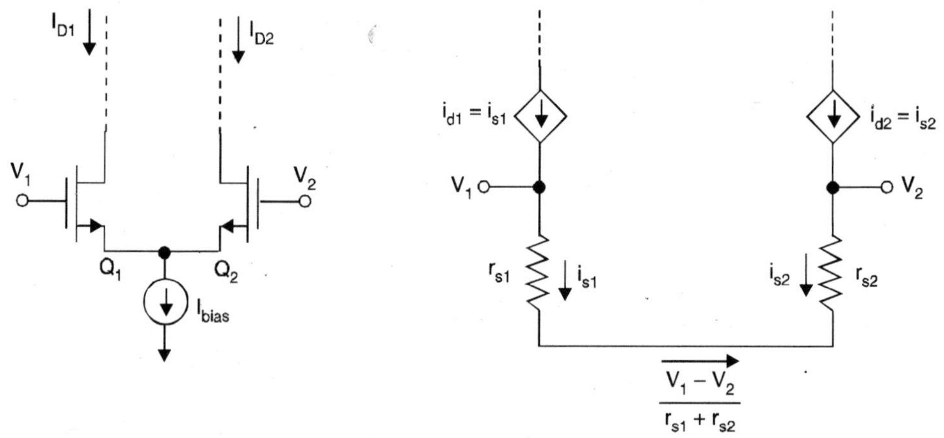

(*a*) A MOS differential pair. (*b*) Small signal model.

Fig. 8.6.1.

Ignoring the output impedance of the transistors differential input voltage $V_{in} = V_d = V_1 - V_2$.

$$i_{d1} = i_{s1} = \frac{V_{in}}{r_{s1} + r_{s2}} = \frac{V_{in}}{1/g_{m1} + 1/g_{m2}}$$

As both Q_1 and Q_2 have the same bias currents, $g_{m1} = g_{m2}$

$$\therefore\ i_{d1} = \frac{g_{m1}}{2} V_{in}$$

Also, $$i_{d2} = i_{s2} = -\,i_{d1} \text{ and } i_{d2} = \frac{-g_{m1}}{2} V_{in}$$

Now differential output current $I_{out} = i_{d1} - i_{d2} = g_{m1} V_{in}$. Next as shown in Fig. 8.6.1 (*c*) current mirror circuit is used as an active load for the differential pair. This circuit is used as the first gain stage in two stage integrated op-amp.

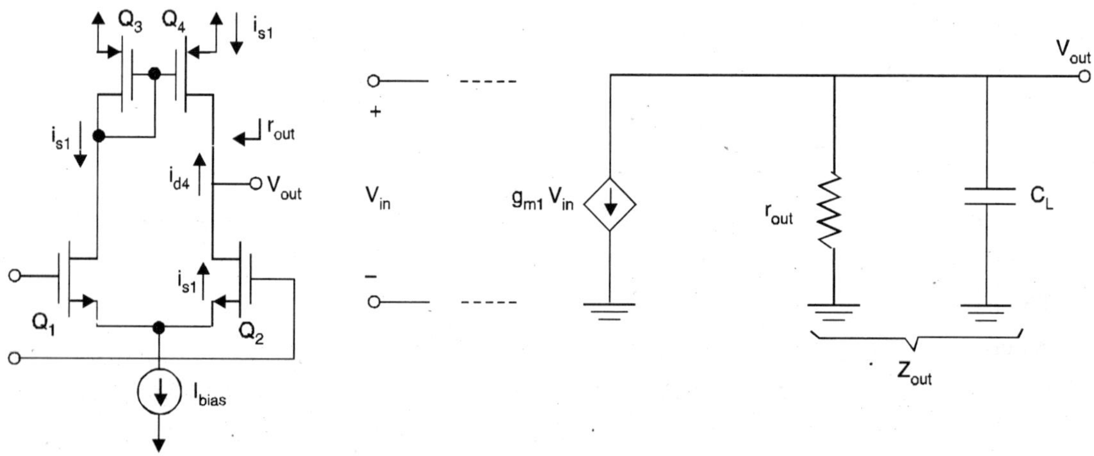

(*c*) A differential input, single-ended output MOS gain stage. (*d*) Small signal model.

Fig. 8.6.1.

In Fig. The differential pair is realised using n-channel transistors and active load (current mirror) is realised using p-channel transistors. From the small signal analysis of the differential pair,

$$i_{d1} = i_{s1} = \frac{g_{m1}}{2} V_{in}$$

If the finite output impedance of current mirror is neglected, $i_{d4} = i_{d3} = -i_{s1}$

Now voltage at node V_{out} is

$$V_{out} = (i_{s1} - i_{d4})\, r_{out}$$

$$= 2i_{s1}\, r_{out} \text{ (Since } i_{d4} = -i_{s1})$$

$$V_{out} = g_{m1}\, r_{out}\, V_{in}$$

Note from the above equation that output impedance is purely resistive. If a capacitive load is used

$$V_{out} = g_{m1} Z_{out} V_{in} \quad \text{and} \quad \text{gain } A_v = g_{m1} Z_{out}$$

Where $Z_{out} = r_{out} \,||\, 1/sC_L$

In the above model, it is assumed that the time constant at the output node is much larger than the time constant due to the parasitic capacitance of the sources Q_1 and Q_2.

However, when high-frequency effects are important, a different model as shown in Fig. 8.6.1 (*e*) should be taken for analysis. In this model an external voltage of V_x is applied to the output node.

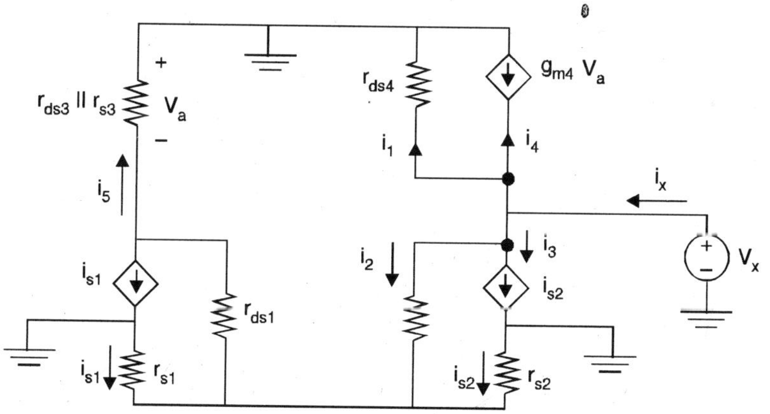

Fig. 8.6.1 (*e*) Small signal model for calculation of output impedance.

The output impedance $r_{out} = V_x/i_x$

Where $i_x = i_1 + i_2 + i_3 + i_4$

$$i_1 = \frac{V_x}{r_{ds4}} \quad i_2 = \frac{V_x}{r_{ds2}} \qquad \text{(since } r_{s1} \,||\, r_{s2} << r_{ds2})$$

Assuming $r_{s1} = r_{s2}$ and ignoring r_{ds1}, it is seen that i_2 splits equally between i_{s1} and i_{s2}

$$\therefore \quad i_{s1} = i_{s2} = \frac{i_2}{2} = -\frac{V_x}{2r_{ds2}}$$

In current mirror circuit $i_4 = i_5$ (as $g_{m4} = 1/r_{s4} = 1/r_{s3}$ and $r_{ds3} >> r_{s3}$)

Now $\quad i_4 = -i_3$ (but $i_3 = i_{s2}$)

$$\therefore \quad i_4 = -i_{s2} = -i_{s1}$$

In otherwords, it can be said that when current equally splits between r_{s1} and r_{s2}, the current mirror causes i_3 and i_4 to cancel each other.

Finally, the output resistance r_{out} is given by

$$r_{out} = \frac{V_x}{i_1 + i_2 + i_3 + i_4} = \frac{V_x}{(V_x / r_{ds4}) + (V_x / r_{ds2})}$$

$$= r_{ds4} \,||\, r_{ds2} \qquad \therefore \; A_v = g_{m1} (r_{ds2} \,||\, r_{ds4})$$

8.6.2. Bipolar Diffrential Pair

Before looking into this let us gain a brief knowledge on Bipolar current mirrors.

Bipolar current mirrors. This is similar to a MOS current mirror and is shown in Fig. 8.6.2 (*a*). Neglecting the finite base current of Q_1 and Q_2 the output current in current mirror is equal to its input current. When the base currents are taken into account $I_{out} = \left(\frac{1}{1+2/\beta}\right) I_{in}$

Where $\quad \beta \rightarrow$ Transistor current gain

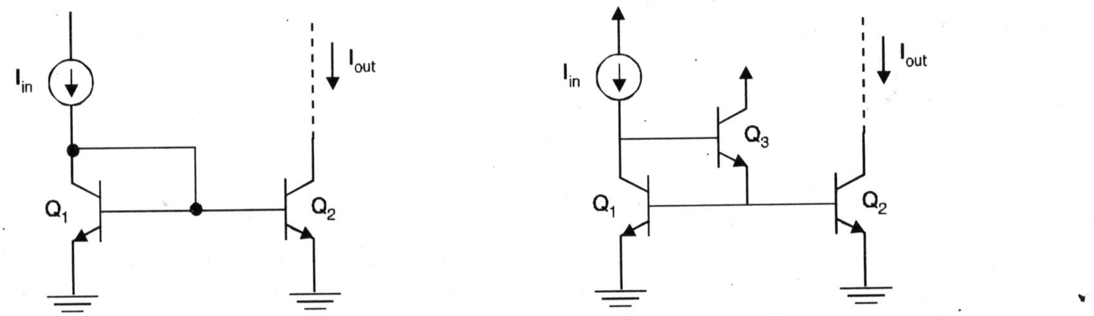

(*a*) Simple bipolar current mirror. (*b*) Current mirror with fewer inaccuracies caused by finite base currents.

Fig. 8.6.2.

In order to minimize the errors due to finite base currents an additional transistors Q_3 (which an emitter follower buffer) is added as shown in Fig. 8.6.2 (*b*). For achieving larger output impedance with minimum errors, emitter degeneration is added. That is a degenerating resistor R_e is connected between emitters of Q_1, Q_2 and ground. The other high output impedance current mirrors are (*i*) cascode and (*ii*) Wilson current mirrors.

Bipolar differential Pair

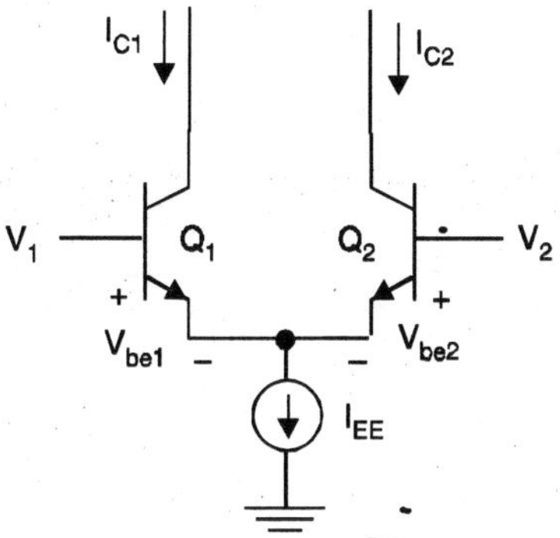

Fig. 8.6.2 (*c*). A bipolar differential pair.

***(i)* Large Signal Behaviour.** Fig. 8.6.2 (*c*) shows a bipolar differential pair. We know the collector current is exponentially related to the base emitter voltage by the relationship.

$$I_c \cong I_s\, e^{V_{be}/V_T}$$

Where I_s = scale current and is proportional to the area of the Emitter base junction.

$$V_T = \frac{KT}{q}$$

K – Boltzmann constant

T – Temperature in Kelvin

q – charge in coulomb

From this Base-emitter voltages can be written as

$$V_{be1} = V_T \ln\left(\frac{I_{c1}}{I_{s1}}\right) \quad \text{and} \quad V_{be2} = V_T \ln\left(\frac{I_{c2}}{I_{s2}}\right)$$

Also from Fig. 8.6.2 (*c*)

$$V_1 - V_{be1} + V_{be2} - V_2 = 0$$

From the above equations, we can write

$$I_{c1}/I_{c2} = e^{(V_1 - V_2)V_T} = e^{V_{in}/V_T}$$

Where $\quad V_{in} = V_1 - V_2$

Also $\quad \alpha\, I_{EE} = I_{c1} + I_{c2}$

Where $\quad \alpha = \beta/(\beta + 1)$

Finally $\quad I_{c1} = \dfrac{\alpha I_{EE}}{1 + e^{-V_{in}/V_T}}$

and $\quad I_{c2} = \dfrac{\alpha I_{EE}}{1 + e^{-V_{in}/V_T}}$

The difference of above equations results in a hyperbolic tangent function and bipolar differential pair V-I relationship is commonly called as tan *h* relationship.

(*ii*) Small signal behaviour

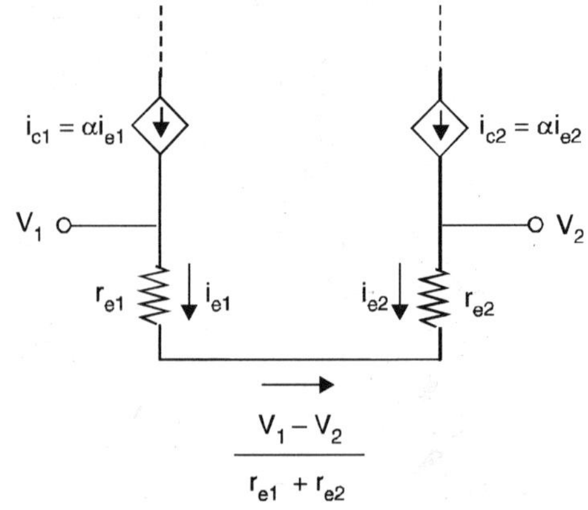

Fig. 8.6.2 (*d*) Small signal model of a bipolar differential pair.

Fig. 8.6.2 (*d*) shows small signal model of bipolar differential pair.

Define $V_{in} = V_1 - V_2$

We have $i_{c1} = \alpha\, i_{e1} = \dfrac{\alpha V_{in}}{r_{e1} + r_{e2}} = \dfrac{\alpha V_{in}}{(\alpha / g_{m1}) + (\alpha / g_{m2})}$

When both transistors have same bias currents through then $g_{m1} = g_{m2}$ and i_{c1} becomes

$$i_{c1} = \frac{g_{m1}}{2} V_{in} = i_{c2}$$

Input impedance of differential pair

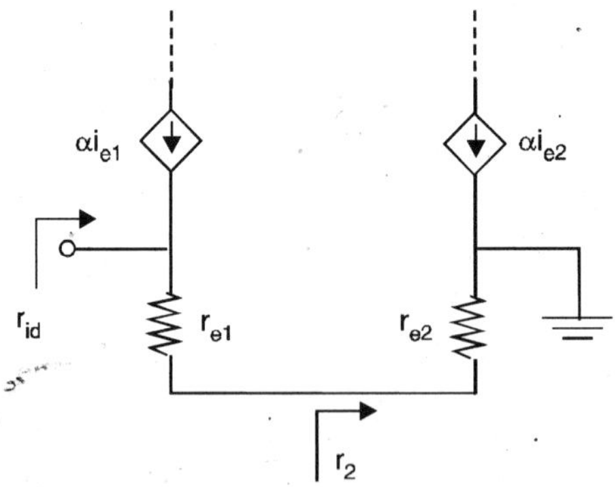

Fig. 8.6.2 (*e*) Finding the input impedance of a bipolar differential pair.

Consider Fig. 8.6.2 (*e*) V_2 is grounded. The impedance, r_2, seen looking into the emitter of Q_2 is r_{e2}. Now the total emitter resistance to the ground of Q_1 is $r_{e1}+r_{e2}$. $\therefore$ the impedance, r_{in}, seen looking into the base of Q_1 is equal to $\beta + 1$ times the emitter resistance.

$$V_{in} = (\beta + 1)\,(r_{e1} + r_{e2})$$

8.7. TWO STAGE C-MOS OP-AMP

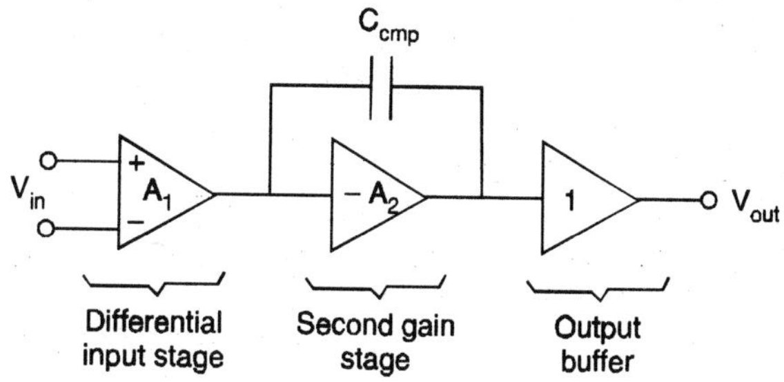

Fig. 8.7 (*a*) A block diagram of a two stage op-amp.

The Fig. 8.7 (*a*) shows block diagram of two stage CMOS OP-amp. It has two numbers of gain stages. The first stage is a differential input stage where complementary circuitry is used. The second stage is a common source stage. This can itself be used as output stage for capacitive loads. But for resistive loads a separate output buffer stage should be used. The capacitor C_c often called as miller capacitance is used to ensure stability when op-amp is used with feedback.

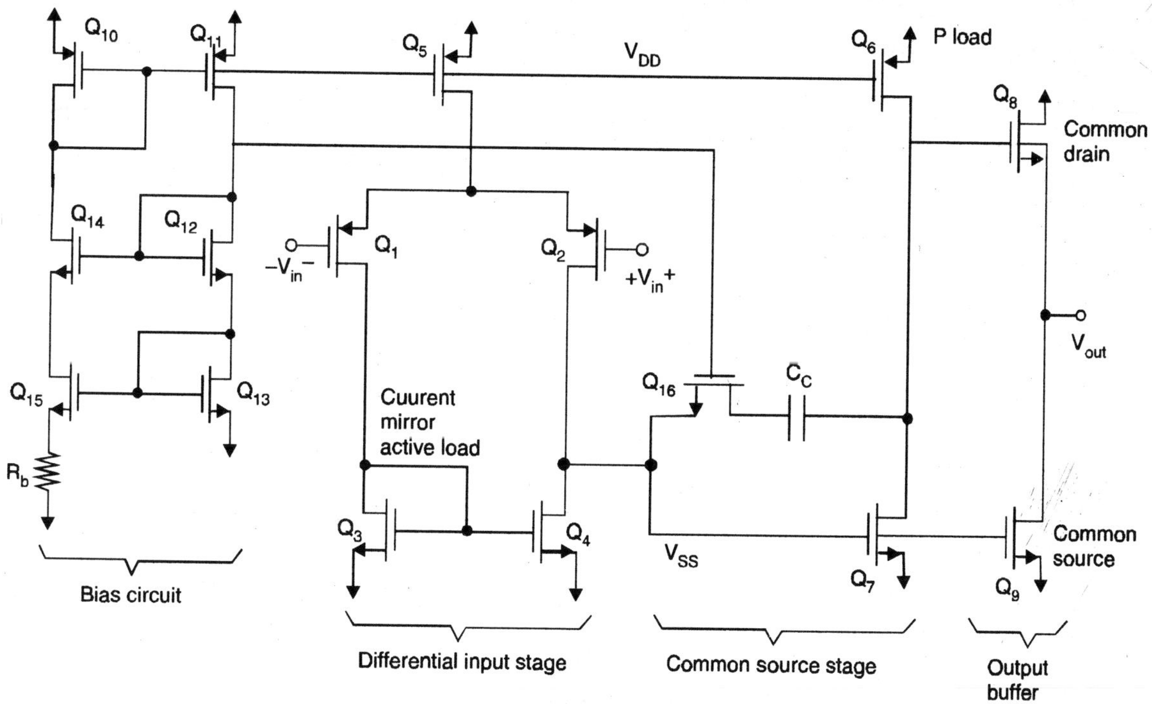

Fig. 8.7 (*b*) A CMOS realisation of a two stage amplifier.

A practical CMOS two stage, op-amp is shown in Fig. 8.7 (*b*) transistors Q_{10}, Q_{11}, Q_{12}, Q_{13}, Q_{14} and Q_{15} form the bias circuit. The first gain stage is a complementary differential stage. P channel transistors Q_1 and Q_2 form the differential input pair with n-channel transistors Q_3 and Q_4 as current mirror active load.

OP-amp gain

The gain of the first stage, which is a differential stage is given by

$$A_{v1} = g_{m1}\,(r_{ds2} \,||\, r_{ds4}) \qquad \text{(Refer MOS differential amplifier 8.6.1)}$$

Also $$g_{m1} = \sqrt{2\mu_p C_{ox}\left[\frac{W}{L}\right] I_{DI}} = \sqrt{2\mu_p C_{ox}\left[\frac{W}{L}\right]\frac{I_{bias}}{2}}$$

The second stage is a common source gain stage formed by Q_{16} with Q_6 a *p*-channel transistor acting as load. The gain of second stage is given by

$$A_{v2} = -\,g_{m7}\,(r_{d6} \,||\, r_{ds7}) \qquad \text{(refer common source amplifer – 8.5)}$$

The third stage is a common-drain buffer stage. This stage is often called a source follower, because the source voltage follows the gate voltage of Q_8. In this substrate of Q_8 is connected to the source to eliminate gain degradations due to body effect. The gain of the source follower stage is given by

$$A_{v3} \cong \frac{g_{m8}}{G_L + g_{m8} + g_{ds8} + g_{ds9}}$$

Where G_L = Load conductance.

Frequency response

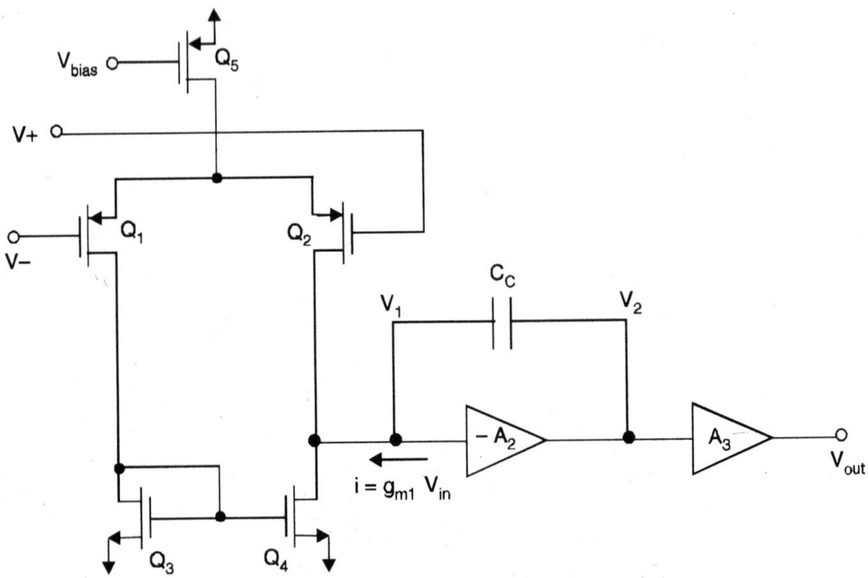

Fig. 8.7 (*c*) Op-amp used to find the mid band frequency response.

The simplified circuit for analysis is shown in Fig. 8.7 (*c*) here all the capacitors except the C_c – compensation capacitor are ignored. Because this C_c always dominated at all frequencies except around the unity gain frequency of the op-amp. Also the transistor Q_{16} is neglected because it has an effect only around the unity gain frequency of the op-amp. The second stage introduces a capacitive load on the first stage due to the compensation capacitor C_c.

Using miller's theorem, C_{eq} at node V_1 is,

$$C_{eq} = C_c(1 + A_2) \approx C_C A_2$$

The gain in the first stage is now given by, see Fig. 8.7c

$$i = -g_{m1} V_{in}$$

$$V_{in} = \frac{-i}{g_{m1}}$$

$$A_1 = V_1/V_{in}$$

Where V_1= Output voltage of first stage

$$\therefore \quad A_1 = V_1/V_{in} = \frac{-g_{mi} - V_i}{i}$$

$$A_1 = -g_{m1} Z_{out1}$$

Where Z_{out1} is output impedance of 1st stage

$$Z_{out1} = r_{ds2} \,||\, r_{ds4} \,||\, 1/SC_{eq}$$

For mid band frequencies, the C_{eq} dominates

$$\therefore \quad Z_{out} \cong \frac{1}{SC_{eq}} \cong \frac{1}{sC_c A_2}$$

The over all gain

$$A_v = \frac{V_{out}}{V_{in}} = A_3 A_2 A_1 \cong A_3 A_2 \frac{g_{m1}}{sC_c A_2}$$

If A_3 is assumed to be one, we can write

$$A_v = \frac{g_{m1}}{sC_c}$$

The C_c causes the magnitude of the gain to decrease at frequencies well below the unity gain frequency of the op-amp. Therefore this op-amp is operated in mid-band frequencies in many applications. The unity gain frequency is given by,

$$\omega_t = \frac{g_{m1}}{C_c}$$

Note this unity gain frequency is directly proportional to g_{m1} and inversely proportional to C_c.

Slew rate. The slew rate is the maximum rate at which the output changes when input signals are large.

$$S_R = \frac{dV_{out}}{d_t}\Big|_{max}$$

For op-amp $SR = \dfrac{2I_{DI}}{C_c}$

or $\quad SR = V_{eff1}\, \omega_t.$

Where $\quad I_{D1}$ = Bias current of Q_1 when no signals are present.

$$V_{eff1} = \sqrt{\frac{2I_{DI}}{\mu_p C_x (W/L)}}$$

From the above equation, it is seen that the slew rate of two stage op-amp can be improved by increasing V_{eff1} and ω_t. Slew rate is inversely proportional to mobility. So *p*-type transistor is chosen at the input differential stage.

Systematic offset voltage. When input to differential input amplifier is zero, there should be no output voltage. Any detection of output voltage in this situation is said to be an offset voltage. While designing the two-stage CMOS op-amp, one should see that no inherent input offset voltage is present. The necessary condition to ensure the above one is

$$\frac{(W/L)_7}{(W/L)_4} = \frac{2(W/L)_6}{(W/L)_5}$$

Where W – Width of transistor

L – Length of transistor

4, 5, 6, and 7 denotes Q_4, Q_5, Q_6 and Q_7 respectively.

Choice of input stage

S.No.	*Parameter*	*P-channel as input*	*n-channel as input*
1.	Overall gain	Good	Poor
2.	Transconductance	High and best frequency response is achieved	Low Moderate frequency response
3.	Voltage drop at input	High	Low
4.	Noise (*i*) White noise	Less	More
	(*ii*) Thermal noise	High and require wide band width	Low

8.8. BJT OP-AMP

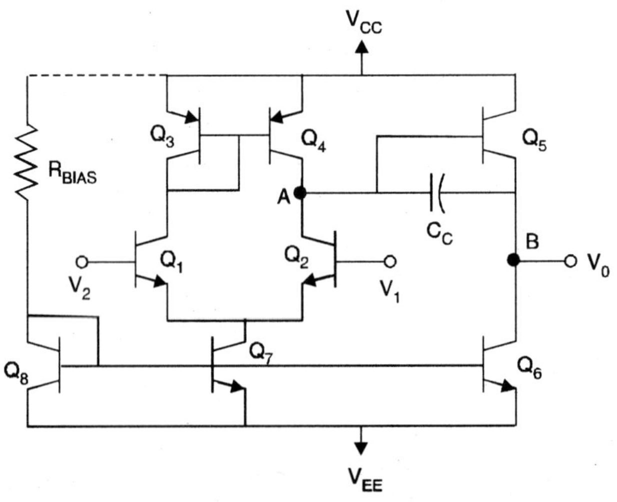

Fig. 8.8 (*a*) Two stage BJT op-amp.

Fig. 8.8 (*a*) shows the Two stage BJT op-amp.

This is also similar to a CMOS op-amp. The first stage is a differential stage formed by Q_1 and Q_2 both *n*-type transistors. Here also active load is a current mirror formed by two *p*-type transistors Q_3 and Q_4 Q_6 acts as second stage amplifier connected in common emitter configuration. C_c is again a compensating capacitor connected between first and second stage.

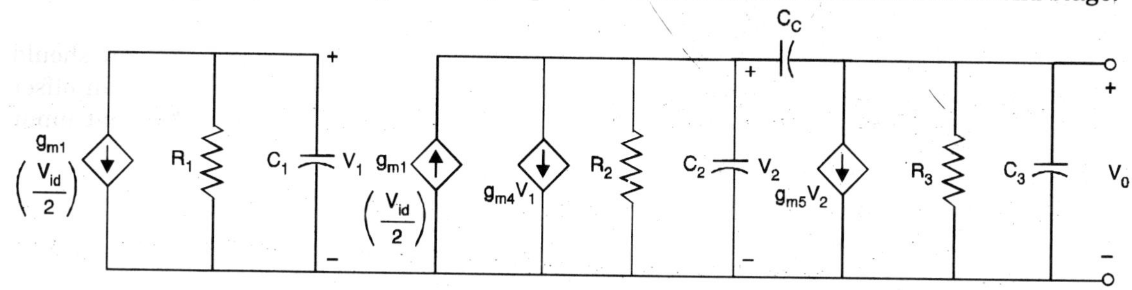

Fig. 8.8.1 (*a*).Small signal.

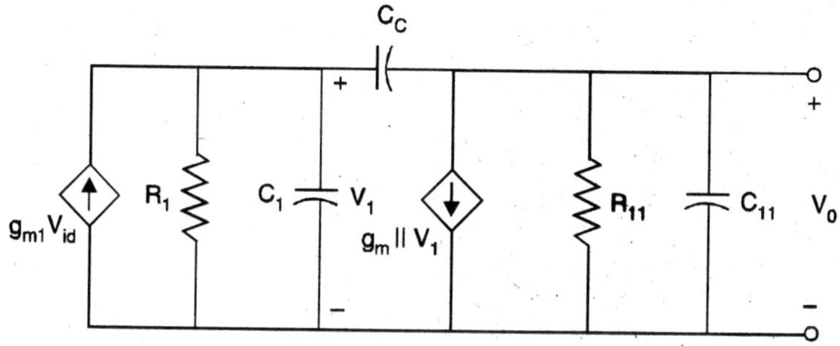

Fig. 8.8.1 (*b*) Simplified model.

Fig. 8.8.1 (*a*) shows the small signal model for analysis and its simplified version is given in Fig. 8.8.1 (*b*) assuming $g_{m3} \simeq g_{m4}$.

$$V_{in} = V_1 - V_{out}$$

In Fig.
$$g_m = g_{m1} = g_{m2}$$
$$R = R_2$$
$$C = C_2$$
$$G'_m = g_{m5}$$
$$R' = R_3$$
and
$$C' = C_3$$

The compensation capacitance C_c is not included in C or C'. The Miller capacitor C_C, splits the poles and accomplishes the dominant pole compensation. If C_c is zero, then

$$\frac{V_0(s)}{V_{in}(S)} = \frac{gm\, R\, gm'\, R'\, \omega'\omega''}{(S + \omega')(S + \omega'')} = \frac{A_0\omega'\omega''}{(S + \omega')(S + \omega'')}$$

Where $\omega' = 1/RC$

and $\omega'' = 1/R'C'$

where ω' and ω'' are poles of BJT op-amp when C_c is zero.

When C_c is not equal to zero we get

$$V_{0(S)}/V_{in}(s) = A_0\,[1-(sC_c/g'_m)]\,/\,\{1+s[R(C+C')+R'(C'+C_c)+g'_m RR'C_c]$$
$$+\,s^2RR'\,[CC'+C_c(C+C')]\}$$

Now the poles are at

$$\omega' = -1/g'_m RR'C_c$$

and $$\omega'' = \frac{-g'_m C_c}{CC'+C'C_c+CC_c}$$

Also a zero is introduced by C_c at $Z = g'_m/C_c$

Summarising

1. Low frequency open-loop gain is $A_o = g_m g'_m RR'$
2. The dominant pole is at $1/\,|\,\omega'\,| = \dfrac{1}{g'_m RR'C_c}$
3. The unity gain band width

   $\omega_t = g'_m C_c$ which is similar to that of a MOS op-amp
4. Slew rate : $SR = I/C_c$ where I is smaller of 2Ic1 or Ic5.

   (It determines how fast C_c can be charged and discharged)

For good margin the following conditions are to be met

(*i*) $g'_m > g_m$

(*ii*) $C' < g'_m C_c/g_m$

For best performance and largest signal swing, both transistors should be in forward active region.

8.9. VOLTAGE AND CURRENT REFERENCES

Circuits that yield a precise dc voltage or current independent of external influences are called voltage or current references. The primary external influences are power supply and temperature variations. So we should try to minimize the dependence of voltage and current on power supply and temperature.

Let V_{ref} – Voltage reference

$S(V_{ref}/V_{ps})$ Sensitivity of V_{ref} to changes in power supply V_{ps}.

$$S(V_{ref}/V_{ps}) = \lim_{\Delta V_{ps}\to 0} \frac{\Delta V_{ref}/V_{ref}}{\Delta V_{ps}/V_{ps}} = \frac{\Delta V_{ps}}{\Delta V_{ref}}\left(\frac{\partial V_{ref}}{\partial V_{ps}}\right) \quad \text{... (1)}$$

Once sensitivity is known, then,

$$\frac{\Delta V_{ref}}{V_{ref}} = S(V_{ref}/V_{ps}) \left(\frac{\Delta V_{ps}}{V_{ps}}\right)$$

The above all equations can be made valid for current references by substituting V_{ref} with $I_{ref.}$

Similarly Let, $S(V_{ref}/T)$ denotes sensitivity of V_{ref} to changes in Temperature and Let the fractional temperature co-efficient (TC_F) of V_{ref} be

$$TC_F\,(V_{ref}) = \frac{1}{V_{ref}}\frac{\partial V_{ref}}{\partial T} = \frac{1}{T}\,S(V_{ref}/T)$$

Fig. 8.9.1 shows three simple voltage references.

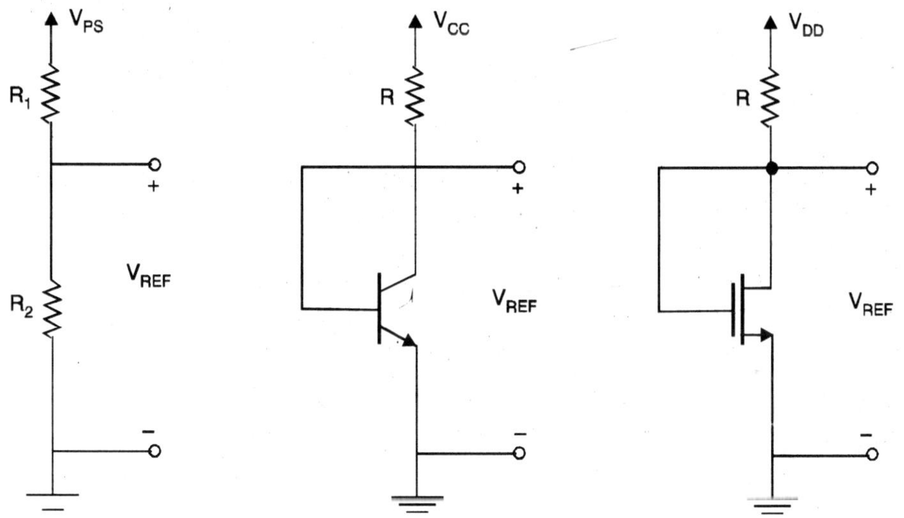

(*a*) Passive voltage divider. (*b*) BJT voltage reference. (*c*) MOS voltage reference.

Fig. 8.9.1. Simple voltage references.

(*i*) Fig (*a*) is simply a voltage divider. From this

$$V_{ref} = V_{ps}\,(R_2/R_1 + R_2)$$

Now

$$\boxed{S(V_{ref}/V_{ps}) = \frac{V_{ps}}{V_{ref}}\frac{\partial V_{ref}}{\partial V_{ps}} = 1}$$

The fractional temperature co-efficient is found to be

$$TC_F\,(V_{ref}) = \frac{1}{V_{ref}}\frac{\partial V_{ref}}{\partial T} = \left(\frac{R_1}{R_2}\frac{V_{ref}}{V_{ps}}\right)\left(\frac{\partial R_2}{R_2\partial T} - \frac{\partial R_1}{R_1\partial T}\right)$$

$$TC_F(V_{ref}) = \left(\frac{R_1}{R_2}\frac{V_{ref}}{V_{ps}}\right)[TC_F(R_2) - TC_F(R_1) \quad \text{... (A)}$$

Where the resistors are assumed to be temperature dependent and V_{ps} is temperature independent.

From the above equation (A) we can understand that if fractional temperature co-efficient of R_1 and R_2 are made equal, then, the fractional temperature co-efficient of V_{ref} is zero, which means temperature independence can be achieved.

Fig. 8.9.1 (*b*) shows a voltage reference formed using a BJT and a resistor. The reference voltage V_{ref} is given by

$$V_{ref} = V_t \ln\left(\frac{V_{cc} - V_{ref}}{RI_s}\right) \quad \text{... (2)}$$

Where $V_t = KT/q$

I_S = saturation current

If $V_{CC} >> V_{ref}$ then we can write

$$V_{ref} \approx V_t \ln(V_{CC}/RIs)$$

Now the sensitivity of V_{ref} with respect to V_{CC} can be found by differentiating the above equations and substituting it in equation (1).

$$S(V_{ref}/V_{CC}) \simeq 1/\ln(V_{CC}/RIs)$$

In this case it is possible to achieve sensitivities less than unity.

To find fractional temperature co-efficient of BJT. In this case V_t, R and I_s all vary with temperature. Hence differentiating (2) with respect to temperature yields.

$$\frac{\partial V_{ref}}{\partial T} = \frac{V_{ref}}{T} - V_t\left(\frac{1}{Is}\frac{\partial Is}{\partial T} + \frac{1}{R}\frac{\partial R}{\partial T}\right) \quad \text{... (3)}$$

Let $Is = KT^3 \exp(-V_{GO}/V_t)$

where V_{GO} is band gap voltage and is temperature independent.

Now
$$\frac{\partial I_s}{\partial T} = \frac{3I_s}{T} + \frac{I_s}{T}\frac{V_{GO}}{T} \quad \text{... (4)}$$

Substituting (4) into (3) and dividing by V_{ref} gives

$$\boxed{TC_F(V_{ref}) = \frac{V_{ref} - 3V_t - V_{GO}}{V_{ref}T} - \frac{V_t}{V_{ref}}\left(\frac{1}{R}\frac{\partial R}{\partial T}\right)}$$

Fig. 8.9.1 (*c*) shows a voltage reference implemented through MOS. In this gate is connected to drain. The device is in saturation for $V_{GS} \geq V_T$ and V_{ref} is given by

$$V_{ref} = V_T + \sqrt{\frac{2(V_{dd} - V_{ref})}{RK'(W/L)}} \qquad \text{... (5)}$$

Where K' = transconductance parameter = μC_{ox}

V_T = Threshold voltage

Now sensitivity of the circuit with respect to V_{DD} for $V_{dd} >> V_{ref}$ is

$$S(V_{ref}/V_{dd}) = \frac{0.5}{1 + V_T\sqrt{RK'W/2V_{dd}L}}$$

To find fractional temperature co-efficient

In MOS device K' and V_T are temperature dependent parameters. They can be modeled as

$$K' = K'_0\,(T/T_0)^{1.5} \qquad \text{... (6)}$$

and
$$V_T = V_{TO} - \alpha(T - T_0) \qquad \text{... (7)}$$

where K'_0 and V_{TO} are values of K' and V_T at T_0 and α is approximately 2.3 mv/c

Assuming R, K'_N and V_T are temperature dependent and differentiating equations (5) (6) and (7) gives

$$\frac{\partial V_{ref}}{\partial T} = \frac{\partial V_T}{\partial T} - \frac{1}{2}\sqrt{\frac{2V_{dd}}{K'R(W/L)}}\left(\frac{1}{R}\frac{\partial R}{\partial T}\right) - \frac{1}{2}\sqrt{\frac{2V_{dd}}{K'R(W/L)}}\left(\frac{1}{K'}\frac{\partial K'}{\partial T}\right) \qquad \text{... (7a)}$$

$$\frac{\partial K'}{\partial T} = -\frac{1.5}{T}K' \qquad \text{... (8)}$$

and
$$\frac{\partial V_T}{\partial T} = -\alpha \qquad \text{... (9)}$$

substituting equations (8) and (9) in (7a) and dividing by V_{ref} gives

$$TC_F(V_{ref}) = \frac{-1}{V_{ref}}\left[\alpha + \frac{1}{2}\sqrt{\frac{2V_{dd}}{K'R(W/L)}}\left(\frac{1}{R}\frac{\partial R}{\partial T} - \frac{1.5}{T}\right)\right]$$

The values of voltage references can be increased by cascading active devices or by the method as illustrated in Fig. 8.9.1 (*d*).

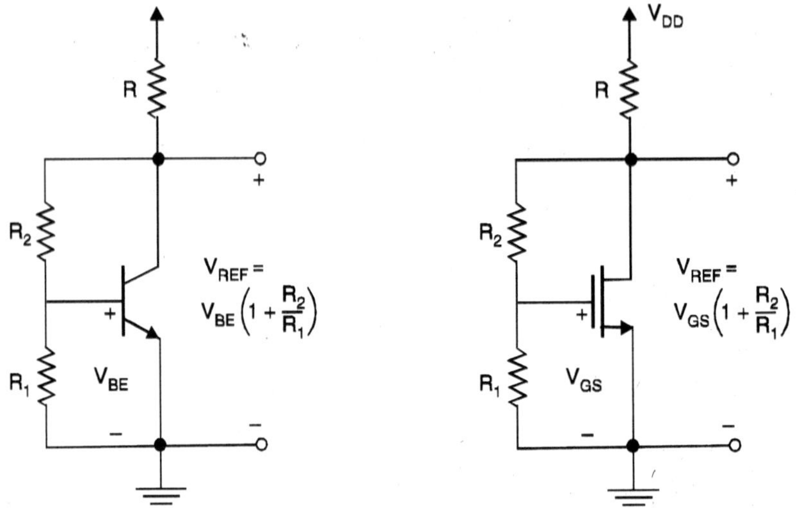

Fig. 8.9.1 (*d*) A method of increasing the value of V_{REF}.

Current references

Fig. 8.9.2 (*a*) and (*b*) shows the current references using B_{JT} and Mos. They are in fact current mirrors where I is given by voltage across R. Assuming ideal current mirrors.

$$I_{ref} = \frac{V_{cc} - V_{be}}{R} \qquad \text{... (10)}$$

and

$$I_{ref} = \frac{V_{dd} - V_{gs}}{R} \qquad \text{... (11)}$$

Assuming $V_{cc} >> V_{be}$ and $V_{dd} >> V_{gs}$, the sensitivity of I_{ref} with respect to power supply is unity for both references. The fractional temperature co-efficient of B_{JT} current reference can be found by differentiating Eq. (10) with respect to temperature and dividing by I_{ref}.

$$TC_F(I_{ref}) = \frac{1}{I_{ref}}\left[\frac{\partial V_{be}}{R\partial T} + I_{ref}\left(\frac{1}{R}\frac{\partial R}{\partial T}\right)\right]$$

Which can be written as

$$TC_F(I_{ref}) \approx \frac{1}{I_{ref}}\left[\frac{V_{go} + 3V_t - V_{be}}{RT}\right] - \left(\frac{1}{R}\frac{\partial R}{\partial T}\right)$$

Similarly the fractional temperature co-efficient of MOS current reference is found by differentiating Eq. (11) with respect to temperature and dividing by $I_{ref.}$

$$TC_F(I_{ref}) = \frac{-1}{I_{ref}}\left[\frac{\partial V_{gs}}{R\partial T} + \frac{I_{ref}}{R}\cdot\frac{\partial R}{\partial T}\right]$$

Which can be written as

$$TC_F(I_{ref}) = \frac{1}{I_{ref}}\left[\frac{\alpha}{R} + \frac{1}{R}\sqrt{\frac{V_{dd}}{2K'R(W/L)}}\left(\frac{1}{R}\frac{\partial R}{\partial T} - \frac{3}{4T}\right)\right] - \left(\frac{1}{R}\frac{\partial R}{\partial T}\right)$$

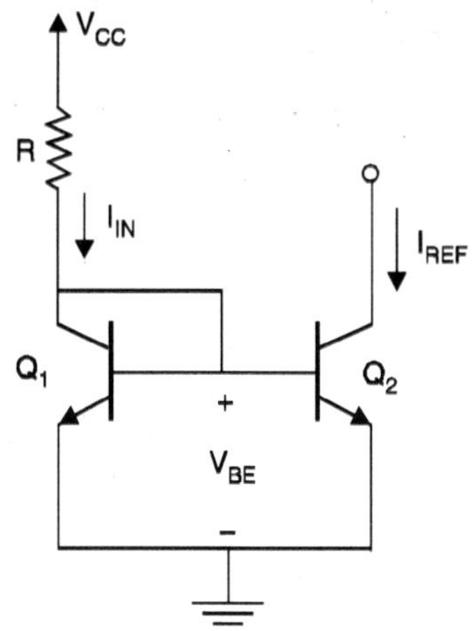

(a) A simple BJT current reference.

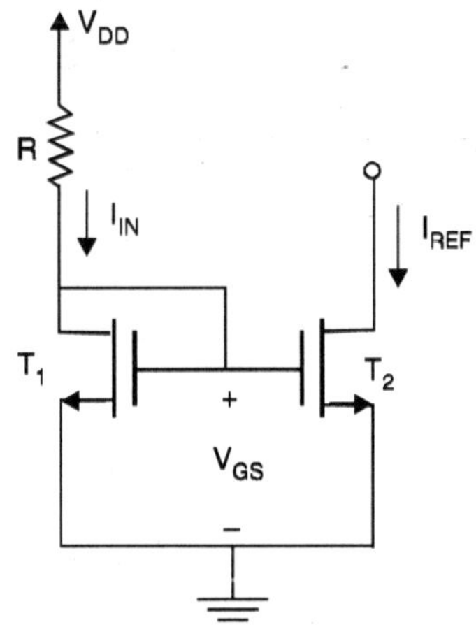

(b) A Simple MOS current reference.

Fig. 8.9.2. Current references.

To achieve a reduction of the dependence of the current on the supply voltage, V_{be} drop of forward biased *pn* junction is used. Neglecting the base current, I_{ref} is expressed as

$$I_{ref} = \frac{V_{be1}}{R_2} = \text{from which it can be shown that}$$

$$\boxed{S\,(I_{ref}/V_{cc}) \approx \frac{V_t}{I_{ref}\,R_2}}$$

Apart from the above there have been many approaches to realize voltage references in Ics. These include

1. Making use of a Zener diode that breaks down at a known voltage when reverse biased.
2. Making use of the difference in the threshold voltage between an enhancement transistor and depletion transistor.
3. Canceling the negative temperature dependence of a *pn* junction with a positive temperature dependence from a PTAT. (Proportional-to-absolute-temperature) circuit.

This is currently the most popular for bipolar and CMOS technologies. This is also called as 'band gap' voltage references because the band gap reference is based on subtracting the voltage of a forward biased diode (or base emitter junction) having a negative temperature coefficient from a voltage proportional to absolute temperature. (PTAT). This PTAT voltage is realized by amplifying the voltage difference of two forward biased base-emitter (or diode) junctions.

8.10. SUBMICRON TECHNOLOGY AND GaAs VLSI TECHNOLOGY

Transistors having channel length < 1μ are classified as submicron devices. It is known that *p*-transistors have inherently slower performance than similar *n*-transistors, due to lower mobility of holes. The mobility influences much in long channel devices. But, when the channel lengths are scaled down, the effects of velocity saturation only dominates. We know, for MOS transistors (long channel) drain current in sat region is given by

$$I_{ds} = \frac{W}{L}\mu C_{ox}\left[(V_{gs} - V_t)V_{ds} - \frac{Vds^2}{2}\right]$$

Where $C_{ox} = \frac{\varepsilon_{ins}\varepsilon_0}{D}$ = gate / channel capacitance / unit area.

From the above it is understood that current is directly proportional to mobility and inversely proportional to channel length. Transconductance g_m is also similarly influenced. When Velocity saturation occurs then the drain current is given by

$$I_{dsat} = W\,C_{ox}\,\upsilon_{sat}\,(V_{gs} - V_t)$$

Where $\upsilon_{sat} \rightarrow$ saturation velocity.

Current is now independent of both mobility and channel length. But dependent on saturation velocity.

In *n*-devices owing to their high mobility, velocity saturation occurs at lower electric field strengths. Hence, as the channel length is reduced, the current in *n*-transistor becomes independent of channel length, but it is not in the case of *p*-transistor until, it run into velocity saturation. In the popular CMOS technology both *n*-transistors and *p*-transistors take active part. Hence for getting faster devices, an alternate technology based on gallium arsenide, has emerged.

GaAs. The GaAs VLSI Technology has emerged in order to overcome the speed limitations in silicon logic. GaAs is used in conjunction with silicon to satisfy the need for Very High Speed Integrated Circuit (VHSIC) Technology.

The compound Gallium Arsenide was realized to be a high speed semiconductor in 1960s. Comparing with Silicon the electron mobility of GaAs is high. Also it has good opto electrical properties. The power dissipation is low. Because of the above said and many other advantages possessed by GaAs, it finds its place in VLSI Technology. The compound GaAs is a semiconductor made of two elements. By using proper dopant, both *n*-type and *p*-type material can be realized from GaAs. The orientation of the crystal influences the properties of GaAs Field effect transistor. Many of the GaAs wafers with < 100 > orientation are currently employed. The superior electron mobility characteristic of GaAs is due to its energy band structure. GaAs is a direct gap material with valence band maximum and conduction band minimum occurring at the same wave vector, which means little momentum change is necessary for the transition of an electron from the conduction band to the valence band. When electric field E across GaAs is increased, the charge carriers gain energy and drift velocity of the charge carrier increases. However, at some point, the energy gained from the field becomes equal to the energy gained from the field becomes equal to the energy lost as a result of collisions. This results in the drift velocity approaching a limiting value referred to as the saturation velocity. Comparing with silicon, GaAs has much higher mobility in low electric field. But in high electric field regions the mobility or silicon increases and is some what comparable with that of GaAs. Fig. Illustrates this.

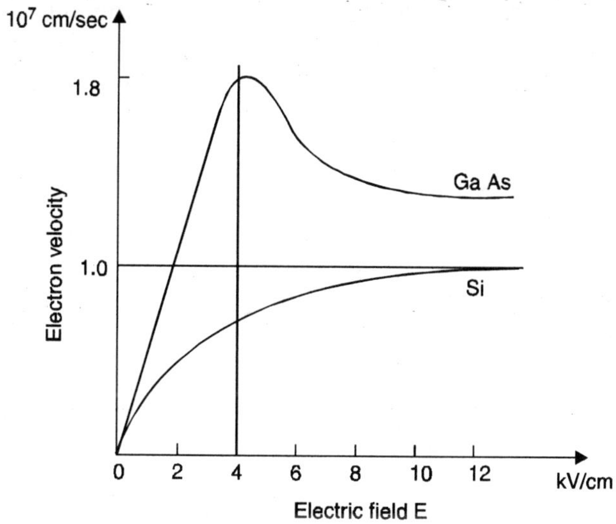

Fig. 8.10. Comparison of Mobility of Si and GaAs in Low and high electric field regions.

Technology development. During last few years considerable progress has been made in GaAs Integrated Circuitry. GaAs finds its application as

1. ON and OFF devices
2. Four inch diameter wafers

It is also suitable where clock rates in the range 1–2 GHz are used.

Also GaAs offers

(*i*) less than 1μ gate geometry

(*ii*) less than 2μ metal pitch

(*iii*) upto four layer metal

Salient features of GaAs Technology

* Comparing with silicon, GaAs possess 6 to 7 times electron mobility therefore electron transit time is very high.
* Saturation drift velocity for GaAs and silicon are approximately equal. But, for GaAs, saturation velocity occurs at a lower electric field than for silicon. See Fig. 8.10
* GaAs has large energy band gap. This minimizes parasitic capacitances and provides good isolation between multiple devices in a single substrate.
* As the gate oxide to trap charges are negligible radiation resistance is high.
* Larger band gap permits wider operating temperature range (–200° to + 200°C)
* Direct band gap of GaAs allows efficient radiative recombination of electrons and holes, this means forward biased *pn* junctions can be used as light emitters. Thus, efficient integration of electronics and optics becomes possible.
* Upto 70% reduction in power dissipation can be obtained over the fastest of the silicon technology such as ECL.

Table 8.10. Comparison between CMOS, bipolar and GaAs technologies

CMOS	*Bipolar*	*GaAs*
Low dissipation	High dissipation	Medium dissipation
High I/P impedance–low drive current	Low I/P impedance–low drive current	High I/P impedance–low drive current
High noise margin	Medium noise margin	Low noise margin
Medium speed–high voltage swing	High speed –Low voltage swing	Very High speed–low voltage swing
High packing density	Low packing density	High packing density
High delay sensitivity to load –fan out.	Low delay sensitivity to load –fan out	High delay sensitivity to load –fan in and fan out
Low Output drive	High Output drive	Low Output drive
$g_m V_{in}$	$g_m e^{V_{in}}$	$g_m V_{in}$
Bidirectional	Undirectional	Bidirectional possible
Ideal switching device	Not Ideal switching device	Reasonable switching device
Medium f_t	High f_t at low current	Very High f_t
Indirect gap	Indirect gap	Direct gap good light -emitter
Mask levels 12 to 16	Mask levels 12 to 20	Mask levels 6 to 10

The speed of operation of a semiconductor depends on (*i*) carrier mobility, (*ii*) carrier saturation velocity and (*iii*) existence of semi insulating substrate. All the above criteria are very well satisfied by GaAs. Apart from this it provides moderate power dissipation. Hence, it plays key role in current generation circuits and subsystems.

Examples of GaAs devices includes the following :

(*i*) First Generation Circuits

- Depletion mode metal semiconductor FET [D-MESFET]
- Enhancement mode metal semiconductor FET [E-MESFET]
- Enhancement mode junction FET [E-JFET] and
- Complementary Enhancement mode JFET [CE-JFET]

First generation GaAs gates have exhibited switching delays as low as 70 to 80 psec for a power dissipation in the order of 1.5 mw to 150 μW.

(*ii*) Second Generation Devices

- High electron mobility transistors, HEMT
- Hetero junction bipolar transistor, HBT

Electron mobility in second generation devices are 5 times greater than that of first generation.

Hence very high speed circuits can be realized.

9

Field Programmable Gate Arrays

INTRODUCTION

Digital Circuit design has gone through several evolutions in the last few decades. The dramatic change in technology has radically changed the design process. Circuit components have evolved from individual transistors to very large scale integrated circuits. (VLSI). In addition to rapid change in the technologies and techniques of VLSI design, the life cycle of modern products is becoming shorter than its design cycle. Thus, the need for rapid prototyping is ever growing.

9.1. DESIGN METHODOLOGIES

Design methodology is the series of the various processes that a designer traverses to implement a design idea onto a silicon chip.

It is the customer whodecides the design methodhlogy, *i.e.,* to say that on the basis of the customers demand the desiner selects design methodology).

The implementation of a digital system is dependent on design cycle. Digital integrated circuits may be realized in different technologies depending on their size and their role in the system. This depends also on the economics of the project. Integrated circuits consist of transistors that are placed on the chip and are interconnected in a such a way as to realize the design. The locations and connectivity of the transistors are defined by several masks. A mask corresponds to one of the silicon compound layers that form the transistors and the interconnect layers. Circuit implementations may be grouped into main categories as follows.

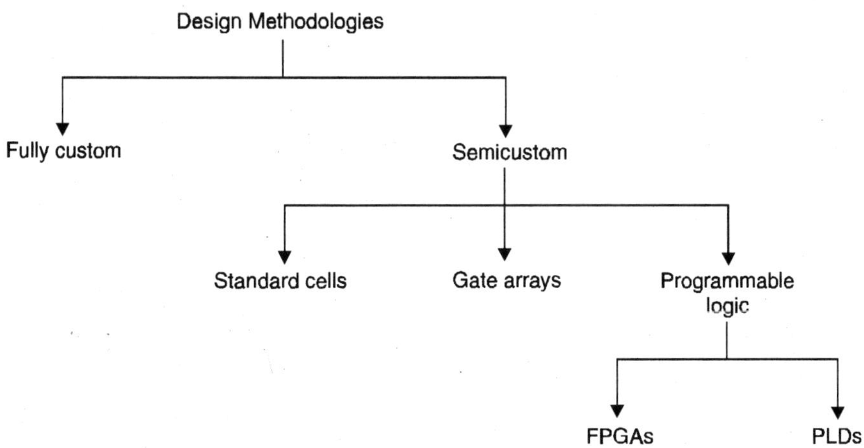

9.1.1. Fully Custom

Each and every component can be designed as per the designers interest (*i.e.*, designers can design (or) implement everything.

Full custom. The full custom approach places no structural restrictions on the designer. Anything can go anywhere, and in many cases logic blocks are interconnected by being abutted directly against each other.

The designer desgins all the circuitry and all the interconnections/communication paths.

The design approach involves the placement of all interconnections, circuits and devices by handcrafted interactive approach using a CAD tool.

Disadvantages :

Increased design time.

Probability of 1st time success very low.

Highest initial cost.

Multiple iterations.

Why go for it. For high volume, cost-sensitive systems that require the most efficient device designs. Several situations can only be best solved with a custom chip.

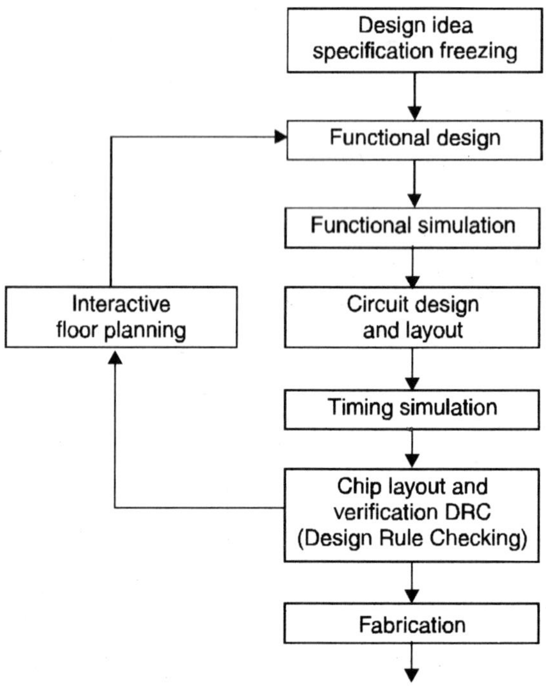

Fig. 9.1.1. Design flow – Full custom.

Semi Custom. The designers can design (or) implement part of the project on their own interest and part of the thing can be taken from the library components.

9.1.2. Standard Cells

The predefined logic cells are called standard cells. The design task is facilitated by the use of predesigned modules *e.g.* AND Gates, OR Gates, Multiplexers, Flip Flops. The Fig. 9.1.2 and 9.1.4 shows the Standard cell and Gate array arrangement respectively.

Features. Cells of standard height and varying widths allow to arrange cells in rows with the voltage and ground buses running across the top and the bottom of each row.

Distinction between circuit and interconnection area

Types

Core Cells – Gates, Flip-Flops, Registers, Decoders, Muxes, Adders etc.

Pad Cells – Input limited, Output limited, Core limited and Normal balanced.

Mega Cells – These are the building block to design complex ASICs solely with a cell based design system.

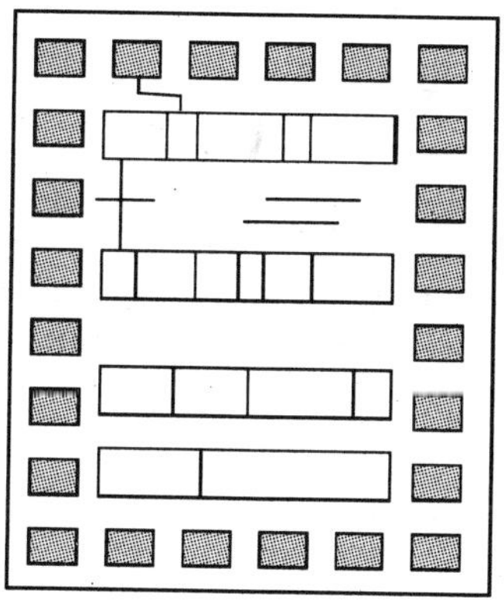

Fig. 9.1.2. Standard cell.

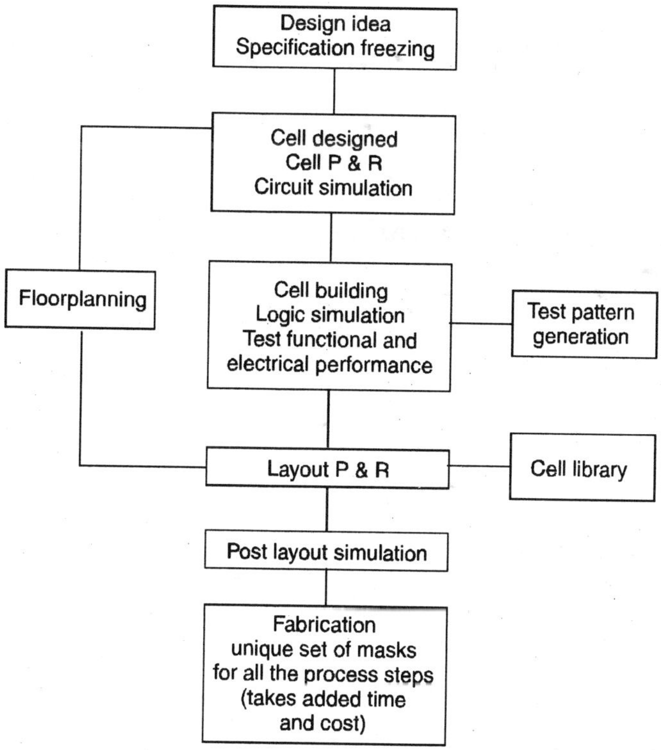

Fig. 9.1.3. Design flow – Standard Cell.

9.1.3. Gate Arrays

In gate arrays a master gate array wafer containing many dice, each with a uniform pattern of n and p transistors, is processed in large volumes upto the last step prior to metallization.

A standard array of many gate circuits diffused into a Silicon chip. The circuit designer provides the semiconductor manufacturer with an interconnection metallisation pattern that converts these basic gates into functional custom circuits.

In other words gate aray or un-committed logic array based design is the one in which standard logic elements are presented for the designer to interconnect and to achieve the desired functional performance.

The core of the chip contains a continuous array of n-and p-transistors *i.e.* the transistors are predefined on the silicon wafer. The interconnect between transistors are defined by the designer using custom masks. Programmability is one way to achieve a wider use of a particular part. In CMOS, the spectrum of programmable devices can be divided in to three areas.

1. Chips with programmable logic structure.
   e.g. PLDs The device is programmed by changing the characteristics of switching elements.
2. Chips with programmable interconnect: Here the routing is programmed.
   e.g. Products from Actel, Quick logic.
3. Chips with reprogrammable Gate arrays: In this gate arrays are programmed.
   e.g. Xilinx FPGAs, Algotronix FPGA..

Features :

- Pre-fabricated or pre-diffused equal-sized *n* and *p* transistors of a fixed number in a fixed layout.
- It is important to remember that the *p* and *n* transistores contained in the logic cell are both of the same size and also the logic cells are all of the same size.
- It results in unbalanced rise and fall time.
- All lithographic levels are standardized except routing.
- Choices of size W/L are limited.
- Transistors can be paralleled for obtaining larger transistors.
- Utilization factor low.
- One level of mask making required for master gate array.

Advantages :

- Wafer pre-fabricated and stock piled using single mask set thus reducing masks makng costs.
- Process data and device characteristics available immediately after routing.

Disadvantages :

- Slower speed.
- Area of die increases.
- Large cell area.
- Cost per chip increases.

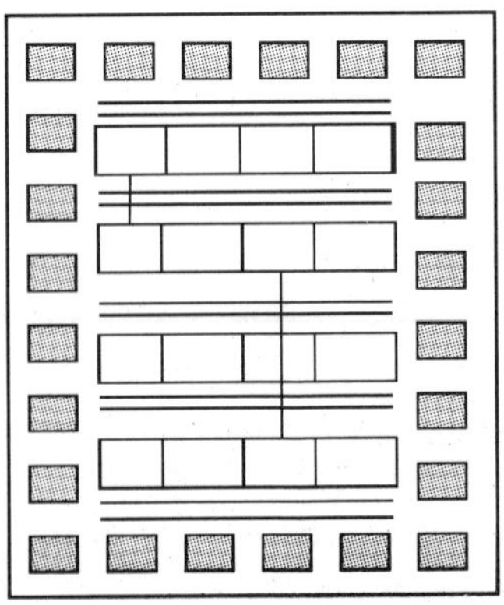

Fig. 9.1.4. Gate array arrangement.

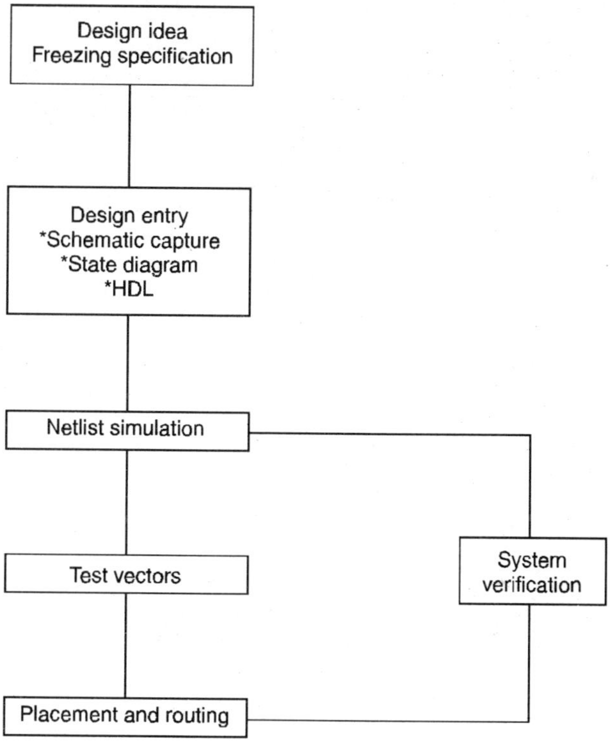

Fig. 9.1.5. Design flow (Gate array).

9.1.4. Field Programmable Devices

Field programmable devices are prefabricated. However, the logic is implemented by electrically programming the interconnects and personalizing the basic cells, typically in the user's laboratory instead of a factory. The field programmable devices have a variety of architectures. Implementing the design in programmable logic devices have the advantage of fast turnaround but limits the design flexibility. Development time and costs are significantly lower than those of any other IC implementation. According to their architectures, we can distinguish two main categories of user programmable logic devices.

(*i*) PLDs – Programmable Logic Devices.

(*ii*) FPGAs – Field Programmable Gate Arrays.

9.1.5. PLDS

As the name suggests, PLDs are logic circuits that can be programmmed by the user to perform specific logic functions, *i.e.*, the device when purchased is unprogrammed like a blank EPROM and in a process similar to programming an EPROM, the user programs the PLDs to implement the desired logic function.

Since the logic function definition as well as the logic function interconnections are defined by the data contained in the on-board RAM cells, logic changes can be implemented by loading a new RAM data file.

It consists of programmable AND arrays (Product terms) and fixed fanin programmable OR gates that are followed by flip-flops. The output of the flip-flops can be fed back as input lines in the product terms. The product line can be connected to any combination of inputs. The connections are programmed by users to implement their designs. The connecting device may be a fuse (in the case of bipolar chips) or a transistor. The transistor can be chosen to act as an open connection or to function normally as a switch. PLDs are at the low-density end of field programmable logic devices. Their densities range from 1000 up to 10,000 gates. Utilization varies with applications, but it is typically low because of the rigid AND/OR architecture. PLDS are used to be fabricated with bipolar technology. However, complementary metal-oxide semiconductors (CMOS) are now more popular.

Features :

The two fundamental sections of a PLD are :

Logic array – user programmable consisting of AND gates, OR gates and inverters. The input signals are routed and operated upon in this section.

Macrocell – It allows the user to customize the PLD output structure like the user can configure each output outside of the logic array to get the desired polarity.

- 100% routability is guaranteed. Routing is provided through the array.
- **PALs** – Programmable two level logic (sum-of-products) arrays with programmable 'AND' plane and fixed 'OR' plane.
- **PLAs** – Programmable two level logic (sum-of-products) arrays with both programmable 'AND' plane and programmable 'OR' plane.

9.2. FIELD PROGRAMMABLE GATE ARRAYS (FPGA)

FPGAs – An unprogrammed FPGA has transistors with missing connections (but plenty of interconnected resources and programmable interconnection points are included on a chip).

FPGA – As the name suggests, Field Programmable Gate Arrays (FPGA) (uncommitted **logic** array) the standard logic elements are available for the designer and he has only to interconnect these elements to achieve the desired functional performance.

It combines the versatility of gate arrays with the programmability of PLDs. It is usually larger and more complex than a PLD. Interconnects can be programmed, through which we can implement combinational as well as sequential logic. The FPGA core is a regular array of programmable basic logic cells. It is surrounded by programmable interconnect. Further programmable I/O cells surround the core. The design turn around time is few hours.

In this design the designer is only best designing the interconnections and communications. The user enters her/his schematic, simulates the design and then with the help of software turns the connections into a string of bits. The programmable equipment makes necessary connections in accordance with the configuration file and the FPGA is ready to use. It is a silicon chip that a user can program himself/herself (electrically with a suitable hardware and software). No masks making or IC processing required.

Elements :

The basic constituents of an field Programmable Gate Array are :

Configurable Logic Blocks (CLBs).

Configurable Input/Output Blocks (IOBs).

2 layer of metal network of vertical and horizontal lines for interconnecting the CLBs and IOBs.

Types. Many kinds of programmable logic devices are called FPGAs, it includes devices with internal structures similar to gate arrays, devices with internal structures similar to a collection of PLDs. The term FPGA is generaly used for devices of first type.

The devices of second type are more commonly called as complex PLDs (CPLDs) or Programmable multilevel Devices (PMDs).

We can categorize FPGAs according to their combination of programming technology and device architecture as :

- CPLDs – Complex Prgrammable Logic Devices
- SRAM FPGAs
- Antifuse FPGAs

CPLDs – Complex PLDs : It provides designers with the capability to implement complex design in a single chip. User creates logic and interconnection by programming EPROM (EEPROM) transistors to form wide fan-in gates. It consists of a few function blocks. Each function block contains a PLD AND-Array that feeds its macrocells. The AND-ARRAY consists of a number of product terms. A macro cell consists of OR-gate/registers and I/O pads/additional EPROM cells to control multiplexers.

SRAM FPGAs : In this FPGA the static memory cells hold the programming. Logic is implemented as look-up tables made from memory cells. A look-up table of 2^n memory cells implements function of n inputs look-up table + flip-flop = CLBs (Configurable Logic Blocks).

Antifuse FPGAs : It's a 2 terminal device that when exposed to a high voltage forms a permanent short circuit between the nodes on either side. The individual antifuse are small so an antifuse based FPGA architecturecain have millions of antifuse. It consists of rows of configurable logic elements with interconnect channels between them. The logic block is a simple gate level network which the user programs by connecting its input pins to fixed values or to interconnect nets. There are antifuses to every wire to pin intersection point in the channel and also at all wire to wire intersection points where channels intersect. The pins on the logic blocks extend into the channel to form wire to pin intersection points in the channel.

Elements of FPGA :

(*i*) **CLBs –** The configurable logic block (CLB) which is RAM based or PLD based is the basic logic cell. It consists of registers (memory), muxes and combinatorial function unit. An array of CLBs are embedded within a set of vertical and horizontal channels that contain routing which can be personalized to interconnect CLBs.

(*ii*) **I/O Logic Cells –** CLBs and routing channels are surrounded by a set of programmable I/Os which is an arrangement of transistors for configurable I/O drivers.

(*iii*) **Programmable interconnects –** These are unprogrammed interconnection resources on the chip which have channeled routing with fuse links. Programmable highly interconnected matrix is available. In this case, the design is that of the interconnections and communications only.

(*iv*) **Software/CAD support –** Software tool to design and programme the FPGA are available on Personal Computers rather than large workstations.

Advantages :

Design cycle significantly reduced. A user can program an FPGA design in a few minutes or seconds rather than weeks or months required for mask programmed parts.

No custom masks tooling is required saving thousand of dollars.

Low risk design style where the price of a logic error is small both in terms of money and project delay.

Rapid product development and prototyping.

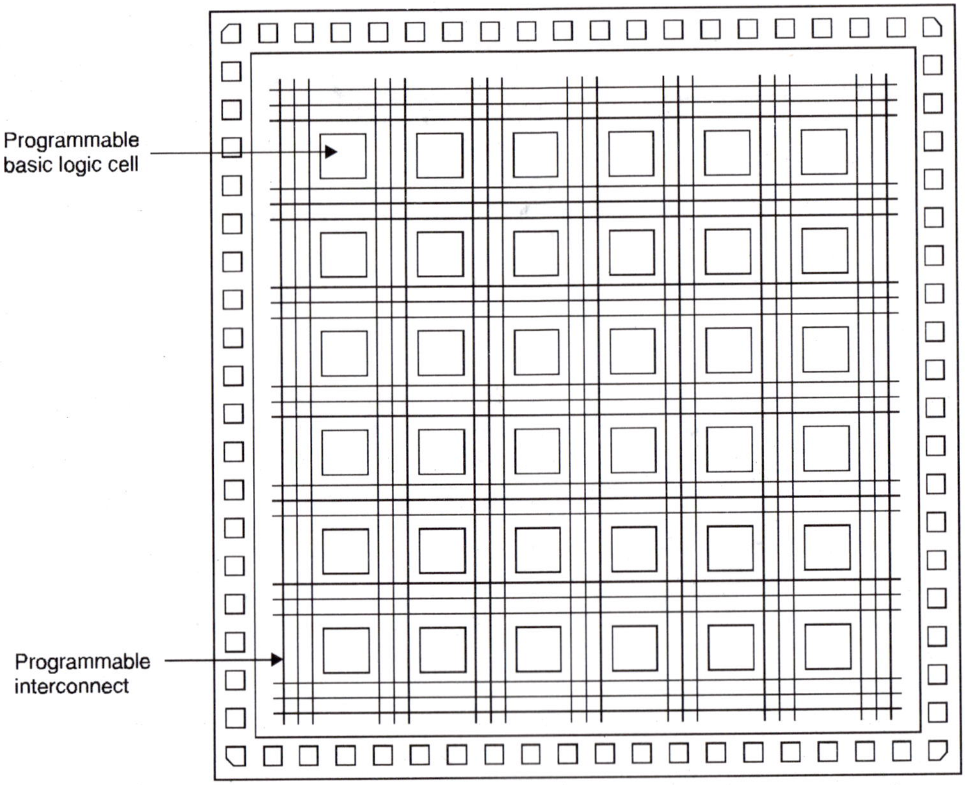

Fig. 9.2 (*a*) FPGA.

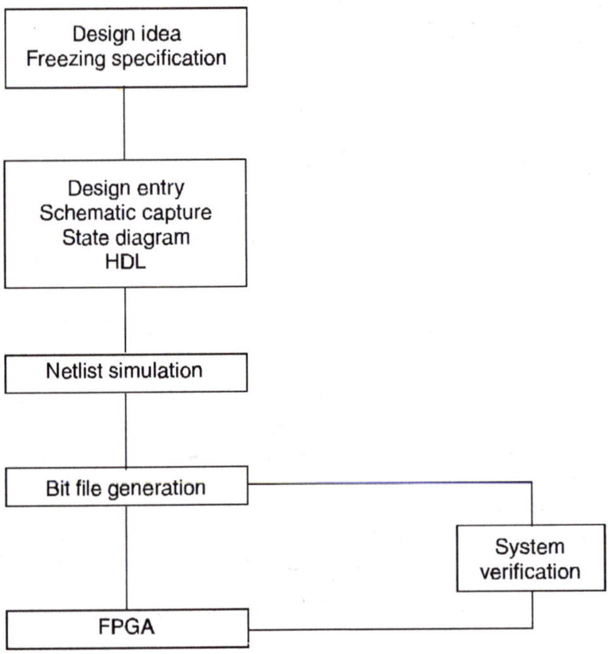

Fig. 9.2 (*b*) FPGA design flow – simplified.

The logic cells or logic block consists of some universal gates, that can be programmed to represent any function., (MUX, RAM). The connectivity between blocks is programmed via different types of devices (SRAM, EEPROM, antifuse). The architecture of the chip depends on the fashion in which the blocks are arranged.

(*i*) *Matrix based architecture* : They form islands in a matrix with horizontal and vertical channels.

(*ii*) *Row based architecture* : They form rows separated by routing channels.

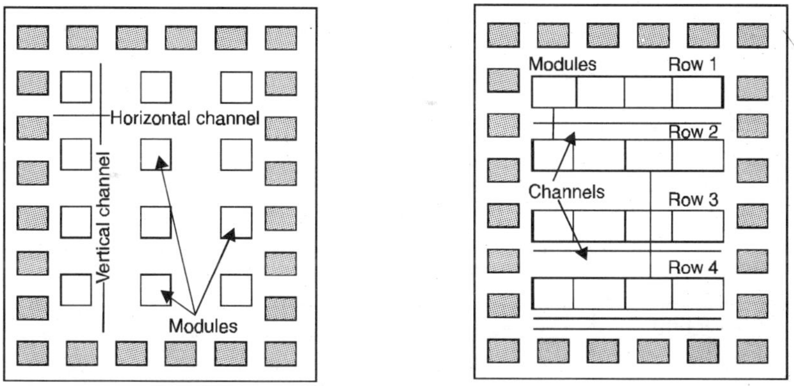

(*a*) Matrix based (*b*) Row based

Fig. 9.2 (*c*) Architectures of FPGA.

Advantage of FPGAs

* High gate density *i.e.* it offers large gate counts. Compared with PLDs they are less dense.
* Short design cycle.
* Low cost.
* Can replace currently used SSI and MSI chips.
* Availability of parts **Off the shelf.**
* Rapid turn around.
* Low risk and highly flexible.
* Reprogrammability for some FPGAs (design can be altered easily).
* Suitable for **Prototyping.**

Prototyping. It is a way of checking the correctness of an integrated circuit without actually fabricating it. *i.e.*, building the system to be designed from discrete components rather than one or a few integrated circuits.

For example **Bread boarding** – It is also a form of prototyping. Prototyping using programmable devices such as FPGAs is quite popular now a days and this type is called rapid system prototyping. The prototype is supposed to show the effects of algorithms in real time.

Limitations of FPGAs

* Speed is comparatively less.

* The circuit delay depends on the performance of the design implementation tools.

The mapping of the logic design into the FPGA architecture requires sophisticated design implementation (CAD) tools than PLDs.

9.3. FPGA ARCHITECTURE

The factors that can influence the design and selection of the device for a given application are:

1. Architecture.
2. Gate density.
3. Routing resources.
4. Programming method.

Table 9.1. Example of FPGAs

Product	*Capacity*	*Architecture*	*Basic Cell*	*Programming Method*
Actel	2,000–8,000	Gate array	MUX	Antifuse
Algotronix	5,000	Sea-of-gates	Functional	SRAM
Altera	1,000–5,000	Extended PLA	PLA	EPROM
Concurrent	3,000–5,000	Matrix	XOR, AND	SRAM
Crosspoint	5,000	Gate array	Transistors	Antifuse
Plessey	2,000–40,000	Sea-of-gates	NAND	SRAM
Quick Logic	1,200–1,800	Matrix	MUX	Antifuse
Xilinx	2,000–10,000	Matrix	RAM block	SRAM

The Xilinx Logic cell array. As an example of FPGA, we will focus on the architectures of Xilinx 3000 and Xilinx 4000 series.

Table. 9.2. Comparison between xilinx devices

Feature	*XC 3000*	*XC 4000*
Equivalent gates	2,000 – 9,000	2,000 – 10,000
Number of CLBs	64 – 320	64 – 400
Inputs per CLB	5	9
Flip-flops per CLB	2	2
I/O Blocks	64 – 144	64 – 240
Maximum RAM bits	0	2,000 – 20,000
Functions per CLB	1 of 5 variable 2 of 4 variables	1 of 5 variables 1 of 4 variables 1 of 9 variables

9.3.1. Xilinx 3000 Series FPGAS

Fig. 9.3.1 (*a*) shows the logic cell array architecture of Xilinx XC 3000. It consists of fixed array of CLBs (Configurable Logic Blocks) input/output blocks (IOBs) as in figure. The blocks can be connected by horizontal and vertical wire segments joined by the switch matrices. The programming elements (configuration memory cells) inside the switch matrices establish the connections between the wire segments. This Xilinx 3000 chip can be reprogrammed since its programmability is based on the SRAM technology.

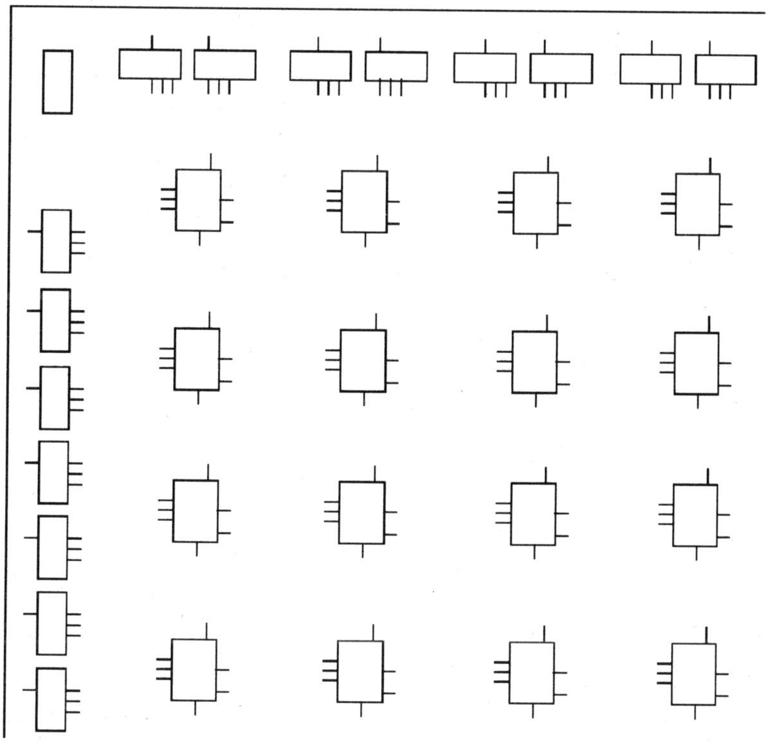

Fig. 9.3.1 (*a*) LCA architecture of XC 3000.

Configuration memory cells. The Fig. 9.3.1 (*b*) shows the configuration memory cells. These are programmed after power has been applied to the LCA. The programmed logic functions and interconnections are retained until the power is turned off. When a write signal is applied to the pass transistor, Data is stored in the cell. Each connection point in the LCA has an associated memory cell, and the data stored in that cell determines whether the connection is made or not.

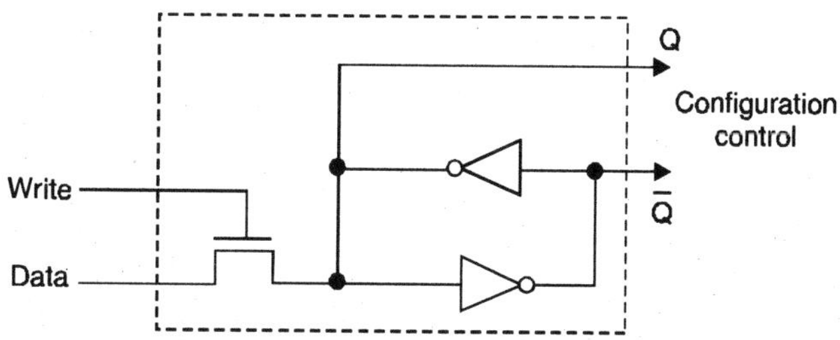

Fig. 9.3.1 (*b*) Configuration memory cells.

Fig. 9.3.1 (*c*) shows Xilinx 3000 series logic cell. It consists of two flip flops, a five input look up table (or combinational function block) and several internal multiplexers a data input (DI) a clock input (k), a clock enable (EC), a direct reset (RD) and two O/Ps X and Y. The multiplexers can be programmed to select one of the inputs. For example, the X O/P can come either from the upper flip flop (QX) or from the 'F' O/P of the "Combinational Function" block. Likewise, the Y O/P can come either from the lower flipflop (QY) or from G. Each M represents a configuration memory cell, and the data in the cell determines which MUX input is selected. The look up table is a block of 32 bit RAM cell. The I/P variables (A,B,C,D,E) are applied to the address lines of the RAM cells. The RAM cells can be programmed to realize any single function of five variables or any two functions of four variables.

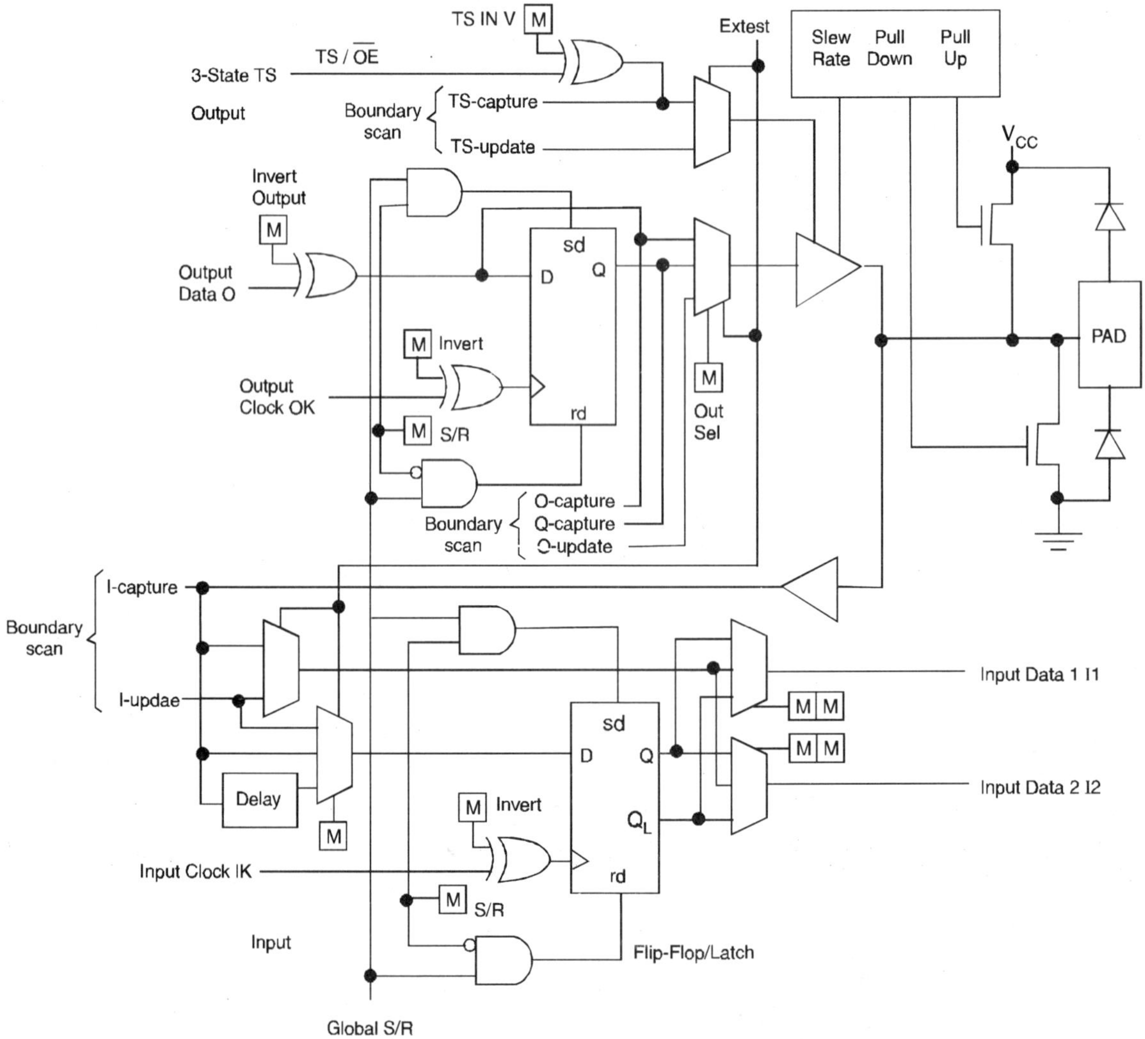

Fig. 9.3.1 (*c*) Xilinx 3000 series logic cell.

Fig. 9.3.1 (*d*) shows the 3 different modes of operation for this block. The multiplexers can be programmed to select one of its inputs.

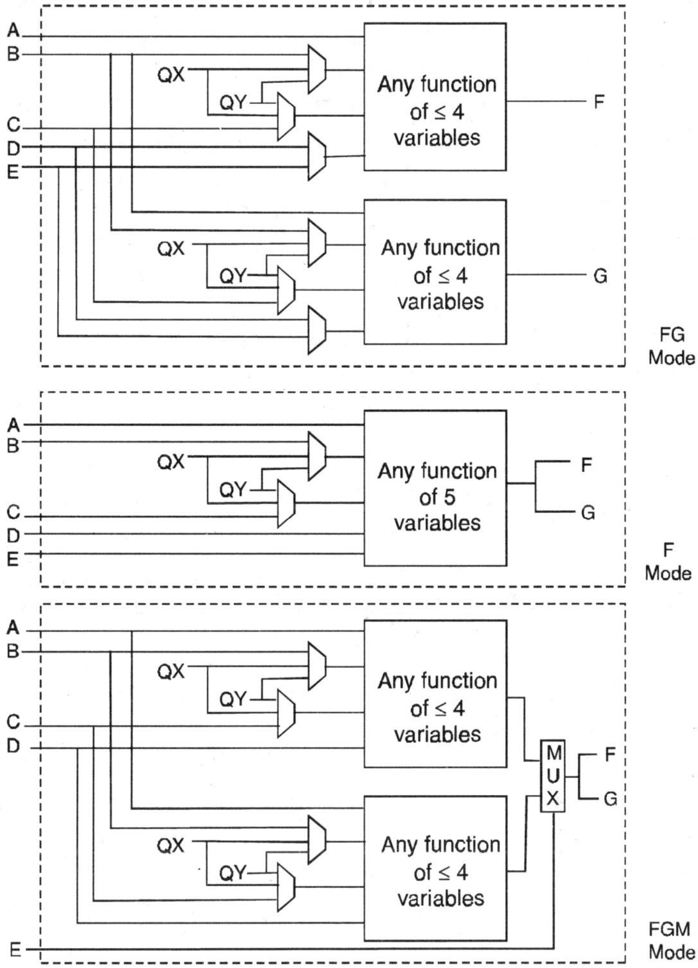

Fig. 9.3.1 (*d*) Different modes of logic cell block.

***(a)* FG Mode.** It generates two functions of four variables each. In this variable 'A' is selected as common to both functions. *i.e.* it is available in both the functions.

For function F the input variables are :

(*i*) Variable A

(*ii*) D or E (selected using MUX)

(*i*) B or QX or QY (selected using MUX)

(*ii*) C or QX or QY (selected using MUX)

Likewise for function G also the 4 variables are selected as above.

For example $G = A'B' + QX.D$ and $F = AC + QY.E$

If QX and QY are not available, then for function F and G the four variables are selected as follows :

(*i*) A (*ii*) B (*iii*) C (*iv*) either D or E

(*b*) F Mode. It can generate one function of five variables as follows :

Three variables A, D, E are chosen as compulsion. The remaining two are chosen from B, C, QX, QY. Using this we can realize simple functions like OR (F = G = A + B + C + D + E) and also any complex function which may consist of several terms in its SOP (Sum of Product)

(*c*) FGM Mode. It also generates two four variable functions. The input variables are A, B and any two from B, C, QX, QY. A Mux is used to select one function from the two available. The input E serves as a control input to MUX in selecting one function. In this mode, even, function with 6 or 7 variables can be realized.

e.g., F = G = E(A′B + QY.D) + E′ (AC + QX.D)

For the two flip flops the D input comes from one MUX. The MUX selects one from F or G or DI or QX or QY. The two FFs are connected to common clock (K) and can be programmed to change state either on rising or falling edge of the clock. The clock can also be controlled by enable clock (EC) input. The MUX connected to the D input of each FF is used to effectively disable the clock.

If EC = 0 Q O/P is fedback to D input and hence $Q^{n+1} = Q^n$. In this condition the FF will never change state even when clock is high.

If EC = 1 The D input is connected to F, G or DIN and now the FF will change state in response to clock. The D FF and MUX combination is equivalent to a D FF with an enable clock (EC).

9.3.2. Parallel Adder Subtractor Implementation in XC 3020

This can be implemented by operating combinational logic block in FG mode. Fig. 9.3.2 shows parallel adder subtractor logic cell, with an accumulator. To the input variable B and A, the logic input *bi* and carry from previous cell C_i are fed. We know each logic cell consists of two flip flops. From that one is chosen to serve as accumulator. The O/P from accumulator (or the value stored in accumulator) acts as one of the logic input *ai*. The *ai* output from FF is fed back internally with in the cell to input variable QX. The control bit S_u can be connected to D or E.

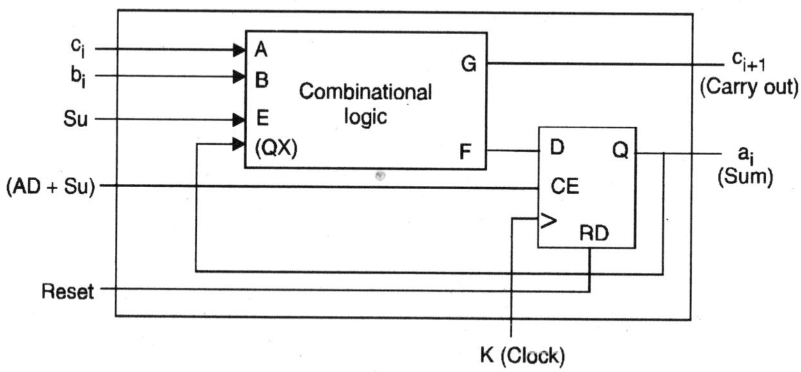

Fig. 9.3.2. Parallel adder subtractor logic cell.

If $A_d = 1$, the B input (*bi*) will be added to the accumulator. If $S_u = 1$, the B input will be subtracted from the accumulator. Subtraction is accomplished by adding 2's complement of B to the accumulator. If $A_D = S_u = 0$ the accumulator will remain unchanged. Hence, the clock enable

EC is connected to $A_d + S_u$ and only when either $A_d = 1$ or $S_u = 1$ it will enable the D-FF and (inturn accumulator will be enabled).

The combinational logic block implements the sum and carry functions as follows :

$$F = \text{Sum} = a_i \oplus (b_i \oplus S_u) \oplus c_i$$

$$G = \text{carryout} = a_i c_i + (a_i + c_i)(b_i \oplus S_u)$$

9.3.3. IOB of XC 3000

Fig. 9.3.3. shows a configurable Input-output block (IOB). The I/O pads are connected to one of the pins on the IC package. Through this, external signal inputs can be given to logic cells and logic cell O/P can be taken to external O/P. To use the cells as an input the tristate buffer should be in high impedance state. To use the cell as an O/P, the tristate buffer must be enabled. The input and output values can be stored in IOB using the flip flops. When direct input or output is desired the flip flops can be bypassed.

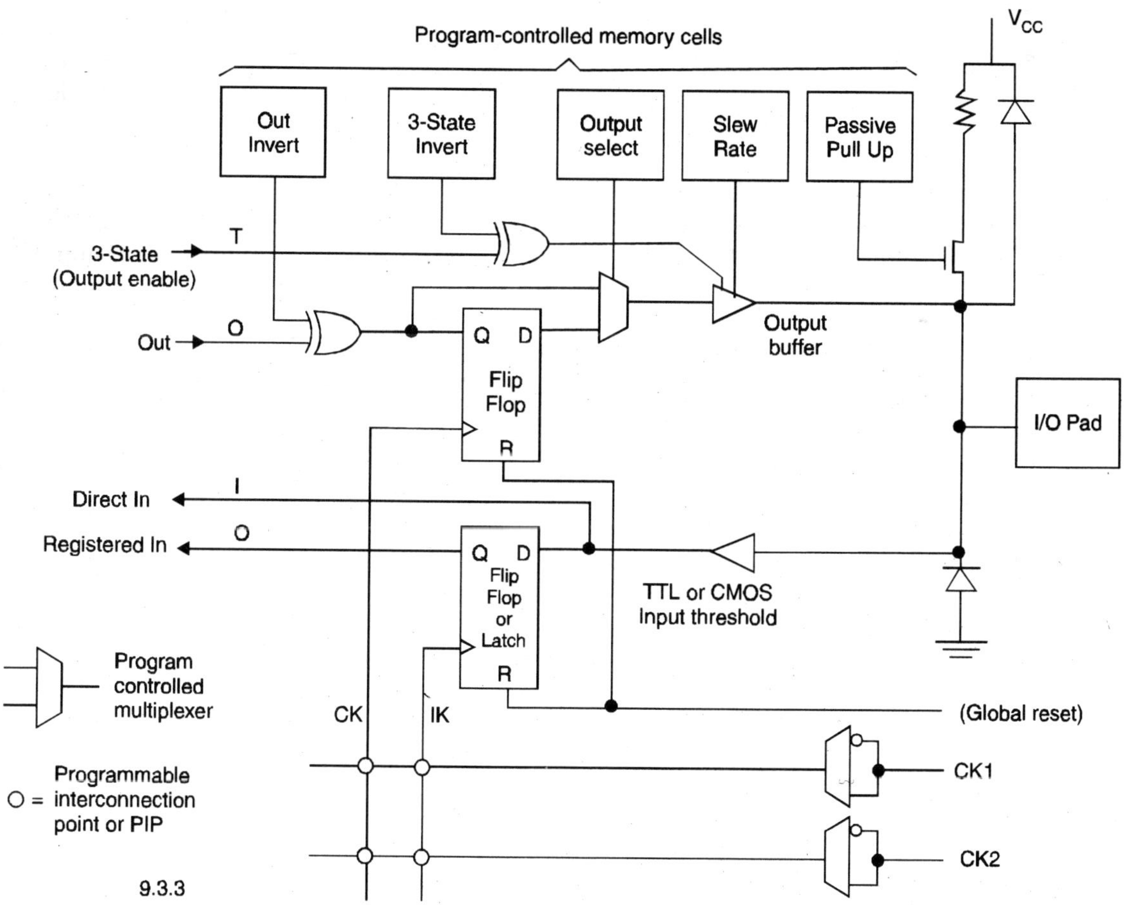

Fig. 9.3.3. Xilinx 3000 series I/O block.

I/O pad as O/P. An OUT signal coming from the logic cell first goes through an EX-OR gate. Depending on out-invert bit, either it is complemented or un complemented. If desired, this out signal can be stored in FF. Either the out signal or FF output is passed to output buffer by the MUX which is controlled by output select bit. If the 3 state signal is '1' and the 3 state invert bit is '0' (or if the 3 state signal is 0 and the 3 state invert bit is '1'), the output buffer has high impedance output. Otherwise the output buffer drives the output signal to the I/O pad.

I/O pad as I/P. To use I/O pad as input, the output buffer should be in high impedance. An external signal coming into the I/O pad goes to the buffer. From the buffer the signal can be directly fed to logic array which is called a Direct IN. The O/P of buffer can also be fed to a D.FF. where the signal can be temporarily stored and from there it can be fed to logic array. This is called a Registered IN.

Slewrate (SR) bit. This controls the rate at which the output signal can change. When the output drives an external device, reduction of the SR is desirable to reduce the induced noise that can occur when the O/P changes rapidly.

Passive pull up bit. When this is set a pull up is connected to the I/O pad. This is used to avoid floating.

9.3.4. Programmable Interconnects

The interconnections between CLBs and IOBs can be made in several ways.

***(i)* General purpose interconnects.** In this signals between one CLB and the other, and between CLBs and IOBs are routed through switch matrices. As shown in Fig. 9.3.4 (*a*), at the center of every 4 CLBs, there is a switch matrix with horizontal and vertical lines on their sides and on their top and bottom for making interconnections.

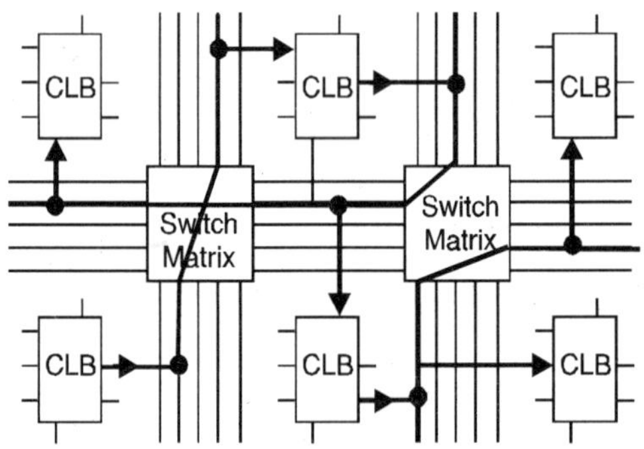

Fig. 9.3.4 (*a*) General purpose interconnects.

***(ii)* Direct interconnects.** Without using switch matrices the interconnections can be made directly between adjacent CLBs as shown in Fig. 9.3.4 (*b*).

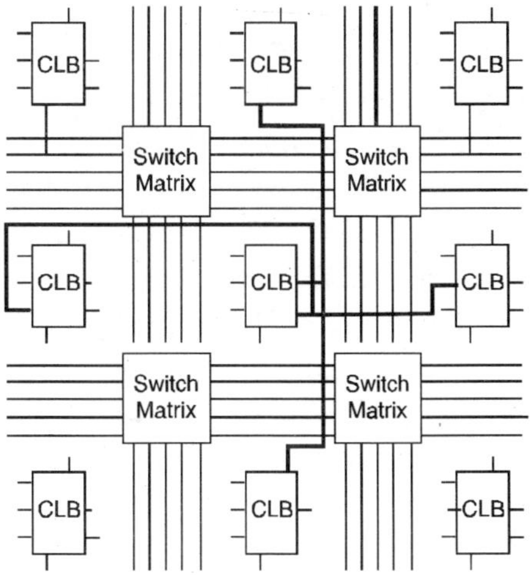

Fig. 9.3.4 (*b*) Direct interconnects.

***(iii)* Long lines.** For connecting CLBs which are far apart long lines are used. As in Fig. 9.3.4 (*c*) between each pair of adjacent columns of CLBs there are four vertical long lines. Out of these 4, two can be used for clocks. In a similar way there are two horizontal long lines between each pair of adjacent rows of CLBs. Each CLB has two tristate buffers connected with the pair of horizontal lines. These tristate buffers are used to multiplex signals onto a horizontal lines. The long lines span the entire length and width of the interconnection area. All the interconnections are programmed by storing bits in internal configuration memory cells with in the logic cell array.

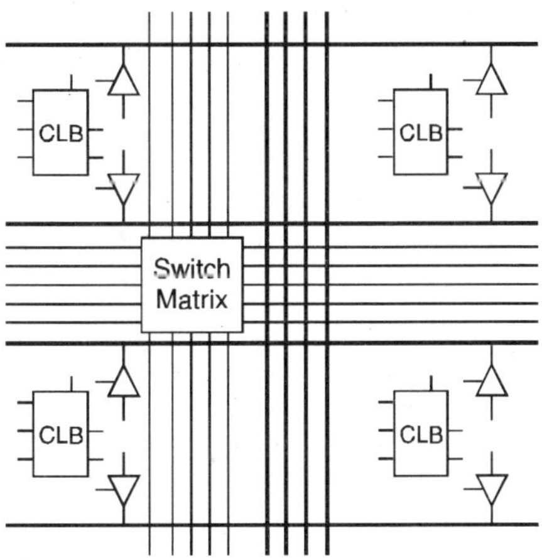

Fig. 9.3.4 (*c*) Vertical and horizontal long lines interconnect.

9.3.5. Xilinx 4000 Series FPGAs

This has more additional features than 3000 series. Fig. 9.3.5 (*a*) shows the block diagram of XC 4000, CLB. It has nine logic inputs and 4 outputs. The inputs are F_1, F_2, F_3, F_4, G_1, G_2, G_3, G_4 and H_1 and the outputs are X, Y, XQ and YQ. It can generate two independent functions of four variables. The logic function block G is a function of G_1, G_2, G_3 and G_4. Similarly the logic functions block F is a function of F_1, F_2, F_3 and F_4. The third logic function block H is a function of F,G and H1. It can generate any function of 6,7,8 and 9 variables. In this also there are two D flip flops with enable clock (EC) whose outputs are YQ and XQ. The other two O/Ps X and Y are obtained from combinational logic function generators. In XC 4000, CLB has no internal feed back. Therefore when feedback is required the FF outputs must be routed externally back to the logic inputs. The CLB has a setreset (S/R) input that can be independently configured to connect to the SD or RD input on each flip flop. Using this S/R input the FF could be set to 1 or 0. There is also a clock input (k) and using this the flip flop can be triggered either on leading edge or trailing edge of the clock. The D input to FF is fed from a 4 to 1 MUX whose inputs are DIN, F_1 G and H. Apart from the above there are several multiplexers for different selections.

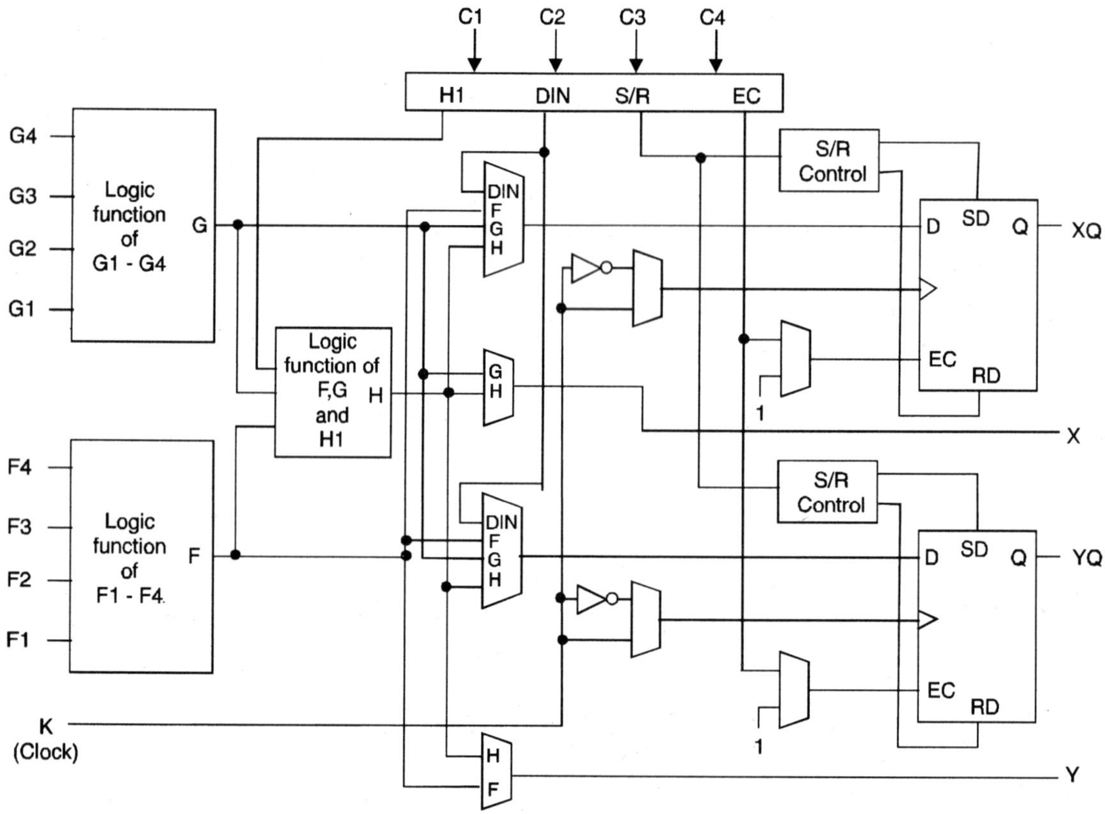

Fig. 9.3.5 (*a*) Block diagram of 4000 series CLB.

Dedicated Carry Logic. The XC 4000 logic cell contains Dedicated carry logic as shown in Fig. 9.3.5 (*b*). This facilitates the implementation of adders, subtractors, incrementers, decrementers and counters. Each cell contains carry logic for two bits, and the function generators F and G can be used to generate the SUM bits. The carry lines between cells are hard wired (permanent connection) and are not programmable. This allows fast propagation of carries. The carries are propagated in vertical direction between the cells. It can be propagated upwards or downwards and the direction is selected using MUX M_1. Multiplexers M_2, M_3 and M_4 allow some of the F and G function generator inputs to come from carries instead of from the normal F and G cell inputs.

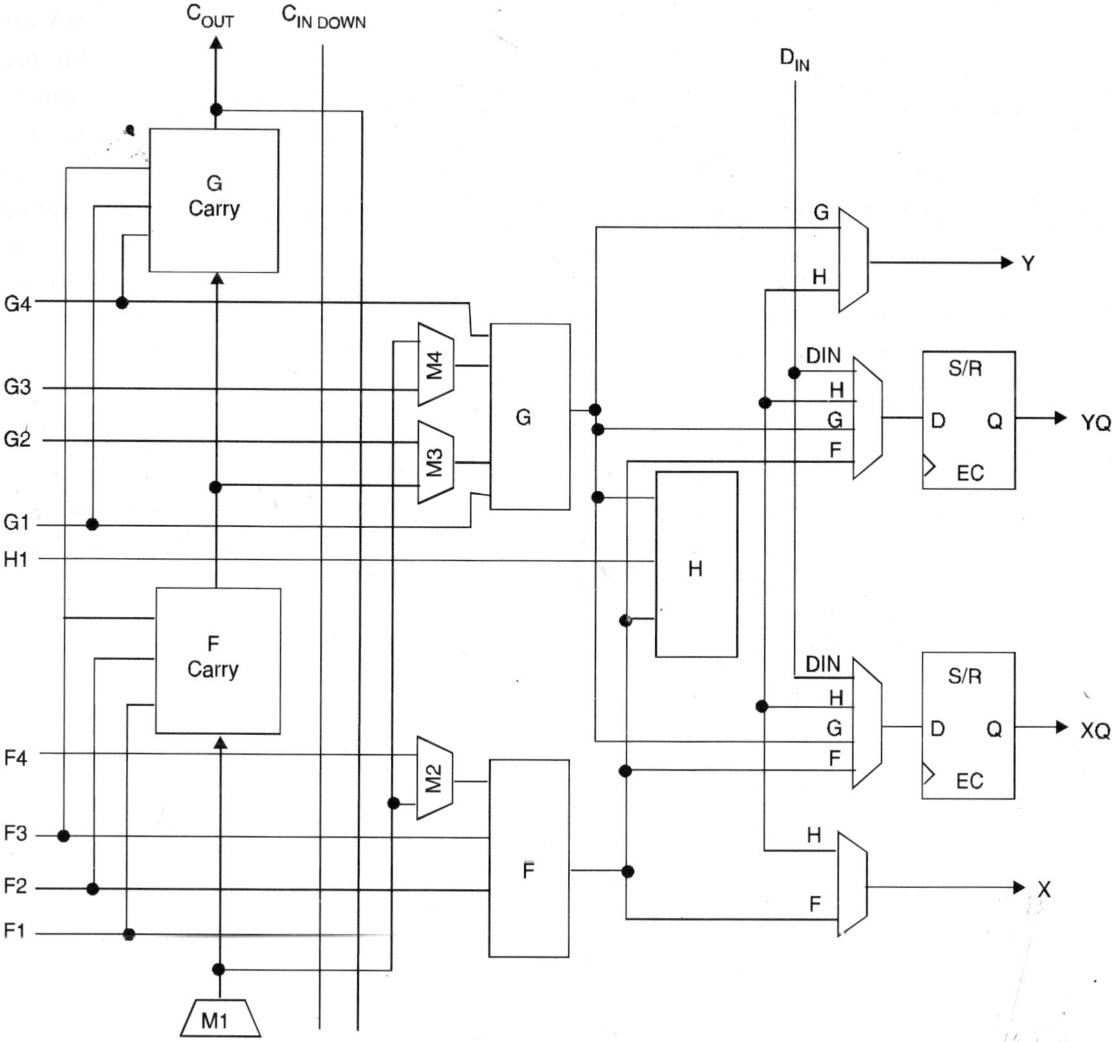

Fig. 9.3.5 (*b*) XC 4000 dedicated carry logic.

Full adder implementation. The Fig. 9.3.5 (*c*) shows how a 2 bit full adder (to add A = a_1a_0 and B = b_1b_0) can be implemented by programming the logic cell. In this configuration a_0 and a_1 come from cell inputs F_1 and F_2. Similarly b_0 and b_1 are obtained from G_1 and G_4. The C_0 input and the C_2 O/P are connected to the hardwired carry propagation lines. Function Generator F generates S_0 (Sum of LSB) and Function Generator G generates S_1 (Sum of MSB). There, using single CLB two bits can be added.

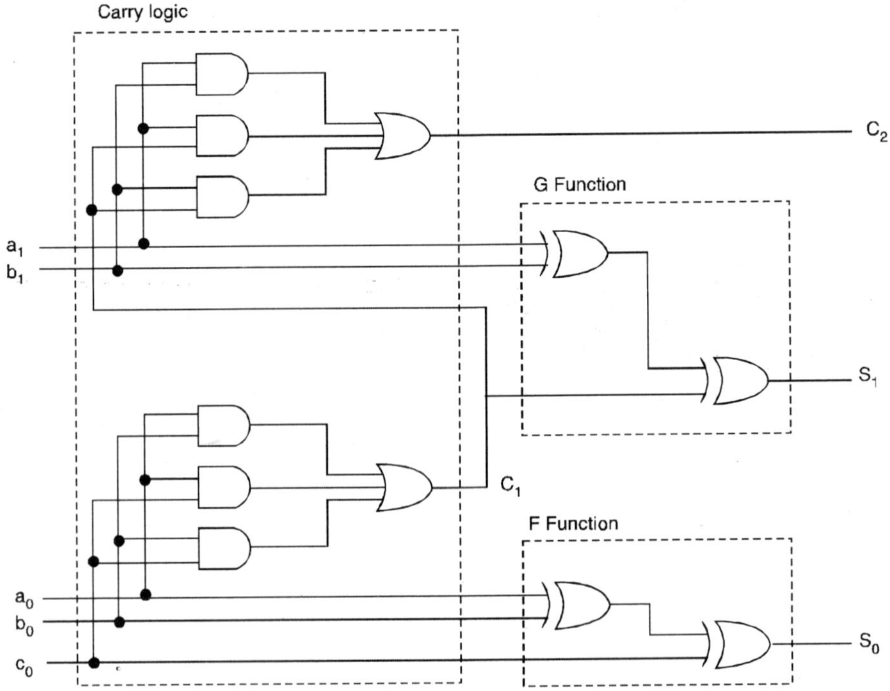

Fig. 9.3.5 (*c*) Full adder implementation.

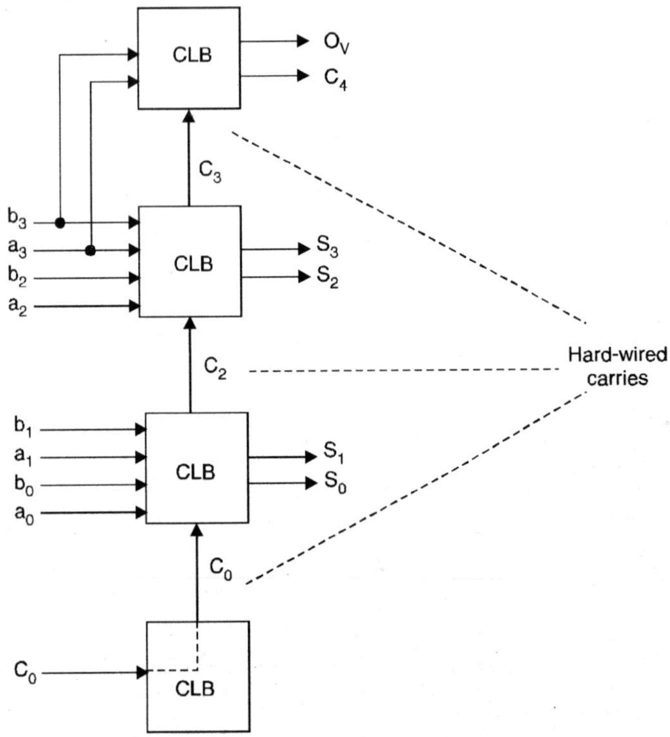

Fig 9.3.5 (*d*) Connections for a four bit adder.

Fig. 9.3.5 (*d*) shows a four bit adder. Obviously it requires two CLBs for performing 4 bit addition. In Fig. the middle two CLBs are used for performing addition. If a carry into the LSB is needed, C_0 must be routed through an additional cell, because no direct connection to the hard wired carry lines is allowed. In a similar way for indicating over flow of the 4 bit adder (*i.e.* carry out from the MSB), another CLB is required as shown in Fig. In This Fig. 9.3.5 (*d*) C_3 is routed to the fourth cell instead of C_4. Also the cell inputs a_3 and b_3 are given to fourth CLB along with C_3. With these inputs it can compute C_4. And finally overflow is detected by ex-or ing C_3 and C_4. This 4 bit adder can be expanded to 8 or 16 bits by using additional CLBs.

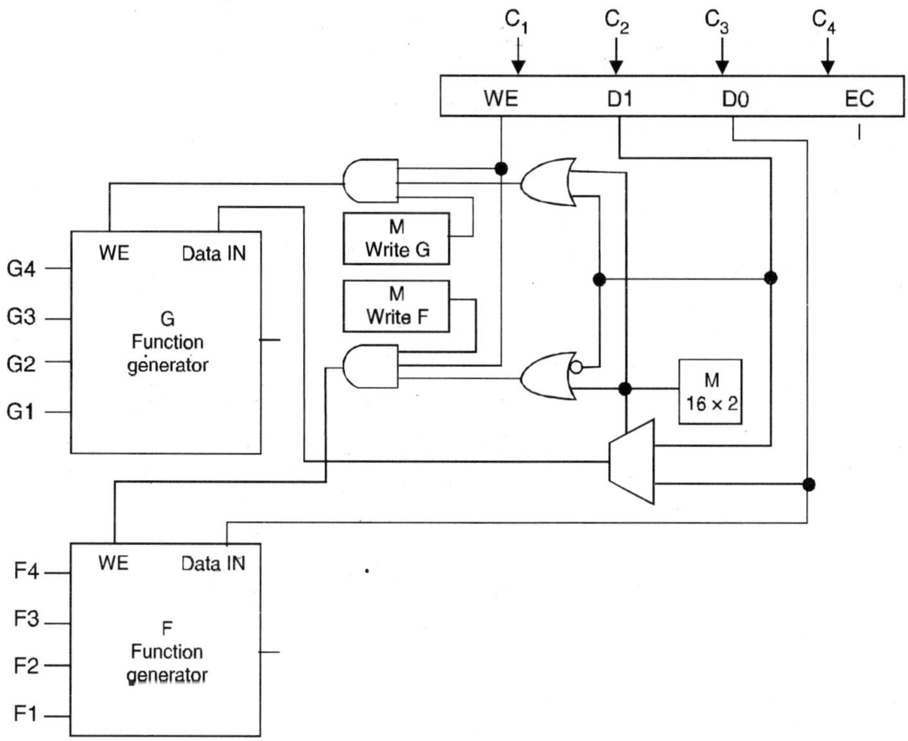

Fig. 9.3.5 (*e*) Read/Write memory cell using CLB.

CLB as a RAM. The CLB of XC 4000 can be used as a Read Write Memory which is shown in Fig. 9.3.5 (*e*). The Function generators F and G can be programmed to serve as RAM memories. Each cell can be configured as a 16 × 2 RAM. (16 words of 2 bit each). The F and G inputs provide a four bit address (Also $F_1 = G_1$, $F_2 = G_2$, $F_3 = G_3$ and $F_4 = G_4$). C1 is used as a write enable line. Data inputs D_1 and D_0 are obtained from C_2 and C_3 using the four address lines 16 locations can be selected in F & G. The contents of 16 locations are available at function generator output. For *e.g.* the content of location 8 can be obtained by making $G_4 = F_4 = 1$ and $G_3 = F_3 = G_2 = F_2 = G_1 = F_1 = 0$. This enables eighth location to be selected. The two bits of the word are stored as 1 bit each in each of the function generator. So at the output of F, LSB of word can be obtained & MSB can be obtained from the output of G.

In other way this CLB can be configured as a 32 × 1 bit RAM, in which case C_2 serves as fifth address line and C_3 is the only data input. The configuration bits labeled WRITE G and WRITE F must be set to allow writing to the Function Generators.

9.3.6. IOB of XC 4000

Fig. 9.3.6 shows a 4000 series IOB. This is similar to 3000 series IOB. External signals are routed into CLB in a similar manner as in 3000 series. Also the signal from CLB are brought to output in a similar way. The M on the diagram represent configuration memory bits. As in the case for 3000, the 4000 IOB can be programmed to invert the input, output, tristate buffer control and clock input. In addition, the global S/R line can be programmed to set or reset each flip flop. The pull up and pull down resistor, both are available as optional in 4000 series.

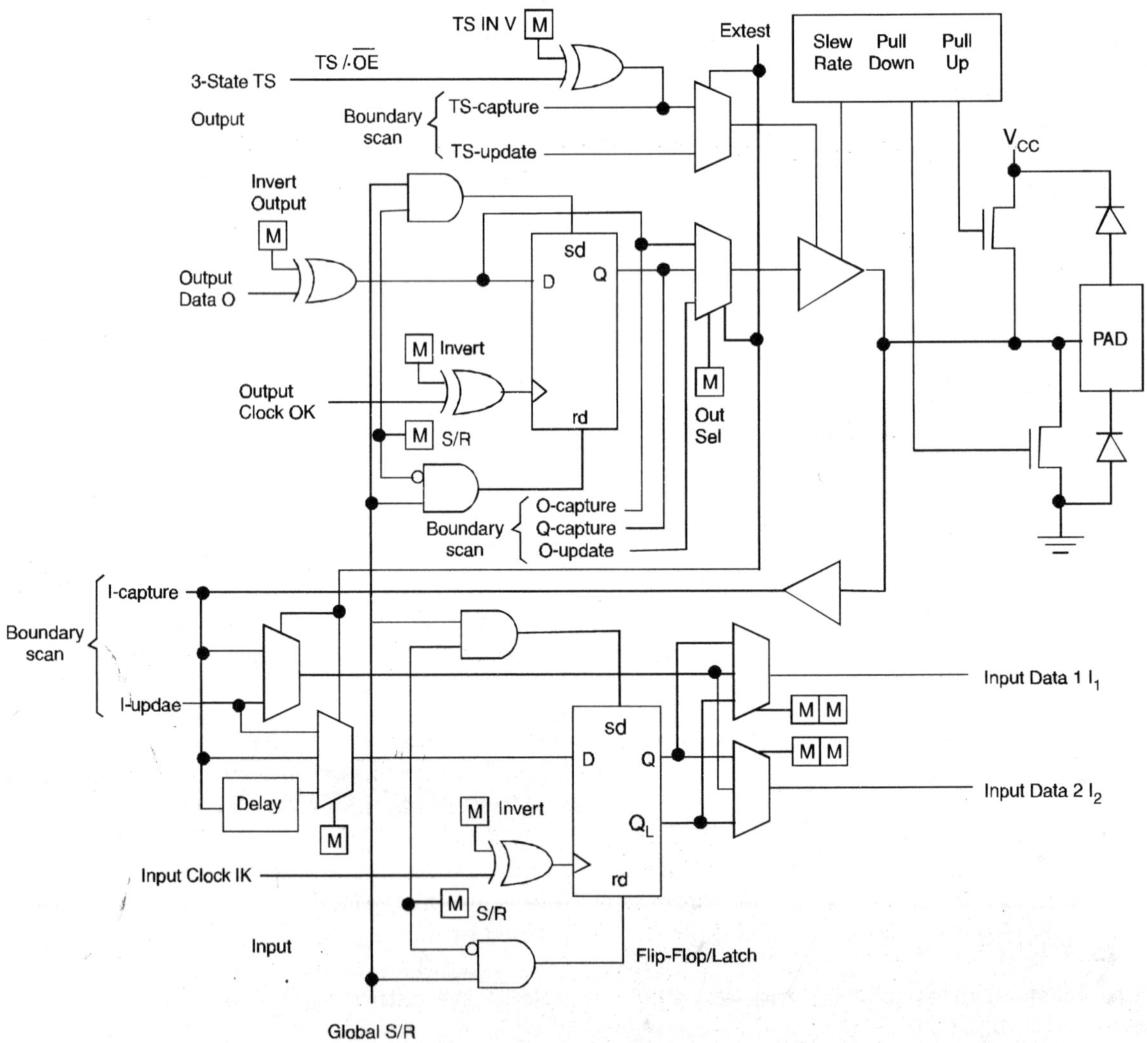

Fig. 9.3.6. XC-4000 series I/O block.

It also contains added logic for Boundary Scan testing (a way of testing). XC 4003 comprises of 100 CLB (10 × 10), 3000 gates, 80 user I/Os, 360 FFs (200 in CLB, 160 in IOB) and 45, 636 configuration data bits. The other member of XC 4000 families have upto 2304 CLBs (48 × 48), 384 user I/Os, 5376 FF and more than 1 million configuration data bits.

9.4. ROUTING

After logic minimization and technology mapping, the design consists of a net list of logical components to be assigned to the physical components of the target FPGA. Communication between the components is established by assigning the netlist to wire segments and programming elements to realize the connections. This physical design process is known as placement and routing. Enough care should be taken while doing this otherwise performance of the design will get affected.

Placement. It is defined as assignment of the logic components to particular physical components on the chip. *i.e.* mapping of logical components with that of physical components on the chip.

Routing. The interconnections between components on integrated circuits are made by depositing metal and joining segments of metal through vias, antifuses, or RAM cells depending on the design methodology. The components may be standard cells or logic blocks in an FPGA device. Interconnecting the physical components by using the available routing resource, wire segments and programming elements, is known as routing. Routing is normally done is two steps

1. Global routing
2. Detailed Routing. The basic FPGA routing terminology is illustrated in Fig. 9.4.1.

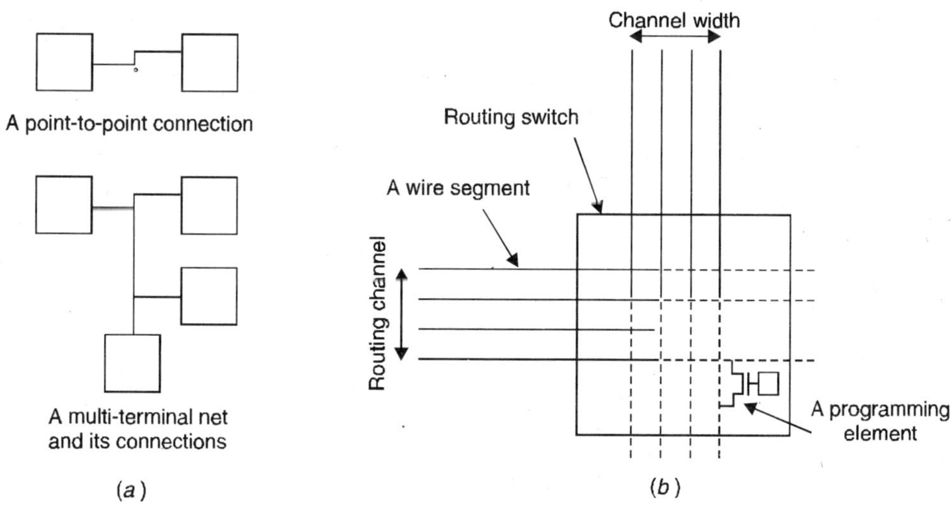

Fig. 9.4.1. Routing terminology.

9.4.1. Global Routing Detailed Routing

Global routing doesn't make any interconnections. It just plans them. Global routers assign "loose paths" to the nets. It determines how they will navigate around the cells. These "loose paths" determine which routing areas a net will traverse to get to its destination(s). The routing areas can be channels, segments of channels, or switchboxes. For multiterminal nets, global routers must also decide in which regions the nets branch to reach different destinations.

The objectives of the assignment might be to shorten the connection lengths, avoid congestion, or use fewer programming elements to decrease delay. After the global routing phase detailed routing is to be done. The detailed router assigns specific tracks to the nets within the routing areas. Detailed routing is heavily influenced by the architecture of the FPGAs.

9.4.2. Segmented Channel Routing

It is applicable to the row-based FPGAs. In this we have two tracks. 1. Horizontal track 2. Vertical track. The horizontal tracks consist of wire segments of possibly varying lengths. The vertical tracks consist of segments connected to the inputs and outputs of the logic or I/O blocks. Programmable devices, antifuses, are located between pairs of adjacent horizontal segments and at the crossing of horizontal and vertical segments.

The approach to segments channel routing depends largely on the number of tracks and their degree of segmentation. There are three cases of possible segmentation. Nonsegmented tracks imply the use of one net per track. In such an arrangement, the number of antifuses is limited (therefore less resistance), but the capacitance per track is larger than for shorter segments. Fully segmented tracks (segments of lengths one) allow maximum flexibility. The left-edge algorithm can be used for this.

Fig. 9.4.2 shows a set of connections a, b, c to be laid on a set of tracks, so each connection spans a set of vertical tracks.

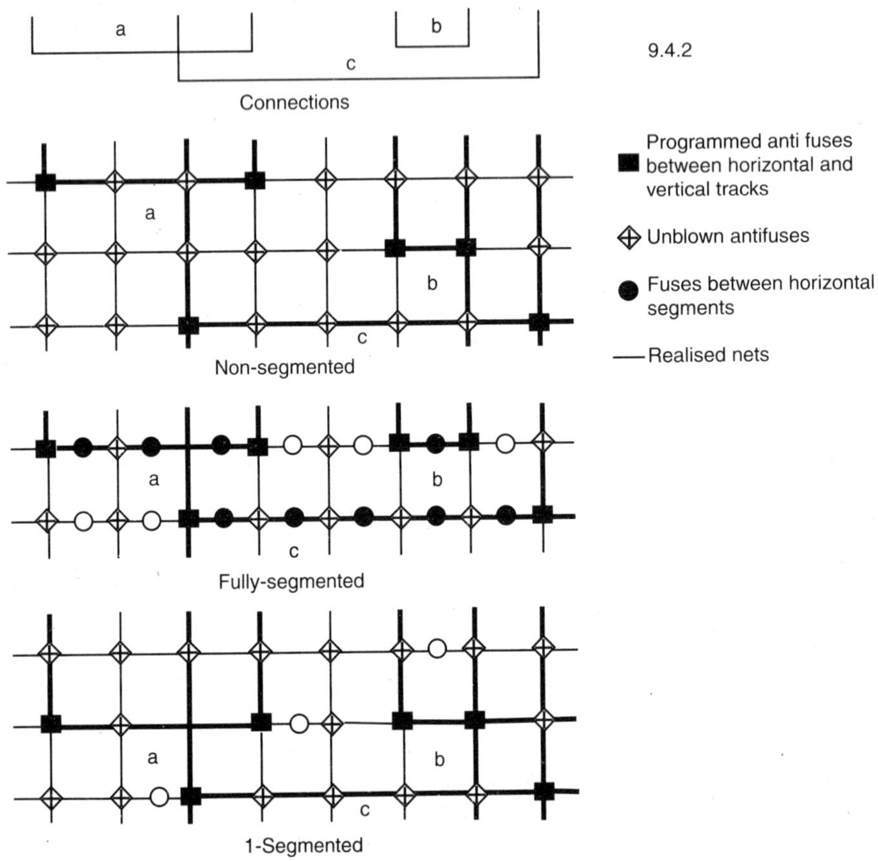

Fig. 9.4.2. Segmented routing.

Steps in LEA. 1. Sort the connections by increasing order of their left edges From Fig.(*a*) we see that the left edge of net '*a*' is left most followed by *b* and *c*. Hence '*a*' is allowed to span left most vertical track as shown in Fig. 9.4.2.

The remaining connections are assigned as follows. Find the next connection whose left edge is to the right of the right edge of the last net assigned to the current track being processed. In Fig., '*b*' is to the right edge of '*a*'. Span '*b*' in same horizontal track where '*a*' is spanned, if possible. If the track is now full or no more nets can be assigned to it then choose next horizontal track for subsequent nets.

9.4.3. Maze Routing

In this algorithm the router finds shortest route between two (or more) points subject to the following restriction: The route is along vertical and horizontal paths. The routing area includes a certain configuration of obstacles that forces the router to detour.

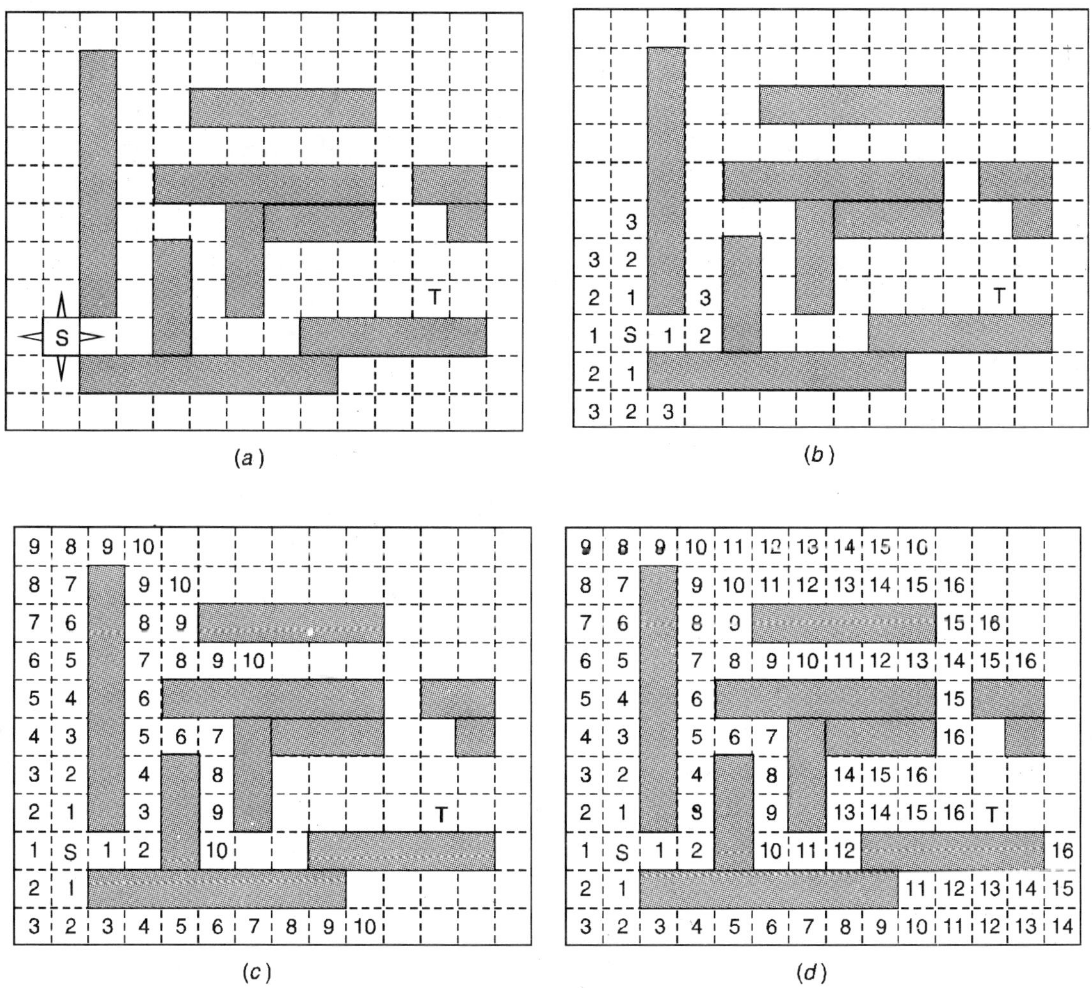

Fig. 9.4.3 (*i*) Maze routing.

(*a*) Starting (S) and Terminal Positions. (*b*) Wave propagation after 3 steps.

(*c*) After 10 steps. (*d*) Different possible routes between S and T.

See Fig. 9.4.3 (*i*), where S is the starting point and T is the target point. The algorithm consists of three phases that are illustrated briefly with the example in Fig. 9.4.3 (*i*). The routing area is organized in a grid and includes the blockages indicated by the gray tiles. The search for the route starts at S (tile 0). From S the next move is to the four neighboring cells up, right, down, and left. These cells are labeled "1" if they are empty. All empty neighbor cells of all 1-cells are then labeled "2". The order of labeling is always in the same sense, for example: up, right, down and left. The process is repeated for all 2-cells, and so on until the target, T is reached. Algorithmic wise we are performing a breadth-first search resulting in a wave propagation from S until T is reached. The search scheme is illustrated in figs 9.4.3 (*i*) (*b*) to (*d*). The labels of the neighbors of T indicate the length of all possible shortest paths between the two points.

To determine the route, the next phase traces back from T starting with the smallest label adjacent to T and repeatedly selecting a neighbor with a smaller label until S is reached. Tracing back the shortest path we obtain the route as shown in Fig. 9.4.3 (*ii*). The last phase in the algorithm is the sweeping: all cells included in the route between S and T are blocked so that one can proceed with routing another set of points (not shown in Fig.).

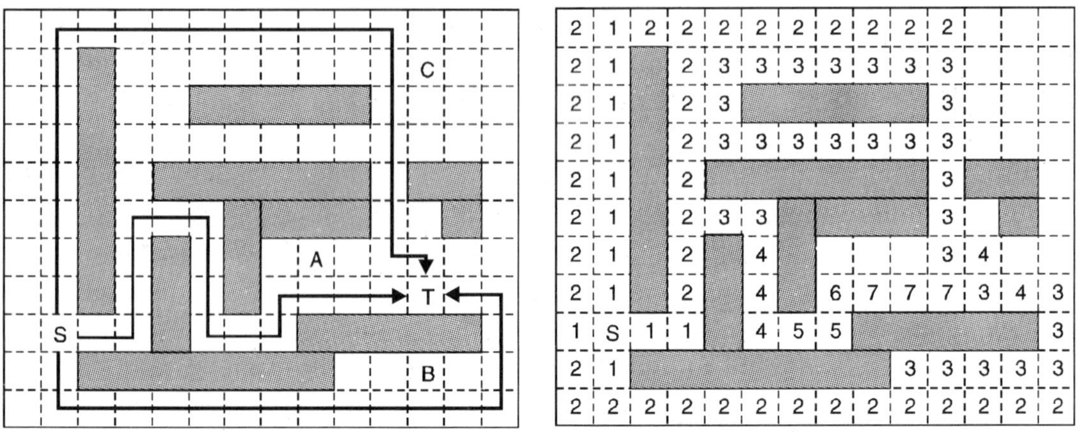

(*a*) (*b*)

Fig. 9.4.3 (*ii*) Shortest path (*a*) Shortest Route A, another route B, and a third route C, (*b*) Path cost based on the number of bends.

It is important to minimize the wiring lengths to improve performance of the design. But, routing delays are also a function of the number of programming elements included in the connection path. Every time a route changes direction, a programming element has to be included in the route to connect the vertical and horizontal wire segments that are on two different metal layers. Three routes in our example are shown in Fig. 9.4.3 (*ii*) as A,B,C. Notice that route A is the shortest (16), but it changes directions six times. As for route B, it is only two cells longer than route A, but it includes only three changes in direction. Note that route C also has only three changes in direction, but it is substantially longer than B. By using proper algorithm the final route can be chosen. In the above example route B is the best one.

Floor planning : Floor planning is the exercise of arranging blocks of layout within a chip to minimize area or maximize speed. Floor plan editors provide graphical feedback about the size and placement of modules without showing internal layout details. Also, the editors show connectivity information between modules in the form of wiring diagrams, where the connected ports of modules are connected by straight lines. These kinds of diagrams indicate the relative density of wiring.

The below figure shows that module D can be flipped around the Y-axis to improve the routing.

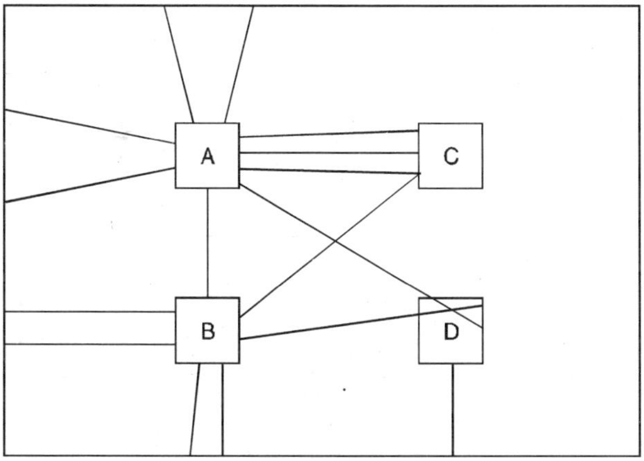

Fig. 9.4.4. A floorplan example.

9.5. THE FPGA DESIGN CYCLE

The design process for using FPGAs generally requires several steps:

1. Entering the design in the form of schematic, netlist, logic expressions, or hardware description languages (Logic minimisation can be done for efficiency)
2. Simulate the design for functional verification then debug the HDL code, and make necessary corrections.
3. Bring out detailed logic design of the system using gates, FFs, registers, counters, adders, etc., Then simulate & debug logic design.
4. Mapping the design into the FPGA architecture: Run a partitioning program to break the logic design into pieces that will fit into the CLBs.
5. Placing and routing the FPGA design: This will place the logic blocks in appropriate places in the FPGA and will route the interconnections between the logic blocks.
6. Extracting delay parameters of the routed design
7. Resimulating for timing verification
8. Generating the FPGA device configuration format
9. Configuring or programming the device (FPGA)
10. Testing the product for undesirable functional behavior: Download the bit pattern into the internal configuration memory cells in the FPGA and test the operation.

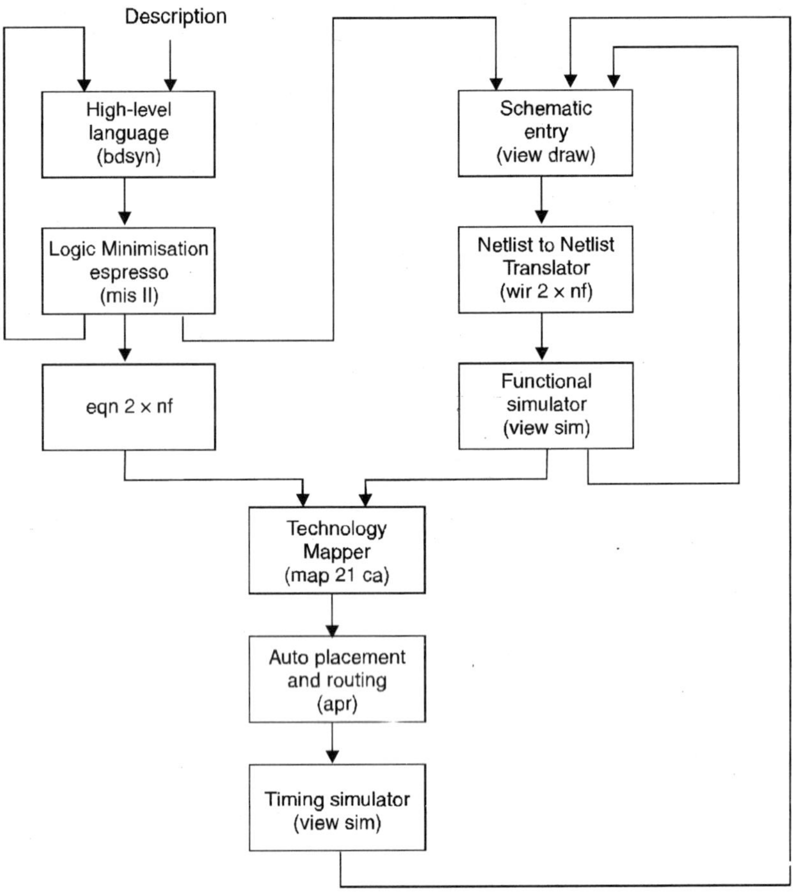

Fig. 9.5.1. Design flow.

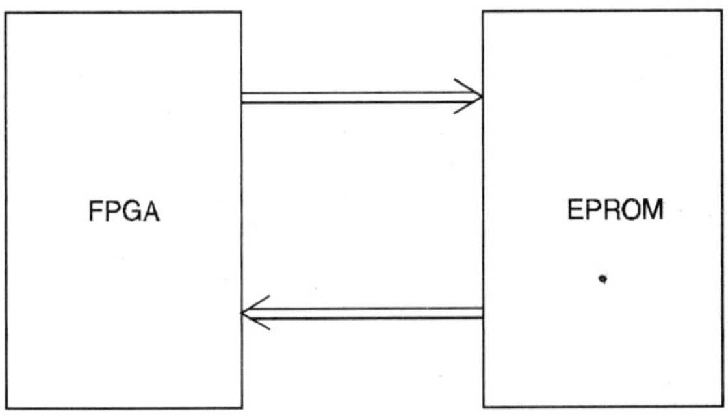

Fig. 9.5.2. Connecting EPROM with EPGA.

When the final system is built, the bit pattern for programming the FPGA is normally stored in an EPROM and automatically loaded into the FPGA when the power is turned on. The Fig. 9.5.2 shows how EPROM is connected to the FPGA. The FPGA resets itself after the power has been applied. Then it reads the configuration data from the EPROM by applying the sequence of addresses to the EPROM inputs and storing the EPROM output data in the FPGA internal configuration memory cells.

Most FPGA vendors allow the entry of the design in a schematic form though it is also possible to enter the design in Boolean expressions. The most versatile design entry however, is some hardware description languages. The most popular languages are verilog HDL and VHDL. Such languages require logic synthesis tools prior to mapping into FPGAs. Automatic synthesis tools are available which can take VHDL description of the system as i/p and generate interconnections of gates and flipflops.

Then the design goes through several iterations of simulation. For every FPGA, the vendor provides design implementation tools to perform steps 3 through 8 and possibly step 6. The front-end design entry (schematic capture or other modes) and simulators may also be part of the tools. Most vendors configure their package with different front end tools to allow the users more choice and flexibility.

Steps 3 and 4 involve several process logic minimization technology mapping placement and routing. Technology mapping binds the technology independent description of the circuit to the basic entities of the target technology. Placement allocates these entities to a specific physical block on the device. Then routing establishes the connections between the different blocks, and is usually done in two stages: *global routing* and *detailed routing*. All FPGA vendors have an automatic placement and routing tool. The placement and routing algorithm has much bearing on the performance of the design.

9.6. VERIFICATION AND TESTING

Testing is an important process in a product life cycle. It is performed at different stages-design, manufacturing, and use. At the design stage, testing is known as design verification. The most common means of verification is simulation. Simulation itself is a multilevel, process, ranging from high level to low level. Formal verification involves proving the properties of a design to guarantee correctness.

Testing after manufacturing is usually known as *digital testing*. This type of testing detects not only failures due to manufacturing defects but also failures due to incorrect design. Often, simulation patterns developed for design verification are complemented with patterns that are generated manually or by an automatic test pattern generator (ATPG) to obtain a complete test set, a test that also verifies the functionality of its logic. The patterns are applied to the circuit using automatic test equipment (ATE). During its lifetime, a product may be tested because it failed or because it is being serviced. The product may also be tested during normal operation. This is known as *online* or *concurrent testing*. Concurrent testing makes use of extra hardware constructs that have been designed exclusively for this purpose. To facilitate digital testing, testing strategies are now an integral part of the design. Designing with testing in mind is called *design for testability (DFT)*.

Computer aided design tools have greatly facilitated the design implementation process. The tools have replaced many of the tedious tasks that a designer used to carry out synthesis (logic minimization, technology mapping, state reduction, and state assignment) design entry and design verification. Done manually these tasks are tedious and time consuming and often error prone.

9.7. SIMULATION

Simulation is a process that imitates the functionality or behavior of the digital design on a computer. It is mainly used to identify possible design errors in the design or timing problems in a circuit. Simulation employs *models* that represent those system's attributes to be imitated; this may include behavioral or timing models. Simulation was originally used in lieu of *prototyping*. Now, simulation is often used to debug a prototype. FPGAs have also been used to prototype designs before their actual fabrication in other technologies, such as standard cells or gate arrays.

There are different aspects of digital circuits that need to be studied before implementation-functionality, timing, effect of certain parameters, etc. There are also a variety of simulations types that are dictated by circuit-level representations. For example, the functionality of a circuit may be simulated at the behavioral level, the gate or logic level, or the circuit level. In the case of FPGAs the functional simulation can be run with zero (no) delays, or one unit delays. In the case of timing simulation, several approaches can be taken. Timing can be checked with normal delays for the technology or with worst case scenarios. More important, timing can be verified using the actual layout of the design on the FPGA. For this, **back annotation** is necessary. The actual delays of the placed and routed design can be extracted and used in timing simulation.

Back annotation

The structural description of a circuit provides a list of leaf cells and their interconnectivity. The generic leaf cells specified by the structural description are mapped to specific library cells. (Leaf cell : The cells at the bottom of the hierarchy tree are called leaf cells). The above information and a specification of the type and placement of the I/O pads on the chip boundary may be used to construct a physical layout for this circuit.

Once a layout has been constructed, the structural connectivity (*i.e.,* the connections between modules) may be reestablished from the geometry by special software. In addition, the actual transistor sizes and physical capacitances may be calculated to a high degree of accuracy. This information may be mapped back to the structural description to place more accurate delays on gates and nets. This is called the back annotation.

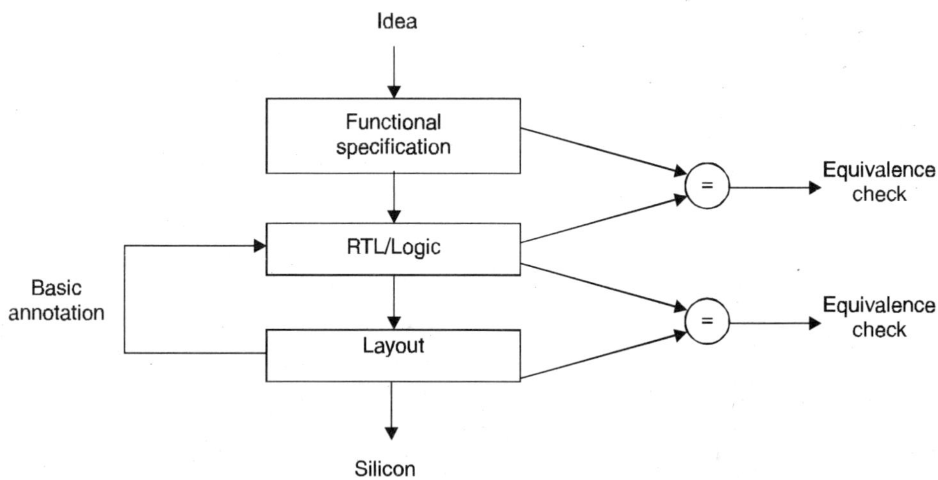

Fig. 9.7. Block diagram illustrating Back annotation.

9.7.1. Logic Simulation

Two types of simulations are used to verify the design; *functional simulation* and *timing simulation*. In functional simulation, the designer may verify the correctness of the design but not at operational speed. That is, no delays or at most nominal delays of the functional units are included. This is usually known as zero (or unit) delay. The main concern are

To check if each block does its intended function.

To modify the design to evaluate alternatives before finalizing the design.

The simulator translates the schematic into a netlist, which is a data structure listing all the gates and their connectivities. The designer prepares a set of inputs, verification patterns, and applies them to the circuit. The output is then examined by the user for correctness. Then if the designer is satisfied with the outcome, the translation should be performed. It is a good practice to use the same set of test vectors on the actual device to confirm that the circuit is a correct implementation of the design entered in the simulator.

9.7.2. Design Validation

The FPGA development system includes validation steps that are performed *before* and *after* the design's netlist is translated to the device configuration format. One would like to think the vendor's design implementation tools faithfully construct the design as specified by the users. However, there are some pitfalls. First, the design implementation tools can detect and report errors and warnings such as

Design exceeds the chip size

A net has no driver

Fanin or fanout is excessive

In general, attention should be given to the error list; otherwise, the design will not be realized. Warnings should also be analyzed carefully since sometimes they may have repercussions in subsequent implementation steps. For example, in some cases, excessive fanin may be just a warning. However, it will eventually affect the circuit performance. Some design implementation tools attempt to "optimize" the design by removing all sub circuits that seem to be redundant, for example, logic modules that have unconnected outputs, two consecutive inverters, and so forth. A report on these actions is given and the designer needs to examine carefully before placement and routing. The major problem with this "optimization" is that the discard subcircuit might affect the functionally or the performance of the design. It is usually not prudent to accept the modifications. Luckily the designer has the option to turn off this optimization stage before it is performed.

Some FPGA development systems have verification tools for validation of the FPGA device configuration during and after configuration, for example, using cyclic redundancy codes (CRC) in the configuration format. After programming the device, a user can even "read back" the configuration format as well as the internal states of an FPGA device for further verification and validation.

9.7.3. Timing Verification

Timing verification can be performed in conjunction with function verification or independently. The delays associated with the different gates are assigned. Usually nominal delays are assigned to the different gates (or functional units in the case of an HDL model). The delays are part of the library used in the design, or they may be modified. In this type of simulation, the net delays are not included, and thus the timing verification is not complete.

It is only when the design is placed and routed that the actual delays can be assigned to the gates and to their interconnects. Also, loading can be evaluated and assigned to the output pads.

9.8. SYNTHESIS

Synthesis is the translation of a design representation to a form that is amendable to minimal realization. For example, the translation of a verilog behavioral model into a register transfer language (RTL) is known as *high-level synthesis*. Mapping the RTL representation into a gate level representation is part of *logic synthesis*. Logic synthesis also involves other processes used in the design of digital circuits such as logic minimization or state reduction and state assignment.

For logic minimization, Karnaugh maps can be used to find a minimal cover for a Boolean expression with a small number of inputs. The Quine-McCluskey method is more amenable to programming on a computer to accelerate and facilitate two-level logic minimization. However, since it is based on exhaustive comparisons and merging, it demands a fair amount of computer resources and time. The same problem arises in conjunction with state assignment for large finite state machines. Given a set of Boolean expressions, the time required to find the minimal cover for logic circuits increases tremendously with the number of terms to be minimized.

9.9. PROGRAMMING METHODS

The main technologies or methods used in the control store of FPGAs and Programmable logic devices (PLDs) are as follows :

(*i*) Static RAM

(*ii*) EPROM

(*iii*) EEPROM

(*iv*) FUSE

(*v*) Antifuse

In memories **(Chapter-6)** itself necessary descriptions on static RAM, EPROM and EEPROM have been given. So a brief description is given under:

Static RAM. It consists of two CMOS inverters to form a bistable device. A new bit can be stored in this device or the state of device can be altered by proper handling of bit and word lines. A five transistor and 6 transistor RAM cell is available. Five transistor RAMs are significantly more dense than 6 transistors RAMs because less area is required to route bit and word line through the array. But six transistors RAM is more stable than 5 transistor RAM. The RAM cell is a storage device only, and in the FPGA application its O/P will control a separate transistor switch.

EPROM. The device is based on special MOSFETs with floating gates. If charge in trapped on the gate the channel will conduct. Charge is injected into the SiO_2 above the transistor channel in a special programming mode by generating high-energy "hot" electrons with in the channel region using higher-than normal voltages. When the high programming voltage is removed, the electrons are trapped in the floating gate. UV Light is used to erase the device by giving the trapped electrons sufficient energy to escape from the gate dielectric. Programming and erasing of EPROM based devices normally take place outside the target system and require special equipment.

EEPROM. This is similar to EPROM, but a second poly silicon gate overlapping the first floating gate is added. EEPROM devices are erased electrically by applying a suitable voltage to this second gate to allow electrons trapped on the floating gate to be discharged by a tunneling mechanism. EEPROM cells are some what less dense than EPROM cells. Most EEPROM technologies have a limit on the number of read/write cycles before the device fails. This limit makes the technology unsuitable for use in computational applications where the device may be reprogrammed thousands of times.

Fuse. This technology relies on simple fuses formed from metal links that are forced open circuit by high programming currents. The fuses are made of metal and have low impedance. So they can support high speed operation. These are used on most low density (Programmable Array Logic) PAL devices based on bipolar technology. They are very rarely used in high density FPGA devices. The reliability of this technology is good.

Antifuse. It is a two terminal circuit element that normally offers a high impedance between its terminals. When a programming voltage higher than the operating voltage is applied, the impedance is reduced to a low value permanently. The behavior is opposite to that of fuse hence the name antifuse. Normally open antifuses are more suitable than fuses for use in gate array like structures.

Advantages of antifuses

(*i*) The metal segments and logic resources of the device can be tested in isolation before any antifuse are programmed.

(*ii*) The antifuses are blown where connection are required on the user design, whereas fuses are blown wherever a potential connection provided by the underlying architecture is not required: this implies that in a general purpose device far fewer antifuses need to be programmed than fuses. There are several antifuse structures in use, of which the most popular is the programmable low-impedance circuit element (PLICE). The PLICE antifuse is bi-directional in its low-impedance state.

9.10. PROGRAMMING ISSUES

The different classes of FPGA device have different goals and requirements for their programming interface. The main issues in the choice of programming interface and control store technology are discussed in the following sections.

Overhead pins. Overhead pins support programming in RAM programmed FPGAs. They increases package and hence part costs. Extra data and address busses are expensive at the board level. However, they allow random access to the control store for rapid reconfiguration and access to the state within the user logic. In some devices, programming and device I/O signals share a physical pad, with a mode control input to select which is active. One disadvantage of this scheme is that if the configuration of the device needs to use the shared I/O pins, control store accesses cannot be made without affecting the computation being implemented by the device.

External support circuitry. With RAM programmed FPGAs it is necessary to load the device configuration in the RAM control store from external non volatile storage, normally EPROM or ROM. It is desirable that this is done with the minimum of support chips. Thus placing the bootstrap logic on the FPGA itself is highly desirable in stand-alone applications. When the FPGA is to be used in a microprocessor-based system, a straightforward memory-like interface can be attractive; depending on the function of the FPGA, it may be possible for the microprocessor or micro controller to load its configuration from the system's nonvolatile storage.

Testability. The choice of programming method affects testability of systems using FPGAs. In RAM based rapidly reprogrammable FPGAs Special FPGA configurations can be developed for use in system tests. These can also be used to test, board traces and even adjacent non-FPGA devices. This technique reduces test time for completed assemblies and there is need for complex board testers and to allow self-test in the field. This method of testing using FPGAs can be successfully integrated with the JTAG boundary-scan methodology. In Write once technologies such as antifuses there are adverse implications for tests, in that the configurations of the devices themselves must be verified. In general, it is not always sufficient to rely on the device programmer to check that fuses read back correctly since a partially blown fuse may pass a static read back test but cause the device to fail at operating speed and temperature.

Reliability. There are reliability concerns associated with the configuration store of all FPGA technologies.

Write once fuse based devices such as PALs can be sensitive to the calibration of the programming equipment. Pin-equivalent devices from different manufacturers often have different programming algorithms. Failure to observe the manufacturer's specifications exactly may cause higher numbers of devices to fail on the programmer.

With RAM-programmed technologies, the most serious problem is that alpha-particle radiation can potentially change the state of one of the control RAM cells. This is called a soft error, since the state of the RAM can be restored by writing it again. Failures are less likely to occur on FPGAs than normal SRAMs because FPGA RAM cells normally have a higher load capacitance and are accessed relatively rarely, have better power supply connection and are designed for stability rather than access speed.

EPROM-based technologies are sensitive to the device the programmer used, as are all fuse-programmed devices. In addition they can be erased by UV light; manufacturer's data gives a figure of about 1 week of exposure to direct sunlight or 3 years exposure to fluorescent room lighting to erase an EPROM if a suitable opaque label has not been fitted. Technologies based on electrically erasable memory may be unreliable if exposed to either radiation or voltage transients on programming pins.

Manufacturing concerns. There are several manufacturing concerns related to the use of write-once FPGA devices: RAM based devices present no special difficulties. The most obvious concern is that the devices must be programmed during manufacture. This has several consequences. Some devices will fail to program correctly in price comparisons between technologies and in determining inventory levels.

Programming the FPGA devices is an additional manufacturing step that must be coordinated with the others and requires its own tooling, which can be relatively expensive. Most complex FPGA antifuse-based devices take a significant amount of time to program and some require that user supplied test vectors be run.

It may be necessary to swap PLDs in the field after design revisions or after a completed board fails the final test. Therefore, it is common practice to use sockets with write-once PLD devices, while other components are normally soldered directly to the board.

Access to internal state. Access to internal state is desirable to support design debugging and in computational applications to access results. There are two main techniques for providing such access :

Special output lines, which can be selectively connected to the outputs of internal gates and are available at dedicated pads is very useful for debugging FPGA designs embedded in board level system.

Mapping cell outputs to control store RAM. This technique is most useful in devices that allow random access to the control store.

Design security. Many write-once PLDs have a “security” fuse which when blown, prevents the fuse array from being read back by a device programmer. This feature is intended to prevent competitors from determining the principles of operation. The security fuse on simple PLDs is of no value. With most complex PLDs based on a PAL architecture, the security fuse would be little more than an inconvenience to the determined. Less obviously, RAM-programmed FPGAs can give design security since the contents of their control store is lost when power is removed. When RAM-programmed FPGAs are used in conjunction with a host microprocessor, extremely secure operation is possible.

Appendix
A VHDL Codings

Note : VHDL is case insensitive.

1. ADDERS AND SUBTRACTORS

(*a*) Half Adder

```
    sum <= a xor b;
library IEEE;
use IEEE.std_logic_1164.all;
entity half is
    port (
            a,b: in STD_LOGIC
            sum, carry: out STD_LOGIC
            );
end half;
architecture half_arch of half is
begin
    sum<=(a and not (b)) or ((not (a) and b);
    carry<=a and b;
end half_arch;
```

(*b*) Full Adder

```
c1<=(a and b) or (b and c) or (a and c);
library IEEE;
use IEEE.std_logic_1164.all;
entity fulladd is
    port(
            a,b,c : in STD_LOGIC;
            s,c1  : out STD_LOGIC
            );
entity fulladd;
architecture fulladd_arch of fulladd is
begin
    s<=a xor b xor c;
```

```
end fulladd_arch
```

(*c*) Half Subtractor

```
library IEEE;
use IEEE.std_logic_1164.all;

entity HSUBTAS is
   port (
         A,B : in STD_LOGIC;
         D,B1 : out STD_LOGIC
         );
end HSUBTAS;

architecture HSUBTAS_arch of HSUBTAS is
begin
   D<= A XOR B;
   B1<= NOT (A) AND B;
End HSUBTAS_arch;
```

(*d*) Full Subtractor

```
library IEEE;
use IEEE.std_logic_1164.all;

entity fsubas is
   port (
         a,b,c : in STD_LOGIC;
         d,b1  : out STD_LOGIC
         );
end fsubas;

architecture fsubas_arch of fsubas is
begin
   process (a,b,c)
   variable x,y,z : std_logic;
begin
      x : = not (a) and not(b) and c;
      y := a and not(b) and c;
      z := a and b and c;
      d <= x or y or z or (a and not(b) and not(c));
      b1 <= x or y or z or (not(a) and b and c);
end process;
end fsubas_arch;
```

(e) Adder/Subtractor

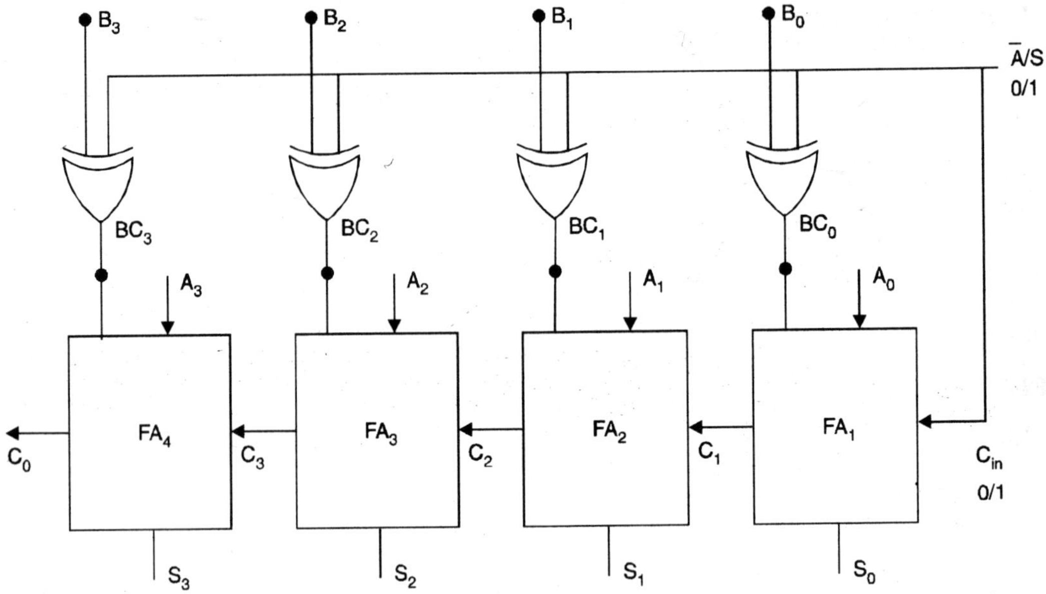

Fig. 1. Binary Adder/Subtracter.

```
library IEEE;
use IEEE.std_logic_1164.all;
entity ADDSUB is
   port (
            A,B : in STD_LOGIC_VECTOR (3 downto 0);
            CIN : in STD_LOGIC;
            C: out STD_LOGIC_VECTOR (3 downto 0);
            S : INOUT STD_LOGIC_VECTOR (3 downto 0);
        CO : OUT STD_LOGIC
        );
end ADDSUB;
architecture ADDSUB_arch of ADDSUB is
SIGNAL BC, C: STD_LOGIC_VECTOR(3 DOWNTO 0);
COMPONENT FULL IS
   PORT (
            A,B,CI : IN STD_LOGIC;
            S,C : OUT STD_LOGIC
            );
END COMPONENT;
begin
   BC(0)<=B(0) XOR CIN;   BC (1)<=B(1) XOR CIN;
   BC(2)<=B(2) XOR CIN;   BC(3)<=B(3) XOR CIN;
```

```
FA1 : FULL PORT MAP (A(0), BC(0), CIN, S(0), C(1));
FA2 : FULL PORT MAP (A(1), BC(1), C(1), S(1), C(2));
FA3 : FULL PORT MAP (A(2), BC(2), C(2), S(2), C(3));
FA4 : FULL PORT MAP(A(3), BC(3), C(3), S(3), CO);
end ADDSUB_arch;
```

A separate coding for full adder is to be written with entity name FULL and ports A, B, CI, S, C and should be called as component here.

PORT MAP. This is called component instantiation. *i.e.,*

• Naming a component within another component as subcomponent. In the above example 'FULL' is named as FA1, FA2, FA3, FA4 under 'ADDSUB'.

2. FLIPFLOPS

RS Flip-Flop

```
library IEEE;
use IEEE.std_logic_1164.all;

entity clkrsffa is
  port(
        s,r,clk : in std_logic;
        q : out std _ logic;
          qp : out std _ logic
       );
end clkrsffa;

architecture clkrsffa _arch of clkrsffa is
  begin
        process (s,r)
        variable t1 : std _logic;
  begin
  if (clk = '1' and clk'event) then
  t1 : = '0';
        if (s='1' and r='0') then
        q <='1';    qb <='0';
        t1 ='1';
        elseif (s='0' and r='1') then
        qp<='1'; q<='0';
        t1 : = '0';
        elseif (s='0' and r='0') then
          q<=t1; qp <=not(t1);
```

```
            end if;
          end if;
       end process;
    end clksffa_arch;
```

T-Flip-Flop

```
elseif (t = '1') then
elseif (t = '0') then
library IEEE;
use IEEE.std_logic_1164.all;

entity tffas is
    port (
          t,clk : in STD_LOGIC;
          q,qb : inout STD_LOGIC
          );
end tffas;

architecture tffas_arch of tffas is
begin
            process(t)
            begin
            if(clk = '0' and clk'event) then
            q<='0'; qb<='1';
            elseif (t = '1') then
            q<=not q;        qb<=q;
            elseif(t = '0') then
            q<=q;   qb<=qb;
            endif;
            end process;
end tffas_arch;
```

D – Flip-Flop

```
library IEEE;
use IEEE.std_logic_1164.all;
entity d_comp is
          port (
                d, clk : in STD_LOGIC;
                q, qb : out STD_LOGIC
                );
end d_comp;

architecture d_comp_arch of d_comp is
```

```
begin
   process (clk)
   begin
        if (clk='1' and clk'event )then
        q<=d;
        qb<=not(d);
end if;
end process;
end d_comp_arch;
```

JK – Flip-Flop

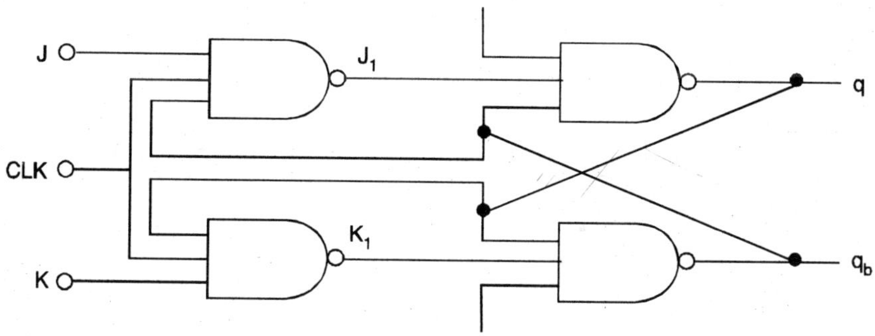

Fig. 2. JK Flip-Flop.

```
library IEEE;
use IEEE.std_logic_1164.all;
entity ff_com is
   port (
        clk, j, k: in STD_LOGIC;
        q,qb : out STD LOGIC
        );
end ff_com;

architecture ff_com_arch of ff _ com is
component nand3
   port (
        a,b,clk : in std_logic;
        d : out std _logic
        );
end component;
signal j1, k1, qt, qbt : std _ logic;
```

```
begin
   q1 : nand3 port map (j, qbt, clk, j1);
   q2 : nand3 port map (k, qt, clk, k1);
   q3 : nand2 port map (j1, qbt, qt);
   q4 : nand2 port map (k1, qt, qbt);
   q<=qt;
   qb<=qbt;
end ff _com_arch;
```

Note : While port mapping remember that the i/p and o/p port names should appear in the order as if it appears in the called component.

Input NAND Gate – Component description for the above diagram.

```
library IEEE;
use IEEE.std_logic_1164.all;
entity nand3 is
   port (
         a,b,clk : in std_logic;
               d : out std_logic
         );
end nand3;
architecture nand3_arch of nand3 is
begin
   process(clk)
   begin
   if (clk='1' and clk'event) then
   d<=(a nand b);
   end if;
end process;
end nand3_arch;
```

3. PARITY GENERATOR/CHECKER

```
library IEEE;
use IEEE.std_logic_1164.all;
entity paritygen is
   port(
         a,b,c,d : in std_logic;
               p : out std_logic
         );
end parity gen;
```

```
architecture paritygen_arch of paritygen is
begin
   PROCESS(a,b,c,d)
   variable s1:std_logic;
   variable s2:std_logic;
   variable s3:std_logic;
   begin
   s1 : = a xor b;
   s2 : = s2 xor b;
   s3 : = s2 xnor d;
   p <= s3;
   end process;
end paritygen_arch;
```

Parity Generator (N-BIT)

```
library IEEE;
use IEEE.std_logic_1164.all;
entity pargen-n  is
   port (
                 a, : in std_logic_vector (3 downto 0) ;
       odd,even : out std_logic
      );
end pargen-n;
architecture gen_arch of pargen-n is
begin
   process(a)
   variable I,t1,t2 : std_logic;
   begin
   t1 : = a(3) xor a(2);
   for i in 1 downto 0 loop
       t2:=t1 xor a(i);
       t1:=t2;
   end loop;
   if (t1 = '1') then
       odd <= '0';   even <= '1';
   else
       odd<='1';  even<='0';
       end if;
end process;
end gen_arch;
```

4. MULTIPLEXER AND DEMULTIPLEXER

(*a*) Multiplexer

```
library IEEE;
use IEEE.std_logic_1164.all;

entity Mux1 is
   port (
          A,B,C,D: in std_logic;
               S: in std_logic_vector(1 downto 0);
                  O :  out std_logic
          );
end Mux1;

architecture Mux1_arch of Mux1 is
begin
   PROCESS(S,A,B,C,D)
   begin
   case S is
   when    "00" => O <= A;
   when    "01" => O <= B;
   when    "10" => O <= C;
   when    "11" => O <= D;

   when others=> O<='X';
   end case;
end process;
end Mux1_arch;
```

(*b*) Demultiplexer

```
library IEEE;
use IEEE.std_logic_1164.all;

entity demux is
   port (
          I,RST : in std_logic;
          S : in std_logic_vector(1 downto 0);
          O1,O2,O3,O4 : out std_logic
          );
end demux;

architecture behave of demux  is
```

```
begin
   process(S,RST)
begin
   if RST = '1' then
     O1 <= '0'; O2 <= '0'; O3 <= '0'; O4 <= '0';
   CASE S IS
   When  "00" => O1 <= I;
   When  "01" => O2 <= I;
   When  "10" => O3 <= I;
   When  "11" => O4 <= I;
   When  others =>  NULL;
END CASE;
end if;
end process;
end behave;
```

5. DECODER AND ENCODEER

(*a*) Encoder

```
library IEEE;
use IEEE.std_logic_1164.all;

entity enco is
   port (
            d: in std_logic_vector(7 downto 0);
         x,y,z: out std_logic
         );
end enco;
architecture enco_arch of enco is
begin
   x <= d(4) or d(5) or d(6) or d(7);
   y <= d(2) or d(3) or d(6) or d(7);
   z <= d(1) or d(3) or d(7) or d(7);
end enco_arch;
```

(*b*) Decoder

```
library IEEE;
use IEEE.std_logic_1164.all;
```

```
entity deco is
   port (
         e : in std_logic;
         s:in std_logic_vector(1 downto  0);
         d0,d1,d2,d3:out std_logic
         );
end deco;
architecture deco_arch of deco is
begin
   process(e,s)
   begin
   if (e='1') then
   case s is
   when   "00" => d0 <= '0';
   when   "01" => d1 <= '0';
   when   "10" => d2 <= '0';
   when   "11" => d3 <= '0';
   when others => null;
   end case;
   else
         d0 <= '1';   d1 <= '1';  d2 <= '1';  d3 <= '1';
   end if  ;
end process;
end deco_arch;
```

6. UP/DOWN COUNTER (WITH LOADABLE CCOUNT)

```
library IEEE;
use IEEE.std_logic_1164.all;
use IEEE.std_logic_unsigned.all;
entity updown is
port (
      clk, rst: in std_logic;
      enable,updown: in std_logic;
      cout:inout std_logic_vector(4 downto 0)
      );
end updown;
architecture updown_arch of updown is
begin
   process(clk)
```

```
    begin
    if (clk= '1' and clk'event)then
    if rst = '1' then
          cout <= "00000";
    else
    if (enable = '1' and updown = '1') then
          cout <= cout + 1;
    elsif enable = '1' and updown = '0' then
          cout <= cout–1;
    end if;
    end if;
    end if;
end process;
end updown_arch;
```

7. SHIFT REGISTERS

(a) Parallel in Serial Out

```
library IEEE;
use IEEE.std_logic_1164.all;

entity piso is
port (
        clk,d1,load : in std_logic;
        din : in std_logic_vector(7 downto 0);
        dout : out std_logic
        );
end piso ;
architecture piso_arch of piso is
signal reg:std_logic_vector( 7 downto 0);
begin
    process(clk)
    begin
    if clk'1' and clk'event then
    if load='1' then
          reg <= din;
    else
          reg <= d1&reg (7 down to 1);
end if;
end if;
```

```
dout <= reg (0);
end process;
end piso_arch;
```

(*b*) Serial in Parallel Out

```
library IEEE;
use IEEE.std_logic_1164.all;
entity sipo is
port (
        clk,din : in std_logic;
        dout : out std_logic_vector(7 downto 0)
        );
end sipo;
architecture sipo_arch of sipo is
signal reg:std_logic_vector(7 downto 0);
begin
    process(clk, reg)
    begin
    if clk='1' and clk'event then
        reg <= din&reg (7 down to 1);
    end if;
    dout <= reg;
end process;
end sipo_arch;
```

8. COMPARATOR

1. Bit Magnitude Comparator – Structural Description

```
library IEEE;
use IEEE.std_logic_1164.all;
entity comp1 is
    port    (
            A,B,G,L,E :in bit;
            GT, LT,EQ: out bit
            );
end comp1;
architecture comp1arch of comp1 is
component nand1
port(
        a,b : in bit;
        c : out bit
        );
```

```
end component;
component nand3
port(
        a,b,c : in bit;
        d : out bit
        );
end component;
component not1
port(
        a : in bit;
        b : out bit
        );
end component;
signalim1,im2,im3,im4,im5,im6,im7,im8,im9,im10: bit;
begin
```

... *a* greater than *b* output

```
  g0 : not1 port map(a,im1);
  g1 : not1 port map(b,im2);
  g3 : nand1 port map(a,im2,im3);
  g4 : nand1 port map(a,g,im4);
  g5 : nand1 port map(im2, g,im5);
  g6 : nand3 port map(im3,im4,im5,gt);
```

... *a* equal *b* output

```
  g7 : nand3 port map(im2,im1,e, im6);
  g8 : nand3 port map(a,b,e,im7);
  g9 : nand1 port map(im6,im7,eq);
```

... lesser than *b* output

```
  g10 : nand1 port map(im1,b,im8);
  g11 : nand1 port map(im1,l,im9);
  g12 : nand1 port map(b,l,im10);
  g13 : nand3 port map(im8,im9,im10,lt);
end comp1arch;
```

2. Input nand gate

```
library IEEE;
use IEEE.std_logic_1164.all;
```

```
entity nand1 is
port (
        a,b: in bit;
        c:out bit
        );
end nand1;

architecture nand1arch of nand1 is
begin
    c < = a nand b;
end nand1arch;
```

3. Input nand gate

```
library IEEE;
use IEEE.std_logic_1164.all;

entity nand3 is
port(
        a,b,c:in bit;
        d: out bit
        );
end nand3;

architecture nand3arch of nand3 is
begin
    d < = not (a and b and c);
end nand3arch;
```

4. Not gate – gate level

```
library IEEE;
use IEEE.std_logic_1164.all;

entity not1 is
port(
        a:in bit;
        b: out bit;
        );
end not1;
architecture not1arch of not1 is
begin
    b < = not a;
end not1arch;
```

5. Comparator – Structural Level – (4 Bit comparator using single bit comparator)

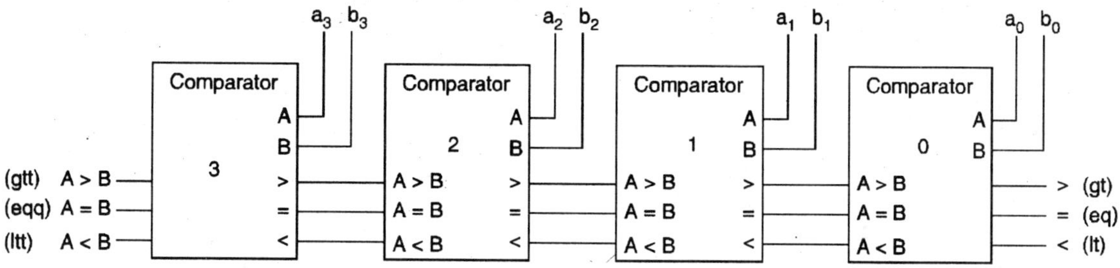

Fig. 3.

```
library IEEE;
use IEEE.std_logic_1164.all;

entity comp4 is
port(
        a,b : in bit_vector (3 downto 0);
        gt,lt,eq : in bit;
        gtt, ltt, eqq : out bit
        );
end comp4;

architecture comp4arch of comp4 is
component comp1
port(
        A,B,G,L,E : in bit;
        GT,LT,EQ: out bit
        );
end component;

signal im:bit_vector(11 downto ));
begin
   c0:comp1 port map (a(0), b(0), gt, lt, eq, im(0), im(1), im(2));
   c1 to c2: for i in 1 to 2 generate
   c:comp1 port map(a(i),b(i),im(i*3-3), im(i*3-2), (im(i*3-1),
   im(i*3+0), im(i*3+1), im(i*3+2));
end generate;
   c3: comp1 port map(a(3),b(3), im(6), im(7), im(8),im(9),im(10), im(11));
   gtt <= im (9);   ltt <= im (10);   eqq <= im (11);
end comp4arch;
```

9. MULTIPLIER – STRUCTURAL LEVEL

```
library IEEE;
use IEEE.std_logic_1164.all;
entity MUL is
port(
        x,y: in std_logic_vector( 3 downto 0);
        p: out std_logic_vector(7 downto 0)
        );
```

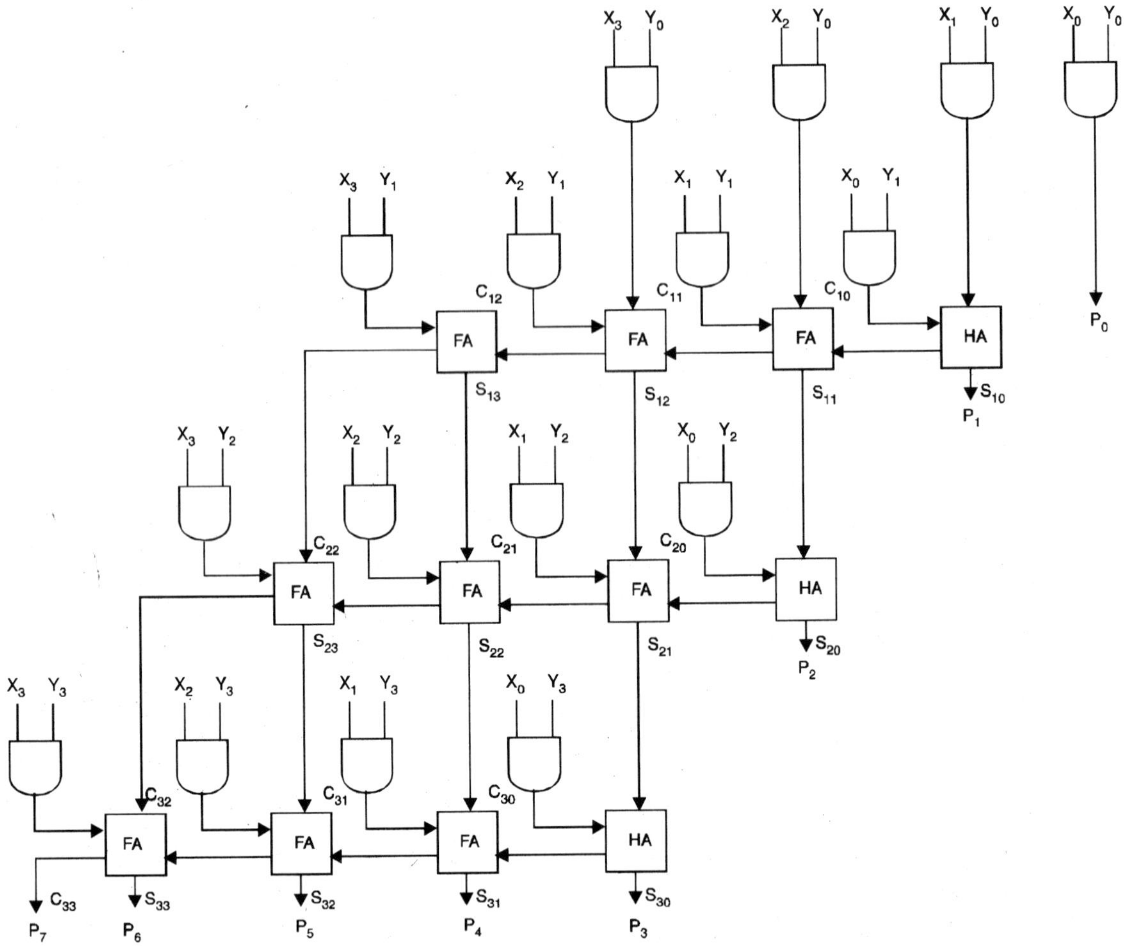

Fig. 4. Block diagram of 4 × 4 Array Multiplier.

```
end MUL;
architecture MUL-arch of MUL is
signal I : std_logic_vector (15 downto 0);
signal s1,s2:std_logic_vector (4 downto 0);
signal c1,c2: std_logic;
```

```
component FADD  is
port(
        a,b : in std_logic_vector(3 downto 0);
         cin: in std_logic;
            s: out std_logic_vector(3 downto 0);
        carry: out std_logic
        );
end component;
component AND2 is
port(
        a,b  : in std_logic;
        c : out std_logic
        );
end component;
begin
i(3) <= '0';
s1(4) <= c1;
s2(4) <= c2;
an1 : AND2 port map(x(0), y(0), p(0));
an2 : AND2 port map(x(1), y(0), i(0));
an3 : AND2 port map(x(2), y(0), i(1));
an4 : AND2 port map(x(3), y(0), i(2));
an5 : AND2 port map(x(0), y(1), i(4));
an6 : AND2 port map(x(1), y(1), i(5));
an7 : AND2 port map(x(2), y(1), i(6));
an8 : AND2 port map(x(3), y(1), i(7));
ad1 : FADD port map(i(7 downto 4), i(3 downto 0), i(3), s1(3 downto 0),c1);
an9 : AND2 port map(x(0), (y(2), i(8));
an10 : AND2 port map(x(1), (y(2), i(9));
an11 : AND2 port map(x(2), y(2), i(10));
an12 : AND2 port map(x(3), y(2), i(11));
ad2 : FADD port map(i(11 downto8), s1(4 downto 1),i(3), s2(3 downto 0),c2);
an13 : AND2 port map(x(0), y(3), i(12));
an14 : AND2 port map(x(1), y(3), i(13));
an15 : AND2 port map(x(2), y(3), i(14));
an16 : AND2 port map(a(3), y(3), i(15));
ad3 : FADD port map(i(15 downto 12), s2(4 downto 1), i(3), p(6 downto3), p(7));
p(1) <= s1(0);     p(2) <= s2(0);
end MUL-arch;
```

Note : The complete description for the components called by the above program (1) AND2 and (2) FADD are given below.

2. Input and gate – gate lev el

```
library IEEE;
use IEEE.std_logic_1164.all;
entity AND2 is
port(
        a,b: in std_logic;
        c: out std_logic
        );
end AND2;
architecture AND2arch of AND2 is
begin
        c < = a and b;
end AND2arch;
```

Full Adder – Gate Level

```
library IEEE;
use IEEE.std_logic_1164.all;
entity FADD is
port(
        a,b : in std_logic_vector( 3 downto 0);
        cin: in std_logic;
        s: out std_logic_vector(3 downto 0);
        carry : out std_logic
        );
end FADD;
architecture FADDarch of FADD is
begin
process(a,b)
begin
        ca(0) < = cin;
for i in 0 to 3 loop
        s(i) < = (a(i) and b(i) or (a(i) and ca(i)) or (b(i) and ca(i));
end loop;
                carry < = ca(4);
end process;
end FADDarch;
```

N X M Multiplier – Behavioural Level

```
library IEEE;
use IEEE.std_logic_1164.all;
use IEEE.std_logic_unsigned.all;

entity mulb is
generic (n,m :POSITIVE:=12);
port(
        mult1 : in std_logic_vector(n downto 0);
        mult2 : in std_logic_vector(m downto 0);
        result : out std_logic_vector(m+n+1 downto 0)
        );
end mulb;
architecture mulbarch of mulb is
TYPE shift IS ARRAY (m downto 0) OF std_logic_vector(m+n+1 downto 0);
SIGNAL  v :std_logic_vector(m+n+1 downto 0);
begin
   process(mult1, mult2,v)
      variable s: shift;
      variable I : std_logic;
      variable c : std_logic_vector(m+n+1 downto 0);
      variable A : std_logic_vector( m downto 0);
begin
   c := (others = > '0');
   A:= (others = > '0'):
   V < = A&mult1;
for i in 0 to m loop – SHIFT AND ADD METHOD
   if i/ = 0 then
   s(i):= v(m+n+1-i  downto 0) & v(m+n+1 downto n+m+2-i);
   elseif i = 0 then
   s(i) : = v(m+n+1 downto 0);
   end if;
   end loop;
for i in 0 to m loop
   if(MULT2(i) = '1') then
   c:= c+s(i);
   end if;
   end loop;
   result<=c;
   end process;
end mulbarch;
```

10. SUBPROGRAMS

Function

```
library IEEE;
use IEEE.std_logic_1164.all;
use IEEE.std_logic_unsigned.all;
use IEEE.std_logic_arith.all;

entity bin-dec is
port(
        B : in BIT_VECTOR(3 downto 0);
        C: out INTEGER RANGE 0 TO 15
        );
end bin-dec;
architecture bin-dec arch of bin-dec is

FUNCTION BI-DEC(BI:BIT_VECTOR(3 DOWNTO 0)) RETURN INTEGER IS
VARIABLE DEC: INTEGER;
BEGIN
   DEC:=0;
   FOR J IN 0 TO 3 LOOP
   IF BI(J)='1' THEN
   DEC:=DEC+2**J;
   END IF;
   END LOOP;
   RETURN DEC;
END BI-DEC;
begin
   C < = BI-DEC(B);
end bin-dec arc;
```

Procedure

```
library IEEE;
use IEEE.std_logic_1164.all;
entity decbi is
port(
        BYTE : in std_logic_vector(7 downto 0);
        INT    : out integer range 0 to 255
        );
end decbi;

architecture decbin of decbi is
```

```
PROCEDURE BY-INT(BY:IN STD_LOGIC_VECTOR(7 DOWNTO 0)
INTE:OUT INTEGER) IS
VARIABLE RESULT : INTEGER := 0;
BEGIN
   FOR I IN 0 TO 7 LOOP
   IF BY(I) ='1' THEN
         RESULT : = RESULT+2**I;
   END IF;
END LOOP;
INTE: = RESULT;
END BY-INT;

begin
   process (BYTE)
   variable OT: integer;
begin
   BY-INT(BYTE,OT);
   INT<=OT;
   end  process;
end decbin;
```

11. SERIAL ADDER – STRUCTURAL LEVEL

```
library IEEE;
use IEEE.std_logic_1164.all;
entity serialadd is
port (
      clk,load : in std_logic;
      a,b: in std_logic_vector( 7 downto 0);
      do : out std_logic_vector(7 downto 0)
      );
end serialadd;
architecture serialadder of serial add is
component piso
port(
      clk, d1, load: in std_logic;
      din: in std_logic_vector( 7 downto 0);
      dout : out std_logic
      );
end component;
component sipo
```

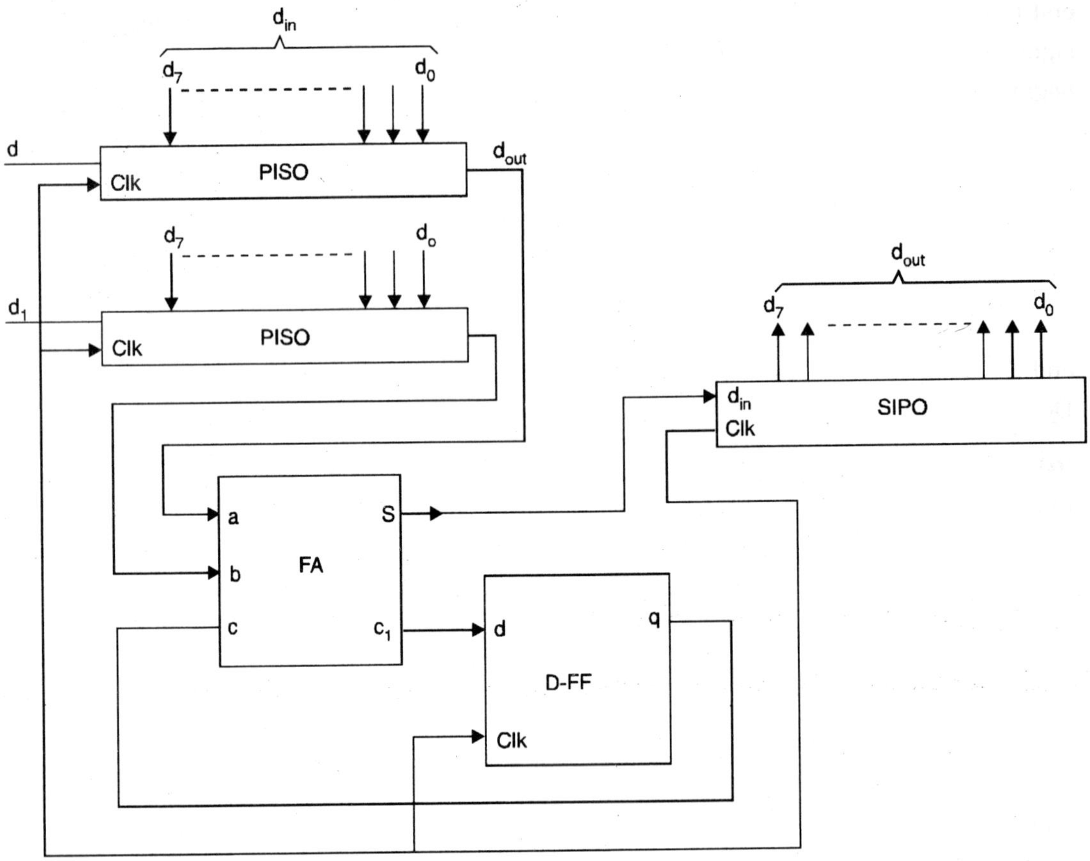

Fig. 5. Serial Adder.

```
port(
        clk, din : std_logic;
        dout:out std_logic_vector(7 downto 0)
        );
end component;
component d_comp
port(
        d,clk : in std_logic;
        q, qb : out std_logic;
        );
end component;
component full_com
port(
        a,b,bin: in std_logic;
        s,c1: out std_logic;
        );
```

```
end component;
signal s1,s2,s3,s4,s5,s6,s7,s8,s9: std_logic;
begin
        s6<='0';
        sr1:piso port map(clk,a,s6,load,s1);
        sr2:piso port map(clk,b,s6,load,s2);
        sr3:full_com port map(s1,s2,s5,s3,s4);
        sr4:d_comp port map(clk,s4,s5,open);
        sr5:sipo port map(clk,s3,do);
end serialadder;
```

Description or components PISO, SIPO, d-comp, full-com

(*a*) Parallel in Serial Out

```
library IEEE;
use IEEE.std_logic_1164.all;
entity piso is
port(
        clk,d1,load : in std_logic;
        din : in std_logic_vector(7 downto 0);
        dout : out std_logic
        );
end piso ;
architecture piso_arch of piso is
signal reg:std_logic_vector( 7 downto 0);
begin
        process(clk)
    begin
        if clk'1' and clk'event then
        if load='1' then
        reg < = din;
        else
        reg<=d1&reg(7 downto 1);
    end if;
    end if;
        dout < = reg(0);
    end process;
end piso_arch;
```

(*b*)Serial in Parallel out

```
library IEEE;
```

```
use IEEE.std_logic_1164.all;

entity sipo is
port (
        clk,din : in std_logic;
        dout : out std_logic_vector(7 downto 0)
        );
end sipo;

architecture sipo_arch of sipo is
signal reg:std_logic_vector(7 downto 0);
begin
   process(clk, reg)
begin
        if clk = '1' and clk'event then
        reg < = din&reg(7 downto 1);
end if;
        dout < = reg;
end process;
end sipo_arch;
```

Fulladder

```
library IEEE;
use IEEE.std_logic_1164.all;

entity fulladd is
port(
        a,b,bin : in STD_LOGIC;
        S, C1 : out  STD_LOGIC;
        );
entity fulladd;
architecture fulladd_arch of  fulladd is
begin
        S <= a x or b x or bin;
        C1 <= (a and b) or (b and bin) or (a and bin);
end fulladd_arch
```

D – Flip-Flop

```
library IEEE;
use IEEE.std_logic_1164.all;

entity d_comp is
```

```
port (
        d, clk : in STD _ LOGIC;
        q, qb : out STD _LOGIC
        );
end d_comp;
   architecture d_comp _arch of d_comp is
begin
   process (clk)
   begin
   if (clk = '1' and clk'event )then
   q < = d;
   qb < = not(d);
   end if;
end process;
end d_comp_arch;
```

12. BARREL SHIFTER – BEHAVIOURAL LEVEL

```
library IEEE;
use IEEE.std_logic_1164.all;
entity barrel is
port (
        SEL : in STD_LOGIC_VECTOR (3 downto 0);
        B_INPUT : in std_logic_vector(15 downto 0);
        B_output : out std_logic_vector(15 downto 0)
        );
end barrel;

architecture barrel_arch of barrel is
signal SEL_A, SEL_B: std_logic_vector(1 downto 0);
signal C: std_logic_vector(15 downto 0);
begin
SEL_A <= SEL (1 downto 0);
SEL_B <= SEL (3 downto 2);
process (SEL_A, B_INPUT)
      begin
      case SEL_A is
      when "00" => — shift by 0
      C(15) <= B_INPUT;
      When "01" => — shift by 1
```

```
        C(15) <= B_INPUT (0);
        C(14 dwonto 0) <= B_INPUT  (15 downto 1);
        When "10" => — shift by 2
        C(15 dwonto 14) <= B_INPUT  (1 downto 0);
        C(13 dwonto 0) <= B_INPUT  (15 downto 2);
        When "11" => — shift by 3
        C(15 dwonto 13) <= B_INPUT  (2 downto 0);
        C(12 dwonto 0) <= B_INPUT  (15 downto 3);
when others =>
C <= B_INPUT;
end case;
end process;
process (SEL_B,C)
begin
   case SEL_B is
     when "00" =>     — shift  by 0 more
     B_OUTPUT <= C;
     B_OUTPUT(15 downto 12) <= C( 3 downto 0);
     B_OUTPUT(15 downto 0) <= C( 15 downto 4);
     when "10" =>     —shift  by 8 more
     B_OUTPUT(15 downto 8) <= C( 7 downto 0);
     B_OUTPUT(7 downto 0) <= C( 15 downto 8);
     when "11" =>     —shift  by 12 more
     B_OUTPUT(15 downto 4) <= C( 11 downto 0);
     B_OUTPUT( 3 downto 0) <= C( 15 downto 12);
when others =>
     B_OUTPUT <= C;
     end case;
     end process;
end barrel_arch;
```

13. SIGNED ADDER

```
Library ieee;
use ieee.std_logic_1164.all;
use ieee.std_logic_unsigned.all;
use ieee.std_logic_arith.all;
entity addsi is
   generic(n:positive:=8);
```

```
    port(
        a:in BIT_VECTOR (n downto 0);
        b:in BIT_VECTOR (n downto 0);
        s:inout BIT_VECTOR (n downto 0));
end addsi;

architecture addsi_arch of addsi is
signal an,bn:bit_vector(n downto 0);
signal m:bit_vector(1 downto 0);
signal c1:bit;
function addsub(x,y:bit_vector(n-1 downto 0);
    ci:bit) return bit_vector is
variable z:bit_vector(n-1 downto 0);
variable c:bit_vector(n downto 0);
begin
        c(0):= ci;
for  i in 0 to n-1 loop
        z(i):= x(i) xor y(i) xor c(i);
        c(i+1):= ((x(i) and y(i)) or (x(i) and c(i)) or (c(i) and y(i)));
end loop;
return z;
end addsub;
begin
process(a,b,m,an,bn,s)
begin
        m <= a(n) & b(n);
case m is
        when "00" =>
        s(n) <= '0';
        s(n–1 downto 0) <= addsub(b(n–1 downto 0), a(n–1 downto 0),'0');
        when "01" =>
if      a(n–1 downto 0) > b(n–1 downto 0) then
        bn <= not (b);
        s(n–1 downto 0) <= addsub(a(n–1 downto 0), bn(n–1 downto 0),'1');
        s(n) <= '0';
else
        an<=not(a);
        s(n–1 downto 0)<=addsub(b(n–1 downto 0), an(n–1 downto 0),'1');
        s(n)<='1';
```

```
if     s(n–1 downto 0)="00000000" then
       s(n)<='0';
end if;
end if;

       when "10"=>
       if a(n–1 downto 0)>b(n–1 downto 0) then
       bn<=not(b);
       s(n–1 downto 0)<=addsub(a(n–1 downto 0), bn(n–1 downto 0),'1');
       s(n)<='1';
if     s(n–1 downto 0)="0000000"then
       s(n)<='0';
end if;
else
       an<=not(a);
       s(n–1 downto 0 )<= addsub(b(n–1 downto 0), an(n–1 downto 0),'1');
       s(n)<='0';
end if ;
       when "11"=>
       s(n)<='1';
       s(n–1 downto 0)<=addsub(a(n–1 downto 0),b(n–1 downto 0),'0');
if     s(n–1 downto 0)=”00000000"then
       s(n)<='0';
end if;
when others => null;
end case;
end process;
end addsi_arch;
```

14. 8 × 8 WALLACE TREE MULTIPLIER

```
Library ieee;
Use ieee.std_logic_1164.all;
entity mul8×8 is
port (
       a: in BIT_VECTOR (7 downto 0);
       b: in BIT_VECTOR (7 downto 0);
       p: out BIT_VECTOR (15 downto 0)
       );
end mul 8 × 8;
```

```
architecture mul8×8_arch of mul8×8 is
component full
port (a,b,ci:in bit;
s,co:out bit);
end component;
component half
port  (a,b:in bit;
        s,c:out bit);
end component;
signal s0,s1,s2,s3,s4,s5,s6,s7:bit_vector(7 downto 0);
signal x11,y11,x12,y12,x13,y13,x14,y14,x15a,y15a,x15b,y15b,
x16a,y16a,x16b,y16b,x17a,y17a,x17b,y17b:bit;
signal x18a,y18a,x18b,y18b,x19a,y19a,x19b,y19b,x1a,y1a,x1b,y1b,x1c,y1c:bit;
signal x22,y22,x23,y23,x24,y24,x25,y25,x26,y26,x27a,y27a,x27b,y27b:bit;
signal x28a,y28a,x28b,y28b,x29,y29,x2a,y2a,x2b,y2b,x2d,y2d:bit;
signal x33,y33,x34,y34,x35,y35,x36,y36,x37,y37,x38,y38,x39,y39,
x3a,y3a,x3c,y3c:bit;
signal x44,y44,x46,y46,x48,y48,x49,y49,x4a,y4a,x4c,y4c:bit;
signal x55,y55,x57,y57:bit;
signal x66,y66,x77,y77:bit;
signal x88,y88,x89,y89,x8a,y8a,x8b,y8b,x8c,y8c,x8d,y8d,x8e,y8e:bit;
begin
s0(0)<=a(0)and b(0);s0(1) <=a(0)and b(1); s0(2)<=a(0)and b(2); s0(3)<=a(0)and b(3);
s0(4)<=a(0)and b(4);s0(5) <=a(0)and b(5); s0(6)<=a(0)and b(6); s0(7)<=a(0)and b(7);
s1(0)<=a(1)and b(0);s1(1) <=a(1)and b(1); s1(2)<=a(1)and b(2); s1(3)<=a(1)and b(3);
s1(4)<=a(1)and b(4);s1(5) <=a(1)and b(5); s1(6)<=a(1)and b(6); s1(7)<=a(1)and b(7);
s2(0)<=a(2)and b(0);s2(1) <=a(2)and b(1); s2(2)<=a(2)and b(2); s2(3)<=a(2)and b(3);
s2(4)<=a(2)and b(4);s2(5) <=a(2)and b(5); s2(6)<=a(2)and b(6); s2(7)<=a(2)and b(7);
s3(0)<=a(3)and b(0);s3(1) <=a(3)and b(1); s3(2)<=a(3)and b(2); s3(3)<=a(3)and b(3);
s3(4)<=a(3)and b(4);s3(5) <=a(3)and b(5); s3(6)<=a(3)and b(6); s3(7)<=a(3)and b(7);
s4(0)<=a(4)and b(0);s4(1) <=a(4)and b(1); s4(2)<=a(4)and b(2); s4(3)<=a(4)and b(3);
s4(4)<=a(4)and b(4);s4(5) <=a(4)and b(5); s4(6)<=a(4)and b(6); s4(7)<=a(4)and b(7);
s5(0)<=a(5)and b(0);s5(1) <=a(5)and b(1); s5(2)<=a(5)and b(2); s5(3)<=a(5)and b(3);
s5(4)<=a(5)and b(4);s5(5) <=a(5)and b(5); s5(6)<=a(5)and b(6); s5(7)<=a(5)and b(7);
s6(0)<=a(6)and b(0);s6(1) <=a(6)and b(1); s6(2)<=a(6)and b(2); s6(3)<=a(6)and b(3);
s6(4)<=a(6)and b(4);s6(5) <=a(6)and b(5); s6(6)<=a(6)and b(6); s6(7)<=a(6)and b(7);
```

```
s7(0)<=a(7)and b(0);s7(1) <=a(7)and b(1); s7(2)<=a(7)and b(2); s7(3)<=a(7)and b(3);
s7(4)<=a(7)and b(4);s7(5) <=a(7)and b(5); s7(6)<=a(7)and b(6); s7(7)<=a(7)and b(7);
p(0)<=s0(0);
h11:half port map (s0(1),s1(0),p(1),x11);
f12:full port map (s0(2),s1(1),s2(0),y12,x12);
h22:half port map (y12,x11,p(2),x22);
f13:full port map (s0(3),s1(2),s2(1),y13,x13);
f23:full port map (s3(0),y13,x12,y23,x23);
f33:half port map (y23,x22,p(3),x33);
f14:full port map (s0(4),s1(3),s2(2),y14,x14);
f24:full port map (s3(1),y14,x13,y24,x24);
f34:full port map (s4(0),y24,x23,y34,x34);
f44:half port map (y34,x33,p(4),x44);
f15a:full port map (s0(5),s1(4),s2(3),y15a,x15a);
f15b:full port map (s3(2),s4(1),s5(0),y15b,x15b);
f25:full port map (y15b,y15a,x14,y25,x25);
h35:half port map (y25,x24,y35,x35);
f55:full port map (y35,x34,x44, p(5),x55);
f16a:full port map (s0(6),s1(5),s2(4),y16a,x16a);
f16b:full port map (s3(3),s4(2),s5(1),y16b,x16b);
f26:full port map (y16b,y16a,s6(0),y26,x26);
f36:full port map (y26,x15b,x15a,y36,x36);
f46:full port map (y36,x25,x35,y46,x46);
h66:half port map (y46,x55,p(6),x66);
f17a:full port map (s0(7),s1(6),s2(5),y17a,x17a);
f17b:full port map (s3(4),s4(3),s5(2),y17b,x17b);
f27a:full port map (s7(0),x16b,x16a,y27a,x27a);
f27b:full port map (s6(1),y17b,y17a,y27b,x27b);
f37:full port map (y27b,y27a,x26,y37,x37);
f57:full port map (y37,x36,x46,y57,x57);
h77:half port map (y57,x66,p(7),x77);
f18a:full port map (s1(7),s2(6),s3(5),y18a,x18a);
f18b:full port map (s4(4),s5(3),s6(2),y18b,x18b);
h28b:half port map (s7(1),y18b,y28b,x28b);
f28a:full port map (y18a,x17b,x17a,y28a,x28a);
f38:full port map (y28a,x27b,x27a,y38,x38);
f48:full port map (y28b,y38,x37,y48,x48);
```

```
h88:full port map (y48,x57,x77,p(8),x88);
f19a:full port map (s2(7),s3(6),s4(5),y19a,x19a);
f19b:full port map (s5(4),s6(3),s7(2),y19b,x19b);
f29:full port map (y19a,x18b,x18a,y29,x29);
f39:full port map (y29,x28b,x28a,y39,x39);
f49:full port map (y19b,y39,x38,y49,x49);
f89:full port map (y49,x48,x88,p(9),x89);
f1a:full port map (s3(7),s4(6),s5(5),y1a,x1a);
f2a:full port map (y1a,x19b,x19a,y2a,x2a);
f3a:full port map (s6(4),y2a,x29,y3a,x3a);
f4a:full port map (s7(3),y3a,x39,y4a,x4a);
f8a:full port map (y4a,x49,x89,p(10),x8a);
f1b:full port map (s4(7),s5(6),s6(5),y1b,x1b);
f2b:full port map (s7(4),y1b,x1a,y2b,x2b);
f8b:full port map (y4c,x4a,x8a,p(11),x8b);
f1c:full port map (s5(7),s6(6),s7(5),y1c,x1c);
f3c:full port map (y1c,x1b,x2b,y3c,x3c);
f4c:full port map (y2b,x2a,x3a,y4c,x4c);
f8c:full port map (y3c,x4c,x8b,p(12),x8c);
f2d:full port map (s7(6),s6(7),x1c,y2d,x2d);
f8d:full port map (y2d,x3c,x8c,p(13),x8d);
f8e:full port map (s7(7),x2d,x8d,p(14),p(15));
end mul8x8_arch;
```

Universal Shift Register

```
Library IEEE;
Use IEEE.Std_logic_1164.all;
entity univ-shift is
   port
      (
         Clk : in std_logic;
         Se1 : in Std_logic_Vector (1 downto 0);
         SRi : in Std_logic;
         SLi : in Std_logic;
         I : in Std_logic_Vector ( downto 0);
         O : inout Std_logic_Vector ( downto 0)
      );
end univ_shift;
```

```
architecture univ_arch of univ_shift is
component Dff is
   port   (d, clk : in Std_logic;
                  q : out Std_logic);
end component;
component MUX is
   port (Se1, Sel0 : in Std_logic;
           a, b, c, d : in Std_logic;
                  O : out Std_logic);
end component;
Signal d1, d2, d3, d4 : Std_logic;
begin
   df1 = Dff    port map (d1, clk, 0(0));
   df2 = Dff    port map (d2, clk, 0(1));
   df3 = Dff    port map (d3, clk, 0(2));
   df4 = Dff    port map (d4, clk, 0(3));

   MUX1 = MUX  port map (Sel(1), Sel(0), 0(0), SRi, 0(1), I(0), d1);
   MUX2 = MUX  port map (Sel(1), Sel(0), 0(1), 0(0), 0(2), I(1), d2);
   MUX3 = MUX  port map (Sel(1), Sel(0), 0(2), 0(1), 0(3), I(2), d3);
   MUX4 = MUX  port map (Sel(1), Sel(0), 0(3), 0(2), SLi, I(3), d4);
end univ_arch;
```

Component DFF

```
Library IEEE;
Use IEEE.Std_logic_1164.all;

entity DFF is
   port
      (
         d : in Std_logic;
         q : out Std_logic;
         Clk : in Std_logic
      );
end DFF;
architecture DFF_arch of DFF is
begin
   process (clk)
   if (clk = '1' and clk' event) than
```

```
        q <= d;
    end if;
end process;
end DFF_arch;
```

Component MUX

```
Library IEEE;
Use IEEE.Std_logic_1164.all;
entity MUX is
    port
            (Sel : in Std_logic_Vector (1 downto 0);
        a,b,c,d : in Std_logic;
              O : out Std_logic);
    end MUX

architecture MUX_arch of MUX is
begin
    process (a,b,c,d, Se1)
    begin
        Case Sel is
            when "00" => y <= a;
            when "01" => y <= b;
            when "10" => y <= c;
            when "11" => y <= d;
            when others => null;
    end case;
    end process;
end MUX_arch;
```

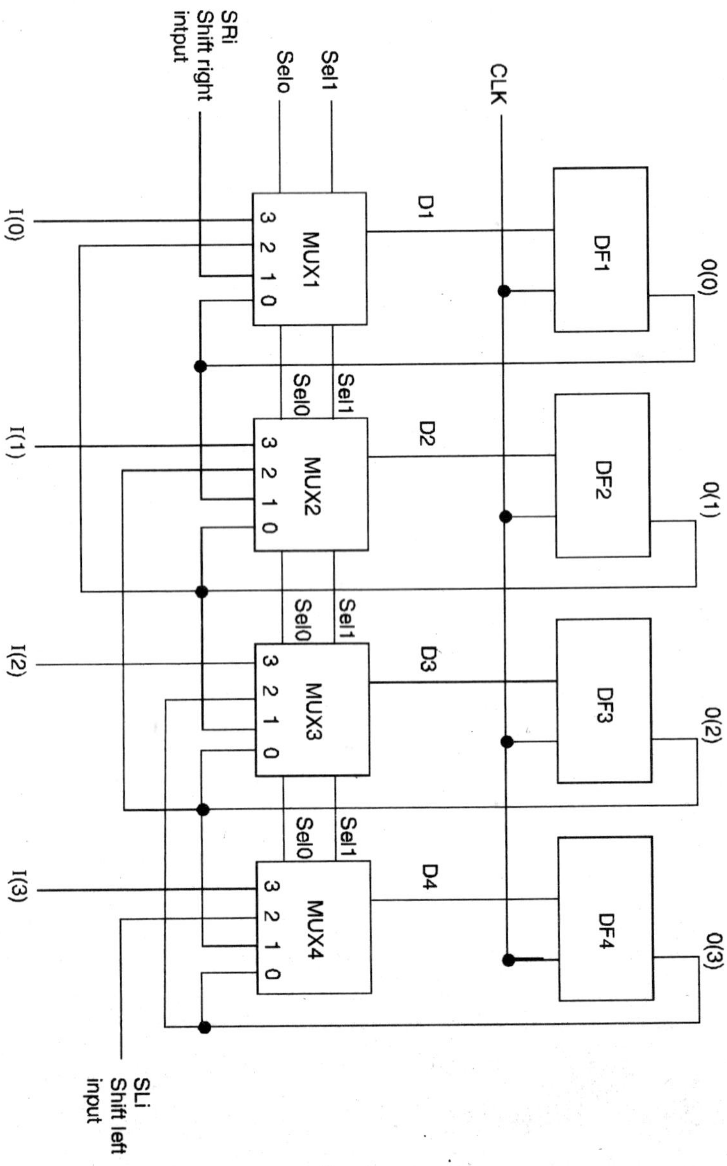

Fig. 6. Universal Shift Register.

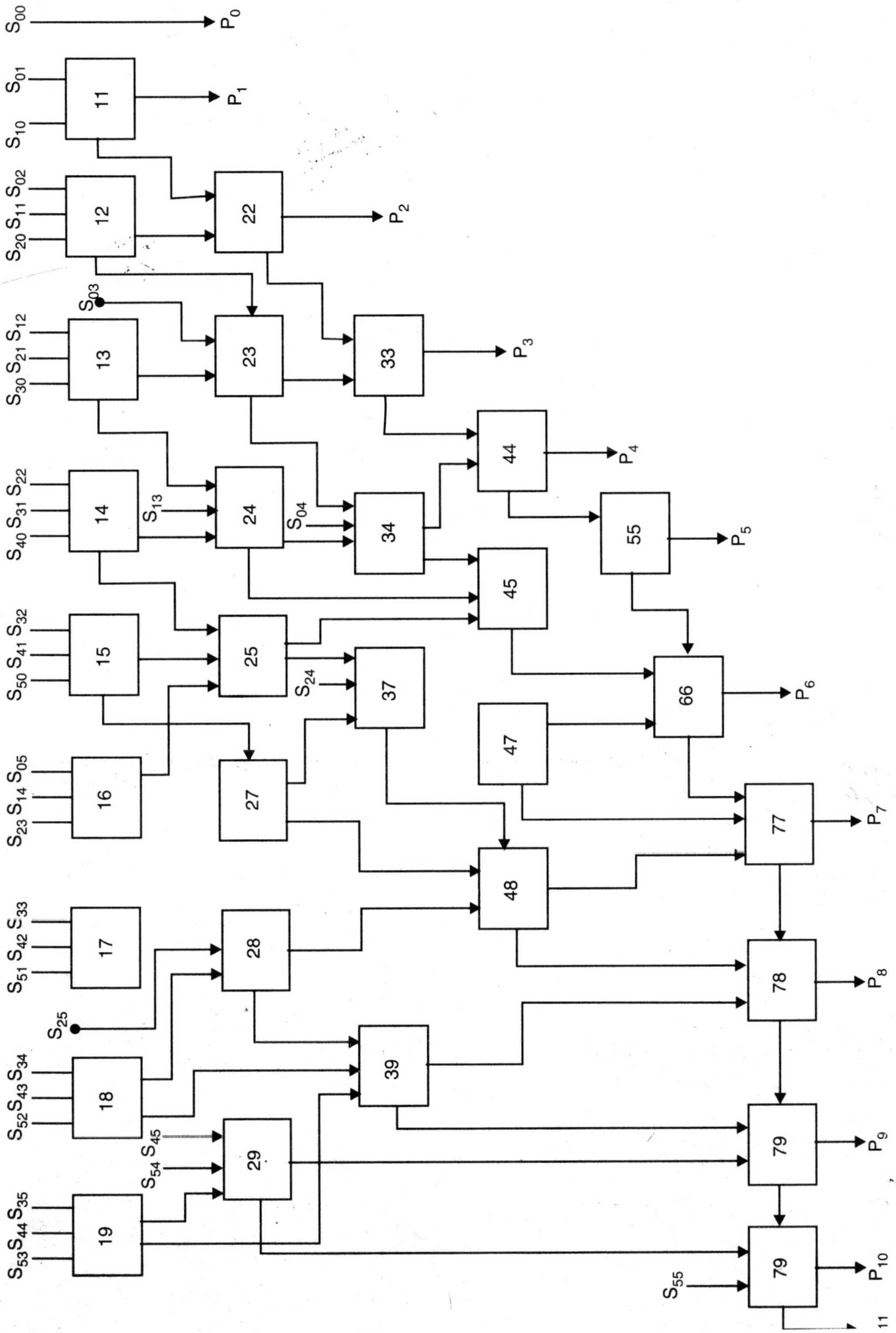

Fig. 7. Wallace Tree Multiplier.

References

Basic VLSI Design Systems and Circuits, Douglas A. Pucknell and Kamran Eshragian, Prentice Hall of India Pvt. Limited.

VLSI Design Techniques for Analog and Digital Circuits, Randal L. Geiger and P.E. Allen, McGraw Hill.

VHDL Analysis and Modeling of Digital System, Navabi. Z, McGraw Hill.

Principles of CMOS VLSI Design, Neil H.E. Weste and Kamran Eshragian, Addison Wesley.

Analog Integrated Circuit Design, David Johns and Ken Martin, John Wiley and Sons Inc.

Fundamentals of Digital Logic with VHDL Design, Stephen Brown and Zvonko Vranesic, Tata McGraw Hill.

Digital Integrated Circuits, Jan, M. Rabaey, Prentice Hall of India Pvt. Limited.

Digital Systems Design using VHDL, Charles H. Roth Jr., Thomson Asia Pvt. Ltd.

Digital Design Field Programmable Gate Arrays, P.K. Chan and Mourad, Prentice Hall of India Pvt. Ltd.

Field Programmable Gate Arrays, John.V. Old Fiels and Richard C. Dorf, John Wiley and Sons Inc.

VHDL Primer, J. Bhaskar, Prentice Hall of India Pvt. Ltd.

Application Specific Integrated Circuits, Smith, M. J., Addison Wesley Press.

Model for pass transistor logic

(i)

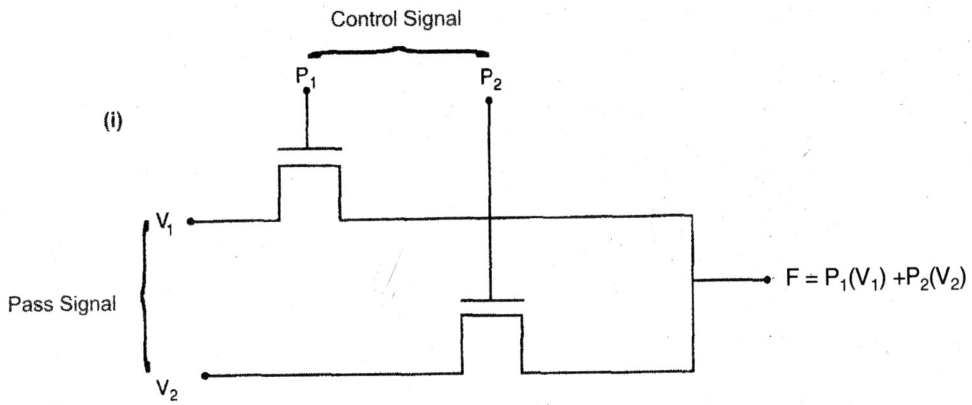

(ii)

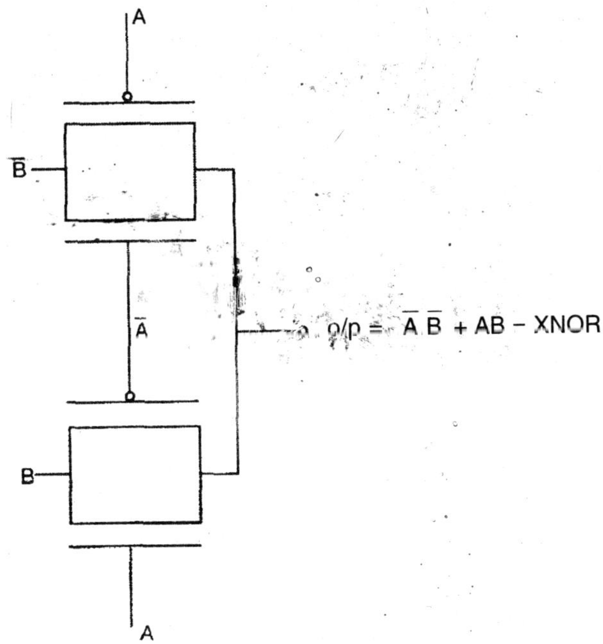